Holomorphy and Calculus
in Normed Spaces

MONOGRAPHS AND TEXTBOOKS IN
PURE AND APPLIED MATHEMATICS

1. *K. Yano*, Integral Formulas in Riemannian Geometry (1970) *(out of print)*
2. *S. Kobayashi*, Hyperbolic Manifolds and Holomorphic Mappings (1970) *(out of print)*
3. *V. S. Vladimirov*, Equations of Mathematical Physics (A. Jeffrey, editor; A. Littlewood, translator) (1970) *(out of print)*
4. *B. N. Pshenichnyi*, Necessary Conditions for an Extremum (L. Neustadt, translation editor; K. Makowski, translator) (1971)
5. *L. Narici, E. Beckenstein, and G. Bachman*, Functional Analysis and Valuation Theory (1971)
6. *D. S. Passman*, Infinite Group Rings (1971)
7. *L. Dornhoff*, Group Representation Theory (in two parts). Part A: Ordinary Representation Theory. Part B: Modular Representation Theory (1971, 1972)
8. *W. Boothby and G. L. Weiss (eds.)*, Symmetric Spaces: Short Courses Presented at Washington University (1972)
9. *Y. Matsushima*, Differentiable Manifolds (E. T. Kobayashi, translator) (1972)
10. *L. E. Ward, Jr.*, Topology: An Outline for a First Course (1972) *(out of print)*
11. *A. Babakhanian*, Cohomological Methods in Group Theory (1972)
12. *R. Gilmer*, Multiplicative Ideal Theory (1972)
13. *J. Yeh*, Stochastic Processes and the Wiener Integral (1973) *(out of print)*
14. *J. Barros-Neto*, Introduction to the Theory of Distributions (1973) *(out of print)*
15. *R. Larsen*, Functional Analysis: An Introduction (1973) *(out of print)*
16. *K. Yano and S. Ishihara*, Tangent and Cotangent Bundles: Differential Geometry (1973) *(out of print)*
17. *C. Procesi*, Rings with Polynomial Identities (1973)
18. *R. Hermann*, Geometry, Physics, and Systems (1973)
19. *N. R. Wallach*, Harmonic Analysis on Homogeneous Spaces (1973) *(out of print)*
20. *J. Dieudonné*, Introduction to the Theory of Formal Groups (1973)
21. *I. Vaisman*, Cohomology and Differential Forms (1973)
22. *B. -Y. Chen*, Geometry of Submanifolds (1973)
23. *M. Marcus*, Finite Dimensional Multilinear Algebra (in two parts) (1973, 1975)
24. *R. Larsen*, Banach Algebras: An Introduction (1973)
25. *R. O. Kujala and A. L. Vitter (eds.)*, Value Distribution Theory: Part A; Part B: Deficit and Bezout Estimates by Wilhelm Stoll (1973)
26. *K. B. Stolarsky*, Algebraic Numbers and Diophantine Approximation (1974)
27. *A. R. Magid*, The Separable Galois Theory of Commutative Rings (1974)
28. *B. R. McDonald*, Finite Rings with Identity (1974)
29. *J. Satake*, Linear Algebra (S. Koh, T. A. Akiba, and S. Ihara, translators) (1975)

67. *J. K. Beem and P. E. Ehrlich*, Global Lorentzian Geometry (1981)
68. *D. L. Armacost*, The Structure of Locally Compact Abelian Groups (1981)
69. *J. W. Brewer and M. K. Smith, eds.*, Emmy Noether: A Tribute to Her Life and Work (1981)
70. *K. H. Kim*, Boolean Matrix Theory and Applications (1982)
71. *T. W. Wieting*, The Mathematical Theory of Chromatic Plane Ornaments (1982)
72. *D. B. Gauld*, Differential Topology: An Introduction (1982)
73. *R. L. Faber*, Foundations of Euclidean and Non-Euclidean Geometry (1983)
74. *M. Carmeli*, Statistical Theory and Random Matrices (1983)
75. *J. H. Carruth, J. A. Hildebrant, and R. J. Koch*, The Theory of Topological Semigroups (1983)
76. *R. L. Faber*, Differential Geometry and Relativity Theory: An Introduction (1983)
77. *S. Barnett*, Polynomials and Linear Control Systems (1983)
78. *G. Karpilovsky*, Commutative Group Algebras (1983)
79. *F. Van Oystaeyen and A. Verschoren*, Relative Invariants of Rings: The Commutative Theory (1983)
80. *I. Vaisman*, A First Course in Differential Geometry (1984)
81. *G. W. Swan*, Applications of Optimal Control Theory in Biomedicine (1984)
82. *T. Petrie and J. D. Randall*, Transformation Groups on Manifolds (1984)
83. *K. Goebel and S. Reich*, Uniform Convexity, Hyperbolic Geometry, and Nonexpansive Mappings (1984)
84. *T. Albu and C. Năstăsescu*, Relative Finiteness in Module Theory (1984)
85. *K. Hrbacek and T. Jech*, Introduction to Set Theory, Second Edition, Revised and Expanded (1984)
86. *F. Van Oystaeyen and A. Verschoren*, Relative Invariants of Rings: The Noncommutative Theory (1984)
87. *B. R. McDonald*, Linear Algebra Over Commutative Rings (1984)
88. *M. Namba*, Geometry of Projective Algebraic Curves (1984)
89. *G. F. Webb*, Theory of Nonlinear Age-Dependent Population Dynamics (1985)
90. *M. R. Bremner, R. V. Moody, and J. Patera*, Tables of Dominant Weight Multiplicities for Representations of Simple Lie Algebras (1985)
91. *A. E. Fekete*, Real Linear Algebra (1985)
92. *S. B. Chae*, Holomorphy and Calculus in Normed Spaces (1985)
93. *A. J. Jerri*, Introduction to Integral Equations with Applications (1985)
94. *G. Karpilovsky*, Projective Representations of Finite Groups (1985)

Other Volumes in Preparation

Holomorphy and Calculus in Normed Spaces

Soo Bong Chae

New College
University of South Florida
Sarasota, Florida

MARCEL DEKKER, Inc.   New York and Basel

Library of Congress Cataloging in Publication Data

Chae, Soo Bong, (date)
    Holomorphy and calculus in normed spaces.

    (Monographs and textbooks in pure and applied
mathematics ; 92)
    Bibliography: p.
    Includes index
    1. Holomorphic mappings. 2. Normed linear
spaces. 3. Calculus, Differential. I. Title
II. Series.
QA331.C439   1985   515   84-25985
ISBN 0-8247-7231-8

MARCEL DEKKER, INC.

270 Madison Avenue, New York, New York 10016

Current printing (last digit):
10 9 8 7 6 5 4 3 2 1

PRINTED IN THE UNITED STATES OF AMERICA

This book is dedicated to

Leopoldo Nachbin

in friendship and admiration

Preface

The study of holomorphic mappings between vector spaces has recently
been the subject of much research by analysts and geometers working
on quite a variety of interesting topics, including foundations,
functional analysis, holomorphic continuation, analytic sets,
symmetric domains, and differential equations. The well-known book
by E. Hille and R. S. Phillips, *Functional Analysis and Semigroups*,
(Amer. Math. Soc., 1955) gives a brief treatment of holomorphy as
developed by M. Zorn in 1945-1946, but since that time the theory
of holomorphic mappings in infinite dimensional spaces (hereafter
referred to as *infinite dimensional holomorphy*) has made vigorous
progress.

   This book is designed to give a thorough exposition of the
elementary theory of holomorphic mappings, taking into account recent
developments, to guide the reader from a rudimentary knowledge of
complex variables to the threshold of infinite dimensional holomorphy,
and also to satisfy the needs of a larger circle of users, in content
as well as didactically.

   To make this book available to a larger group of users I wrote
it at a level that could be accessible, first of all, to advanced
undergraduate students in New College who had successfully completed
the material of an introductory course in metric spaces (or a
rigorous course in real analysis) and complex variables. Thus, the
level of this book could be very comfortable for most graduate

students in mathematics. A rudimentary knowledge of functional analysis, in particular the three principles of functional analysis, Hahn-Banach theorem, open mapping theorem, and the uniform boundedness theorem, is quite helpful for understanding the material, but it is not required of the reader. These fundamental theorems of functional analysis are presented in the first two chapters. The book starts out, whenever possible, from the questions and facts of classical analysis, and tries to enhance the motivation and intuitive understanding of the reader in this abstract theory. The exposition always includes the basic concepts, the essential statements, and the main methods. Results with which it is assumed the reader is familiar are frequently cited without further elaboration; however, in almost all such instances an appropriate reference is given.

The purpose of the book is mainly to present a systematic introduction to the theory of holomorphic mappings in normed spaces which has been scattered throughout the literature. In order to achieve this goal and to present the material coherently, it is necessary to present differential calculus in a modern setting.

Part I of this book provides the necessary, elementary background for all branches of modern mathematics involving differential calculus in higher dimensional spaces; the material in Part I is quite adequate for almost all applications of differential calculus on normed spaces. Part I can be used alone for a course entitled *"Normed Spaces and Calculus,"* and I have given such a course to advanced undergraduates in mathematics and physics as an introduction to functional analysis. One major innovation of my approach to differential calculus on normed spaces is its simplicity of notation; although intrinsically we are studying calculus of infinite variables, my presentation is in the form of calculus of one variable. We also study homogeneous polynomials and general polynomials of order n on normed spaces based on continuous multilinear mappings which are presented in Chapter 3. Polynomials play a central role in the study of holomorphic mappings in Parts II and III. Major theorems in differential calculus are completely presented: Taylor's theorem, mean-value theorem, implicit function theorem, surjection theorem, and Schwarz's symmetric theorem.

Part II begins with a generalization of the classical one-complex-variable theory to vector-valued holomorphic mappings on a complex domain and presents Dunford's theorem showing the equivalence of holomorphy and weak holomorphy; this theorem has been an important tool in the study of Banach algebras and spectral theory. To inform the reader that the theory of vector-valued holomorphic mappings on a complex domain is not a mere generalization of the classical complex analysis, we present Thorp and Whitley's strong maximum modulus theorem in Chapter 10.

Basically, there are two approaches in the theory of holomorphic mappings, one based on convergent power series, and the other based on differentiability. The first is known as the Weierstrass viewpoint, and the second the Cauchy viewpoint. We present both views and show that they are equivalent in Chapter 14. Cauchy integral theorem and Cauchy inequalities are generalized to holomorphic mappings defined on normed spaces. Schwarz's lemma, Cartan's uniqueness theorem, and fixed point theorem of holomorphic mappings are presented in Chapter 13. Some unexpected consequences of infinite dimensional holomorphy are discussed in Chapter 15.

Interplay between functional analysis and theory of holomorphic mappings are well known in finite dimensions. In Part III we study various locally convex topologies on the spaces of holomorphic mappings other than the usual compact open topology; the compact open topology is the most natural topology for finite dimensional cases, but it does not have good topological properties for infinite dimensional cases. The natural topology for the infinite dimensional case turns out to be the Nachbin topology or the Coeuré-Nachbin topology. These topologies are needed in the study of simultaneous holomorphic continuations, i.e., in constructing envelopes of holomorphy and in the study of domains of holomorphy. We also study the Levi problem in Banach spaces. The solution of the Levi problem for the finite dimensional case is considered the highlight of

several complex variables. We complete Part III by presenting the
completeness of the Nachbin topology through interplay between
spaces of holomorphic germs and spaces of holomorphic mappings.

Each chapter ends with a section called *"Exercises"*. Some
exercises are intended to be confidence-builders for beginners,
while others involve theorems not in the text. Routine questions
are imbedded implicitly in the text so that the reader can take an
active part in the flow of the exposition.

To facilitate its use as a reference source and a text, I have
included the original sources in appropriate parts of the text, a
large but certainly not exhaustive bibliography, and an index as
comprehensive as my patience would allow.

The reader will find that the subject is still in a state of
transition and that there are many unanswered problems in the field
to be cultivated. The importance of the study of infinite dimensional holomorphy is addressed by Nachbin and by many others listed
in the bibliography. For a further study of complex analysis in
locally convex spaces, we refer the interested reader to the reference
texts Dineen (1981) and Colombeau (1982). Many nontrivial extensions
of the results in this book to locally convex spaces can be found
in the literature, e.g., Dineen (1981). The historical notes of
A. E. Taylor reprinted in this book are the most comprehensive
account of the development of infinite dimensional holomorphy up to
the mid-1940s; it is reprinted here with the author's permission.
A recent history of the subject is informally given throughout the
text.

Soo Bong Chae

Acknowledgments

Anyone who writes a book of this sort accumulates many outstanding debts which cannot be paid. I owe the greatest debt to Leopoldo Nachbin for seeding in my mind the idea of writing this book and encouraging me throughout the project. I also acknowledge my great debt to Nelson Dunford who read the early version critically and kindly patted my back when I needed it. It is a great privilege to include the historical account written by one of the early pioneers of the subject, A. E. Taylor, as an appendix.

The presentation here has been affected by many students in New College who deserve my thanks for suffering through earlier versions of the manuscript and reviewing my writing critically. These include Rob Gayvert, Tom Schmidt, Dick Canary, and others. I especially recognize Dick for his scrutiny of each statement in the book; I am indeed a lucky one to enjoy these motivated undergraduates in mathematics. I am very grateful to Richard Aron, Yun Sung Choi, and Seán Dineen for reading the final manuscript critically, and to my colleague Mike Frame for proofreading the entire book. Earlier versions of the manuscript were typed by Rob Gayvert, and the final copy was prepared by Phyllis Wren. I would like to thank New College for support throughout the project; Provost Robert Benedetti especially has been most helpful.

Special thanks must go to Sookkyung, Dusan, and Nabin for providing me with their support and for bearing with me through all the hours spent on this.

Contents

PART III.   TOPOLOGIES ON SPACES OF HOLOMORPHIC MAPPINGS

PART I

NORMED SPACES AND DIFFERENTIAL CALCULUS

1

Normed Spaces

1.1. This chapter is devoted to a discussion of a number of elementary facts about normed spaces and Banach spaces. We denote $\mathbb{R}$ and $\mathbb{C}$ the fields of real numbers and complex numbers, respectively. We shall assume that these two fields have their usual metrics and topologies. All of the vector spaces that we consider will be defined over one or the other of these fields. Sometimes it is not necessary to specify over which fields a certain vector space is defined. In that case we shall speak of a vector space over $\mathbb{K}$.

NORMS ON A VECTOR SPACE

1.2 *Norms*. Let E be a vector space over $\mathbb{K}$. A nonnegative real-valued function p on E is said to be a *norm* on E if the following conditions are satisfied:

(a) $p(x + y) \leq p(x) + p(y)$ for all x and y in E;

(b) $p(rx) = |r| p(x)$ for all x in E and all r in $\mathbb{K}$;

(c) $p(x) = 0$ if and only if $x = 0$.

If p satisfies (a) and (b), p is called a *seminorm*. When p is a norm on E, it is customary to denote by $\|x\|$. A *normed space* is a pair $(E, \|\cdot\|)$, where E is a vector space and $\|\cdot\|$ is a norm defined on E. When confusion is unlikely we will denote the normed space $(E, \|\cdot\|)$ by E.

1.3 *Norm Topology*. There is a natural metric associated with a normed space E; we define the distance between any two points x and y of E by

$$d(x,y) = \|x - y\|$$

Then d is a metric defined on E and d generates a metric topology on
E that we call the *norm topology* of E or the topology induced on E
by the norm $\|\cdot\|$.

   We shall use some geometrical terminology with norms. Let a $\in$
E and r > 0. The *open ball* of center a and radius r is the set

$$B_r(a) = \{x \in E: \|x - a\| < r\}$$

and the *closed ball* of center a and radius r is the set

$$\overline{B}_r(a) = \{x \in E: \|x - a\| \leq r\}$$

In particular, $\overline{B}_1(0)$ will be called the *unit ball*.

   A subset U of E is said to be *open* if for any a ε U there exists
r > 0 such that $B_r(a)$ is contained in U. These open sets form the
norm topology on E. Whenever we speak of a convergent sequence in a
normed space E, we mean a sequence of points of E that converges for
the norm topology on E. Similarly, we shall speak of compactness,
connectedness, continuity, etc., on E.

1.4   *Banach Spaces.* A sequence (x_n) in E is called a *Cauchy sequence*
if for any ε > 0 there exists N such that m $\geq$ N and n $\geq$ N imply that
$\|x_m - x_n\| < \varepsilon$. Every convergent sequence is a Cauchy sequence, but
the converse is not true in general.

   A normed space E is said to be *complete* if every Cauchy sequence
in E converges to a point in E; a *Banach space* is a complete normed
space.

EXAMPLES OF NORMED SPACES

In many of the examples below, we shall leave the proof that a certain function is a norm as an exercise.

1.5   *Euclidean n-Space.* Let $\mathbb{K}^n$ be the vector space of all ordered n-tuples $x = (x_1, x_2, \ldots x_n)$ of elements of $\mathbb{K}$. We define a norm on $\mathbb{K}^n$ for each p, $1 \leq p < \infty$, and $p = \infty$, respectively, by

$$\|x\|_p = \sum_{k=1}^{n} |x_k|^p {}^{1/p}$$

$$\|x\|_\infty = \max \{|x_1|, \ldots, |x_n|\}$$

It is easy to see that for each p, $1 \leq p \leq \infty$, $\|\cdot\|_p$ is a norm on $\mathbb{K}^n$. In particular, the norm $\|\cdot\|_2$ is called the *Euclidean norm* on $\mathbb{K}^n$.

For a sequence in $\mathbb{K}^n$ to have a limit $x = (x_1, \ldots, x_n)$, it is necessary and sufficient that for any integer k, $1 \leq k \leq n$, the kth coordinates of the points of the original sequence have the limit x_k. Since $\mathbb{K}$ is complete, by the preceding remark $\mathbb{K}^n$ is also complete; hence it is a Banach space for each of the norms above.

1.6 ℓ^p *Spaces.* Consider the set ℓ^p, $1 \leq p < \infty$, of all sequences $x = (x_n)$ in $\mathbb{K}$ with the property that

$$\sum_{n=1}^{\infty} |x_n|^p < \infty$$

Then ℓ^p is a vector space over $\mathbb{K}$ under coordinatewise addition and scalar multiplication. If we set

$$\|x\|_p = \sum_{n=1}^{\infty} |x_n|^p {}^{1/p}$$

then it becomes a norm on ℓ^p. It is easy to show that ℓ^p with the norm topology is complete; hence it is a Banach space. The norm $\|\cdot\|_p$ is called the ℓ^p-*norm*.

Consider the vector space ℓ^∞ of all bounded sequences $x = (x_n)$ in $\mathbb{K}$ and define a norm as

$$\|x\|_\infty = \sup \{|x_n| : n \in \mathbb{N}\}$$

Then ℓ^∞ is a Banach space.

1.7 *Lebesgue Space* L^p. Let I be an interval in $\mathbb{R}$ and consider the
space $L^p(I)$, $1 \leq p < \infty$, consisting of all scalar valued measurable
functions f defined on I such that

$$\|f\|_p = \left[\int_I |f(x)|^p dx \right]^{1/p}$$

is finite. Then $\|\cdot\|_p$ satisfies all the properties of a norm except
1.2(c). In fact, the equality $\|f\|_p = 0$ does not imply that f is
identically zero; it only implies that f = 0 almost everywhere (i.e.,
with the exception of a set of measure zero). We shall consider two
functions to be equivalent if they are equal almost everywhere; then
if we do not distinguish between equivalent functions, the space
$L^p(I)$ becomes a normed space. (To be pedantic we should say the
elements of $L^p(I)$ are not functions but rather equivalence classes
of functions.) Furthermore, it follows from the theory of Lebesgue
integration that $L^p(I)$ is complete; thus it is a Banach space.

Let $L^\infty(I)$ be the vector space of all scalar valued measurable
functions f defined on I which are bounded almost everywhere. We
define a norm on this space by

$$\|f\|_\infty = \inf\{B: |f(x)| \leq B \text{ almost everywhere on I}\}$$

Then it becomes a Banach space.

1.8 *The Space* $C_b(X;F)$. Let X be a topological space and F a Banach
space over $\mathbb{K}$. A function f: $X \to F$ is said to be *bounded* if

$$\|f\|_X = \sup \{\|f(x)\|: x \in X\}$$

is finite. Then the collection $C_b(X;F)$ of all bounded continuous
functions becomes a vector space over $\mathbb{K}$ with respect to the following
operations: for any f and g in $C_b(X:F)$ and any r in $\mathbb{K}$,

$$(f + g)(x) = f(x) + g(x)$$

$$(rf)(x) = r[f(x)]$$

for all x in X. Then $C_b(X;F)$ is a normed space with respect to the norm $\|f\|_X$. This norm is referred to as the *sup-norm*. Observe that a sequence of continuous functions is convergent for this norm if and only if it is uniformly convergent on X. Hence $C_b(X;F)$ is a Banach space.

FINITE DIMENSIONAL SPACES

1.9 *Equivalent Norms*. It is easy to see that on the vector space $\mathbb{K}^n$, the norms $\|\cdot\|_p$, $1 \leq p \leq \infty$, are mutually distinct. In the study of normed spaces E, we are often concerned with a family of continuous linear maps between E and E. Clearly the set of all such continuous linear maps is determined by the topology on the space and not by some particular norm that induces the topology.

   Two norms p and q on a vector space E are said to be *equivalent* if they induce the same topology. In other words, two norms p and q are equivalent if and only if the identity map $1_E: (E,p) \to (E,q)$ is a homeomorphism. This remark yields easily the following characterization of equivalent norms.

1.10   THEOREM. *Two norms p and q on a vector space E are equivalent if an only if there exist positive numbers a and b such that*

$$ap(x) \leq q(x) \leq bp(x)$$

for all x in E.

1.11 *Finite Dimensional Spaces*. Recall that a vector space E over $\mathbb{K}$ is said to be finite dimensional if E has a finite basis. Choose a basis $\{x_1, x_2, \ldots, x_n\}$ for E for some $n \in \mathbb{N}$. For each x in E then there exists a unique set of scalars $r_1, \ldots, r_n$ such that $x = \Sigma \; r_k x_k$. Now it is easy to see that E can be given a norm. In particular,

$$\|x\|_p = \|(r_1, \ldots, r_n)\|_p$$

are norms on E for each p. It is clear that the coordinate map

$$x \in E \to (r_1, \ldots r_n) \in \mathbb{K}^n$$

is a vector space isomorphism. Furthermore, for each p, $1 \leq p \leq \infty$, the map establishes a homeomorphism between $(E, \|\cdot\|)$ and $(\mathbb{K}^n, \|\cdot\|)$.

1.12   THEOREM. *Any two norms on a finite dimensional vector space are equivalent.*

Proof. It is sufficient to show that every norm p on $\mathbb{K}^n$ is equivalent to the norm $\|\cdot\|_\infty$. Let $\{e_1, \ldots, e_n\}$ be the standard basis for $\mathbb{K}^n$. Then for $x = \Sigma_k x_k e_k$ we obtain

$$p(x) \leq \Sigma_k |x_k| p(e_k) \leq M \|x\|_\infty$$

where $M = \Sigma_k p(e_k)$. This shows one of the desired inequalities and also shows that p is continuous with respect to the norm $\|\cdot\|_\infty$. Since the unit sphere $S = \{x: \|x\|_\infty = 1\}$ is compact (by the Heine-Borel theorem which states that a closed bounded set in $\mathbb{K}^n$ is compact) p takes a minimum on S, say m at y in S, then for any non-zero vector x in E we obtain

$$m \leq p(x/\|x\|_\infty) \qquad \text{or} \qquad m\|x\|_\infty \leq p(x)$$

which establishes the other inequality to conclude that p is indeed equivalent to the norm $\|\cdot\|_\infty$.   □

1.13 *Subspaces.* Let F be a vector subspace of a normed space E. Then the restriction of the norm $\|\cdot\|$ to F is clearly a norm on F. This is called the *subspace norm* on F. Whenever we consider a vector subspace of a normed space we shall always assume, without explicit mention, that it has the subspace norm. When a vector subspace is closed for the topology, we call the subspace a *closed subspace*.

Since a complete subset of a metric space is closed and every finite dimensional space is complete, we conclude that every finite dimensional subspace of a normed space is a closed subspace.

1.14 *Locally Compact Normed Spaces.* A topological space X is said to be *locally compact* if every point has a compact neighborhood. For example, $\mathbb{K}^n$ is locally compact and so is any finite dimensional normed space by the Heine-Borel theorem. It is clear that a normed space is locally compact if and only if the closed unit ball is compact. In fact, if E is locally compact, then the origin O has a compact neighborhood. Therefore, for some $r > 0$, $\overline{B}_r(0)$ is compact; hence $\overline{B}_1(0)$ must be compact since $\overline{B}_r(0) = r\overline{B}_1(0)$.

1.15 THEOREM (F. Riesz). *A normed space is locally compact if and only if it is finite dimensional.*

Proof. Let (E,p) be a locally compact normed space and let B be the closed unit ball of E. Since B is compact, we can find a finite number of points $x_1,\ldots,x_n$ in B such that B is covered by the open balls of radius $1/3$ centered at these points. Let F be the finite dimensional subspace of E generated by these points. We now claim that $F = E$, which obviously shows that E is finite dimensional.

Suppose the contrary and let $x \in E\backslash F$. Since F is closed, the distance $d(x,F)$ between x and F is positive. Recall the formula

$$d(x,F) = \inf \{p(x-y): y \in F\}$$

If we let $a = d(x,F)$, then we can find a point y in F such that

$$a \leq p(x-y) < 2a$$

Denote $z = (x-y)/p(x-y)$. Then z is in B, and hence for some j we have $p(z-x_j) < 1/3$. It follows by simple algebra that

$$x = y + p(x-y)z = y + p(x-y)x_j + p(x-y)(z-x_j)$$

Since $y + p(x-y)x_j$ is in F, we have

$$a \leq p(x-[y+p(x-y)x_j]) = p(x-y)p(z-x_j) < p(x-y)/3$$

Hence, $p(x-y) > 3a$, which contradicts the fact that $p(x-y) < 2a$. □

ABSOLUTELY CONVERGENT SERIES

1.16 *Series.* Let E be a normed space, and let (x_n) be a sequence in E and consider the series

$$\sum_{n=1}^{\infty} x_n$$

For simplicity, we often denote this series by $\sum x_n$ provided that no
confusion can result. We call x_n the nth term of the series, and
the nth partial sum of the series is denoted by

$$S_n = x_1 + \ldots + x_n$$

The series $\sum x_n$ is said to *converge* to a point x is the sequence
(S_n) converges to x.

1.17   THEOREM (Cauchy Criterion). *Let E be a normed space and* (x_n)
a sequence in E. *If the series* $\sum x_n$ *converges, then for any positive
number* ε *there exists a natural number* M *such that if* $n \geq M$ *and*
$p \in \mathbb{N}$ *then*

$$\|S_{n+p} - S_n\| = \|x_{n+1} + \ldots + x_{n+p}\| < \varepsilon$$

The converse is true when E is a Banach space.

The proof of this is omitted since its proof can be a simple
modification of the classical case. For the converse, it is essential to assume that E is complete.

An immediate corollary of the theorem is that if $\sum x_n$ converges,
then $\lim x_n = 0$.

1.18   *Absolutely Convergent Series*. A series $\sum x_n$ in a normed
space E is said to *converge absolutely* if $\sum \|x_n\|$ converges.

THEOREM. *In a Banach space* E, *if* $\sum x_n$ *converges absolutely, then*
$\sum x_n$ *converges and*

$$\|\sum x_n\| \leq \sum \|x_n\|$$

It is essential to assume that E is a Banach space since the Cauchy criterion is required for the proof. We leave the proof to the reader as an exercise.

EXERCISES

1A *Continuity of Vector Operations.* Let E be a normed space. Then

(a)   The addition $(x,y) \in E \times E \to x + y \in E$ is continuous.

(b)   The scalar multiplication $(r,x) \in \mathbb{K} \times E \to rx \in E$ is a homeomorphism if $r \neq 0$.

(c)   The norm $x \in E \to \|x\| \in \mathbb{R}^+$ is continuous.

1B *Closure.* Let E be a normed space. For any subset X of E, let $\overline{X}$ denote the closure of X.

(a)   If F is a vector subspace of E, then $\overline{F}$ is a closed subspace of E.

(b)   If B is any bounded subset of E, then $\overline{B}$ is also bounded.

1C *Sum Sets.* Let E be a normed space and let A and B be subsets of E. We denote by A + B the set of all sums a + b where $(a,b) \in A \times B$.

(a) $B_r(\xi) = \xi + B_r(0)$; $rB_1(0) = B_r(0)$.

(b)   If A is open, then A + B is open for any subset B.

(c)   If both A and B are compact, then A + B is also compact.

(d)   If A is compact and B is closed, then A + B is closed.

(e)   If both A and B are closed and not compact, A + B may not be closed.

(f) $\overline{\xi + A} = \xi + \overline{A}$; $\overline{rA} = r\overline{A}$, $r \neq 0$

(g) $\overline{A} + \overline{B} \subset \overline{A + B}$

1D *Finite Dimensionality.* If a sphere $\{x \in E: \|x\| = r\}$ in a normed space is compact, then the space is finite dimensional.

1E *Complexification.* Let E be a real vector space. The space $E_{\mathbb{C}} = E \times E$, with the coordinatewise addition and scalar multiplication defined by

$$(a + bi)(x,y) = (ax - by, bx + ay)$$

is called the *complexification* of E.

(a) $E_{\mathbb{C}}$ is a vector space over $\mathbb{C}$.

(b) E and E × {0} are algebraically isomorphic.

(c) If E is a normed space, then we can extend the norm on E to a norm on $E_{\mathbb{C}}$ in several ways. But these various norms on $E_{\mathbb{C}}$ are equivalent.

(d) $E_{\mathbb{C}}$ is complete if and only if E is complete.

2

Linear Maps

2.1. Linear maps and linear functionals play a central role in functional analysis and the study of calculus and holomorphy. In this chapter, after defining these concepts and studying the interrelationship between continuity and boundedness, we present the three basic principles of Banach spaces, namely the Hahn-Banach theorem, the open mapping theorem and the uniform boundedness theorem.

CONTINUOUS LINEAR MAPS

2.2 *Linear Maps*. Let E and F be normed spaces over the same field $\mathbb{K}$. A map A: E $\to$ F is called *linear* if

$$A(rx + y) = rA(x) + A(y)$$

for all x, y in E and for all r in $\mathbb{K}$. A linear map may not be continuous in general as shown below.

EXAMPLE. Let E = C(I;$\mathbb{K}$) be the normed space of all continuous $\mathbb{K}$-valued functions on the unit interval I = [0,1] under the sup-norm, and let F = C^1(I;$\mathbb{K}$) be the normed subspace of E consisting of those functions f which have continuous derivative df. Then the linear map f $\to$ df of F into E is not continuous. In fact, if we set $f_n(x)$ = (sin nx)/n, the sequence (f_n) converges to 0, whereas the sequence (df_n) does not converge to 0.

Whenever we work with a linear map between two normed spaces, we assume, without explicit mention, that both spaces are defined over the same field $\mathbb{K}$.

2.3   THEOREM.   *Given* E *and* F *normed spaces and* A: E → F *a linear map,*
the following are equivalent.

(a)   A *is continuous;*
(b)   A *is continuous at* O;
(c)   *There exists* M *such that* $\|A(x)\| \leq M\|x\|$ *for all x in* E;
(d)   A *maps bounded sets onto bounded sets.*

Proof.   The implications (a) → (b)   and (c) ↔ (d) are obvious.

       (b) → (c):   If A is continuous at O, then there exists a closed
ball B in E centered at the origin with radius r such that the image
A(B) is contained in the unit ball of F.   Thus if x ≠ O, then we
have $\|A(rx/\|x\|)\| \leq 1$, and hence

$$\|A(x)\| \leq M\|x\|$$

where M = 1/r.   If x ≠ O, this inequality holds trivially.

       (c) → (a):   If (c) holds, then

$$\|A(x) - A(y)\| \leq M\|x - y\|$$

for every x and y in E; hence A is uniformly continuous on E.   □

2.4   *The Space* L(E;F).   We denote the vector space of all continuous
linear maps between E and F by L(E;F).   Members of this space are
often called by *bounded linear maps* in view of 2.3(c) and 2.3(d).
Given A ∈ L(E;F), we define

$$\|A\| = \inf \{M: \|A(x)\| \leq M\|x\| \text{ for all } x \in E\}$$

By Theorem 2.3, $\|A\|$ is well-defined and

$$\|A(x)\| \leq \|A\|\|x\|$$

for every x in E.   It is easy to show that

$$\|A\| = \sup \{\|A(x)\|: \|x\| \leq 1\} = \sup \{\|A(x)\|: \|x\| = 1\}$$

As defined above, $\|\cdot\|$ makes $L(E;F)$ a normed space. This norm will be called the *usual norm* on $L(E;F)$. If the norms on E and F are replaced by equivalent norms respectively, it is easy to show that the usual norm on $L(E;F)$ is replaced by an equivalent norm. Thus the topology on $L(E;F)$ depends only on the topologies on E and F.

2.5 THEOREM. *If F is a Banach space, then* $L(E;F)$ *is also a Banach space.*

Proof. Let (A_n) be a Cauchy sequence in $L(E;F)$, and let $x \in E$. Then $\|A_m(x) - A_n(x)\| = \|(A_m - A_n)(x)\| \leq \|A_m - A_n\|\|x\|$ so (A_n) is a Cauchy sequence in F. Since F is complete, $(A_n(x))$ converges to an element y in F. If we define $A: E \to F$ by $A(x) = y$, then A is clearly linear. To see that A is continuous, we note that

$$\|A(x)\| = \|\lim_n A_n(x)\| \leq (\sup_n \|A_n\|)\|x\|$$

since (A_n) is a Cauchy sequence, the sup is finite, hence A is continuous by 2.3(c). Now we show that $A_n \to A$ with respect to the usual norm on $L(E;F)$. Let $r > 0$. Choose N such that

$$\|A_m - A_n\| < r$$

whenever $m, n \geq N$. Then

$$\|(A_n - A)(x)\| = \lim_m \|(A_n - A_m)(x)\| < r\|x\|$$

so $\|A_n - A\| \to 0$. $\quad\square$

2.6 *Dual Spaces*. Let E be a normed space over $\mathbb{K}$. Then the Banach space $L(E;\mathbb{K})$ is called the dual space of E or the *conjugate* of E. The dual of E will be denoted by E* and each member of E* will be called a *continuous linear functional* on E. The dual of E* is denoted by E** and is called the *bidual* of E. Clearly, the bidual E** is a Banach space.

2.7 *The Linear Extension Theorem.* Let E be a normed space and F a subspace. Then the closure F^- of F is also a subspace. Indeed, elements in F^- are limits of sequences in F. Thus if

$$x = \lim_n x_n \text{ and } y = \lim_n y_n$$

then

$$x + y = \lim_n (x_n + y_n)$$

$$rx = \lim_n (rx_n)$$

where $r \in \mathbb{K}$. Hence F^- is a subspace of E.

THEOREM. *Let* E *be a normed space,* F *a subspace of* E, *and* G *a Banach space. If* $A: F \to G$ *is a continuous linear map, then there exists a unique extension of* A *to a continuous linear map* $A': F^- \to G$ *such that*

$$\|A'\| = \|A\|$$

Proof. The uniqueness of A' is clear from continuity. We show its existence. Let $x \in F^-$, and let $x = \lim_n x_n$ with x_n in F. Then

$$\|A(x_n) - A(x_m)\| \leq \|A\|\|x_n - x_m\|$$

Hence $(A(x_n))$ is a Cauchy sequence in G, and since G is complete, $(A(x_n))$ has a limit in G which we denote by $A'(x)$. This value $A'(x)$ is well-defined, for if $x = \lim_n y_n$ with $y_n \in F$, then $\lim_n A(x_n) = \lim_n A(y_n)$. Thus we have a well-defined map $A': F^- \to G$. It remains to be shown that A' is a continuous linear map. If $y \in F^-$ and $y = \lim_n y_n$ with $y_n \in F$, then for $r \in \mathbb{K}$,

$$x + y = \lim_n (x_n + y_n) \qquad \text{and} \qquad rx = \lim_n (rx_n)$$

Therefore,

$$A'(x+y) = \lim_n A(x_n+y_n) = \lim_n A(x_n) + \lim_n A(y_n) = A'(x) + A'(y)$$

Similarly,

$$A'(rx) = rA'(x)$$

Hence A' is linear. For $x \in F$, we have $A'(x) = A(x)$ trivially. This shows that A' is a linear extension of A. Finally, we have

$$\|A'(x_n)\| = \lim_n \|A(x_n)\|$$

since the norm is a continuous function. From

$$\|A(x_n)\| \leq \|A\|\|x_n\|$$

it follows that

$$\|A'(x)\| \leq \|A\|\|\lim_n x_n\| = \|A\|\|x\|$$

hence A' is continuous and $\|A'\| \leq \|A\|$. Since $F \subset F^-$ and $A' = A$ on F, we have easily $\|A\| \leq \|A'\|$, which shows that $\|A'\| = \|A\|$.   □

THE HAHN-BANACH THEOREM

2.8. An important part of the study of a normed space E is the investigation of the dual space E* of all continuous linear functionals on E. The Hahn-Banach theorem is, together with the uniform boundedness and open mapping theorems, one of the most important theorems of functional analysis. The theorem assures us that there are always plenty of continuous linear functionals on any normed space. The proof of this depends on Zorn's lemma or its equivalent forms.

ZORN'S LEMMA. *If every totally ordered subset of a partially*
ordered set P *has an upper bound, then* P *has a maximal element.*

2.9 THEOREM (Hahn-Banach). *Let* E *be a vector space over* $\mathbb{K}$ *,* p *a*
seminorm on E, *and* F *a vector subspace of* E. *If* f *is a linear*
functional on F *such that* $\left|f(x)\right| \leq p(x)$ *for all* x $\in$ F, *then there*
exists a linear functional f' *on* E *such that*

   (a)   f´ = f *on* F;
   (b) $\left|f'(x)\right| \leq p(x)$ *for all* x $\in$ E.

We use the following lemma without giving a proof.

LEMMA. *Let* E *be a complex vector space,* f *a linear functional on* E,
and g = Re(f). *Then*

(a)   g *is a real linear functional on* $E_{\mathbb{R}}$, *the space* E *considered as*
      a vector space over $\mathbb{R}$ *instead of* $\mathbb{C}$.

(b)   f(x) = g(x) - ig(ix).

(c)   *If* p *is a seminorm on* E, *then*

$$\sup\{\left|f(x)\right|: p(x) \leq 1\} = \sup\{\left|g(x)\right|: p(x) \leq 1\}$$

Proof. By the preceding lemma, it is sufficient to consider the case
$\mathbb{K} = \mathbb{R}$. In fact, if $\mathbb{K} = \mathbb{C}$, we extend g = Re(f) to g´ on $E_{\mathbb{R}}$, then
by (b) of the lemma, we obtain f´ on E. Since $\left|g´(x)\right| \leq p(x)$ for all
x $\in$ E, $\sup\{\left|g´(x)\right|: p(x) \leq 1\} \leq 1$; hence, $\sup\{\left|f´(x)\right|: p(x) \leq 1\} \leq$
1 by (c). This shows that $\left|f´(x)\right| \leq p(x)$ for all x $\in$ E.

   Let us call (g,G) an *extension* of f if

   (1)   F $\subset$ G, G is a vector subspace of E;

   (2)   f = g on F;

   (3) $\left|g(x)\right| \leq p(x)$ for all x $\in$ G.

Let P be the family of all extensions of f. Clearly P is a non-
empty set since (f,F) $\in$ P. If (g_1, G_1) and (g_2, G_2) are in P, we
define $(g_1, G_1) \leq (g_2, G_2)$ if $G_1 \subset G_2$ and $g_1 = g_2$ on G_1. Let $\{(g_\alpha, G_\alpha)\}$
be a totally ordered subset of the partially ordered set P. Then
$\{(g_\alpha, G_\alpha)\}$ is bounded above by an extension (g,G), where $G = \cup\, G_\alpha$ and

$g = g_\alpha$ on each G_α. Consequently, by Zorn's lemma, P has a maximal element which we call (h,M).

To complete the proof, we must show that M = E. If this were not the case, choose $x \in E \backslash M$ and let

$$L = \{m + ax: m \in M \text{ and } a \in \mathbb{R}\}$$

Then L is clearly a vector subspace of E. To extend h to a linear functional h' on L, we look for a suitably chosen real number b satisfying the following equation:

$$h'(m + ax) = h(m) + ab \tag{i}$$

If this were true, then h' is clearly a linear functional on L satisfying $h' = h$ on M. The main problem is how to choose such a number b so that

$$\left| h'(x) \right| \leq p(x) \tag{ii}$$

for all $x \in L$. Suppose that h' satisfies the relations (i) and (ii). Then it is clear that (h',L) is a nontrivial extension of h, which contradicts the maximality of (h,M).

For a moment let us assume that we found such a number b. Then

$$\left| h(m) + ab \right| \leq p(m + ax)$$
$$\left| h(m/a) + b \right| \leq p(m/a + x) \qquad (a \neq 0)$$
$$\left| h(m) + b \right| \leq p(m + x)$$
$$-p(m + x) - h(m) \leq b \leq p(m + x) - h(m) \tag{iii}$$

for any m in M. The last relation suggests us how to choose such a number. In fact, if m and m' are arbitrary members of M, we have

$$-p(m+x) - h(m) = -p(m+x) - h(m') + h(m') - h(m) = -p(m+x) - h(m')$$
$$+ h(m'-m) \leq -p(m+x) - h(m') + p(m'-m) \leq p(m'+x) + p[-(m+x)]$$
$$- p(m+x) - h(m') = p(m +x) - h(m')$$

This shows that the number b must be an upper bound of the nonempty set A and a lower bound of the nonempty set B defined by

$$A = \{-p(m + x) - h(m) : m \in M\}$$
$$B = \{p(m + x) - h(m) : m \in M\}$$

Therefore, we can choose a number b in the interval [sup A, inf B]. This completes the proof.   □

2.10 THEOREM (Hahn). *Let F be a subspace of a normed space E and let f $\in$ F*. Then f can be extended to f´ $\in$ E* such that*

$$\|f´\| = \|f\| \ .$$

Proof. There is no loss of generality in assuming that $\|f\| = 1$ so that $|f(x)| \leq \|x\|$ for all $x \in F$. Then Theorem 2.9 applied to f and the norm $\|\cdot\|$ on E yields the desired extension f´.   □

2.11 COROLLARY. *For any nonzero x in a normed space E, there exists f $\in$ E* such that $\|f\| = 1$ and $f(x) = \|x\|$.*

Proof. Let F be the one dimensional subspace generated by x and define $g: F \to \mathbb{K}$ to be

$$g(rx) = r\|x\|$$

Then we easily obtain $g \in F^*$ and $\|g\| = 1$. It now follows from the preceding theorem that there exists $f \in E^*$ such that $\|f\| = 1$ and $f = g$ on F; i.e., $f(x) = \|x\|$.   □

2.12 COROLLARY. *For every x in a normed space E,*

$$\|x\| = \sup \{|f(x)| : f \in E^*, \|f\| \leq 1\}$$

Proof. This is an immediate consequence of the preceding corollary.   □

2.13 *The Completion of a Normed Space.* The bidual E** of a normed space E is a Banach space by 2.6. For each fixed x ∈ E define

$$x^{\wedge}(f) = f(x)$$

for all f ∈ E*. Then x^ is clearly a linear functional on E* and

$$\left| x^{\wedge}(f) \right| = \left| f(x) \right| \leq \|f\| \cdot \|x\|$$

and hence x^ ∈ E**. Thus we can define a map Φ: E → E** by letting Φ(x) = x^ for each x in E. Then Φ is a linear isometry; i.e.,

$$\|x^{\wedge}\| = \|x\|$$

for all x ∈ E by Corollary 2.12. We shall call Φ the *natural embedding* of E in E**. Then E becomes a dense subspace of $\overline{\Phi(E)}$ in E**. Since E** is a Banach space, $\overline{\Phi(E)}$ is also a Banach space. Therefore, we have the following theorem.

THEOREM. *Every normed space is isometrically isomorphic to a dense subspace of a Banach space.*

OPEN MAPPING THEOREM

2.14. We turn our attention to another fundamental theorem of functional analysis, the open mapping theorem. The proof of this theorem is an application of the Baire category theorem, which we now recall.

THEOREM (Baire). *A complete metric space cannot be represented as the countable union of closed subsets all having empty interiors.*

2.15 THEOREM (Banach Open Mapping Theorem). *Let E and F be Banach spaces. If A ∈ L(E;F) is a surjection, then A is an open mapping.*

Proof. Let $B_n = \{x \in E: \|x\| < 1/2^n\}$, and let $S_n = A(B_n)$. Since $E = \bigcup_{n \geq 1} nB_1$ and A is a surjection, $F = \bigcup_{n \geq 1} nS_1 = \bigcup_{n \geq 1} \overline{nS_1}$. By the

Baire category theorem, there is an integer k such that $\overline{kS_1}$ has non-empty interior. Thus for some point ξ in F and some a > 0, $D_a(\xi) \subset \overline{kS_1}$, where $D_a(\xi)$ is the open ball in F centered at ξ with radius a. Since the mapping $y \to ky$ is a homeomorphism, $\overline{S_1}$ contains an open ball $D_r(\eta)$ for some r > 0 and $\eta \in \overline{S_1}$. Then

$$D_r(0) \subset \overline{S_1} - \eta \subset \overline{S_1} - \overline{S_1} \subset 2\overline{S_1} = \overline{S_o}$$

hence $D_r(0) \subset \overline{S_o}$ and $D_{r/2}(0) \subset \overline{S_1}$. Inductively, we obtain the following inclusion:

$$D_{\rho(n)}(0) \subset \overline{S_n} \qquad \text{for all n} \qquad\qquad\qquad (i)$$

where $\rho(n) = r/2^n$. Now let $y \in F$ be such that $\|y\| < r/2$. Since y $\in \overline{S_1}$ by (i), there exists $x_1 \in B_1$ with

$$\|y - A(x_1)\| < \rho(2)$$

Then $y - A(x_1) \in \overline{S_2}$, so there exists $x_2 \in B_2$ such that

$$\|y - A(x_1) - A(x_2)\| < \rho(3)$$

Inductively, we can find $x_n \in B_n$ such that

$$\|y - A(x_1) - \cdots - A(x_n)\| < \rho(n+1)$$

Since $\Sigma \|x_n\| < \Sigma 1/2^n = 1$, by 1.18, Σx_n converges to a point in B_o and $A(\Sigma x_n) = \Sigma A(x_n) = y$. Thus $y \in S_o$, which shows that S_o contains $D_{r/2}(0)$. Similarly, we have

$$D_{\rho(n+1)}(0) \subset S_n \qquad\qquad\qquad\qquad (ii)$$

We now show that A is an open mapping. Let U be an open subset of E, x an arbitrary member of U, and y = A(x). Then for some n,

$x + B_n \subseteq U$; hence $y + S_n \subseteq A(U)$. Since S_n contains a neighborhood of O by (ii), $A(U)$ is a neighborhood of y; hence $A(U)$ is open. □

2.16 COROLLARY. *Let E and F be Banach spaces and let* $A \in L(E;F)$ *be a bijection. Then A is a homeomorphism, and hence A is a topological isomorphism.*

Proof. The inverse of A is continuous, so A is a homeomorphism. □

UNIFORM BOUNDEDNESS PRINCIPLE

2.17. If a subset B of $L(E;F)$ is bounded (i.e., $\sup\{\|A\|: A \in B\} \leq M$ for some $M > 0$), then for each $x \in E$ there exists $M_x > 0$ such that

$$\sup\{\|A(x)\|: A \in B\} \leq M_x$$

The surprising fact is that the converse is also true if E is a Banach space. This result is known as the principle of uniform boundedness or the Banach-Steinhaus theorem.

THEOREM (Banach-Steinhaus). *Let E be a Banach space and let F be a normed space. Given a family* $\{A_k\}_{k \in I}$ *in* $L(E;F)$, *the following are equivalent:*

(a) $\{A_k\}_{k \in I}$ *is an equicontinuous family; i.e.,*
   for any $\varepsilon > 0$ *there exists* $\delta > 0$ *such that* $\|x\| < \delta$ *implies* $\|A_k(x)\| < \varepsilon$ *for all* $k \in I$.

(b) *For each* $x \in E$, *there exists* $M_x > 0$ *such that* $\|A_k(x)\| \leq M_x$ *for all* $k \in I$.

(c) *There exists* $M > 0$ *such that* $\|A_k\| \leq M$ *for all* $k \in I$.

Proof. (a) → (b): Assuming the statement (a), we can find a $\delta > 0$ such that $\|x\| < \delta$ implies $\|A_k(x)\| < 1$ for all k in I. If $x \neq O$, then

$$\|A_k(\delta x/\|x\|)\| \leq 1$$

$$\|A_k(x)\| \leq \|x\|/\delta = M_x$$

for all $k \in I$.

(b) $\rightarrow$ (c): For each natural number n, let

$$C_n = \{x \in E: \|A_k(x)\| \leq n \text{ for all } k \in I\}$$

Since each A_k is continuous, it follows that each C_n is closed. By (b), we have $E = \bigcup_{n \geq 1} C_n$. Then the Baire category theorem ensures that some C_n contains a closed ball $\bar{B}_r(\xi)$. We then have

$$\|A_k(x)\| \leq n$$

for all $x \in \bar{B}_r(\xi)$ and $k \in I$. This implies that if $\|x\| \leq r$, then

$$\|A_k(x)\| \leq \|A_k(x + \xi)\| + \|A_k(\xi)\| \leq 2n$$

for all $k \in I$. Therefore, $\|A_k\| \leq 2n/r = M$ for all k.

(c) $\rightarrow$ (a): Let $\varepsilon > 0$ be given and let $\delta = \varepsilon/M$. Then for $\|x\| < \delta$, we have $\|A_k(x)\| \leq M\|x\| < \varepsilon$ for all k, so the family $\{A_k\}_{k \in I}$ is equicontinuous.   □

2.18   THEOREM. *A subset* B *of a normed space* E *is bounded if* f(B) *is bounded for all* $f \in E^*$.

Proof. Consider the set $B^\wedge = \{x^\wedge \in E^{**}: x \in B\}$, where $x^\wedge$ was introduced in (2.13). Then for each $f \in E^*$,

$$\sup \{|x^\wedge(f)|: x \in B\} = \sup \{\|f(x)\|: x \in B\} = M_f < \infty$$

This shows that the set $B^\wedge$ is pointwisely bounded, hence by the Banach-Steinhaus Theorem 2.17 and from the fact

$$\|x\| = \|x^\wedge\|$$

we have

$$\sup \{\|x^\| : x^ \in B^\} = \sup \{\|x\| : x \in B\} < \infty$$

Thus B is a bounded subset of E.   □

2.19 THEOREM. *A linear map* A: E → F *is continuous if for every*
f ∈ F*, *we have the composite map* f o A *in* E*.

Proof. Suppose that A is not continuous, although f o A is continuous for all f ∈ F*. Then there exists a sequence (x_n) in E such
that x_n → O and $\|A(x_n)\|$ > n for all n. Since $(f[A(x_n)])$ is a bounded
sequence for all f ∈ F*, so is the sequence $(A(x_n))$ by the preceding
theorem. This is a contradiction, and hence A must be
continuous.   □

EXERCISES

2A *Additive Map* . Let E and F be normed spaces over ℝ. Let f:
E → F be such that

    (a)   f(x + y) = f(x) + f(y) for all x and y in E;

    (b)   f is bounded on the unit ball of E.

Then f ∈ L(E;F).

2B *Linear Maps on* $\mathbb{K}^n$. Every linear map of a finite dimensional
normed space into another normed space is continuous.

2C *Norm on* L(E;F). Let A ∈ L(E;F). Then

$$\|A\| = \sup \{\|A(x)\| : \|x\| \leq 1\} = \sup \{\|A(x)\| : \|x\| = 1\}$$

2D *Composition of Linear Maps.* Let E, F, and G be normed spaces,
and let f ∈ L(E;F) and g ∈ L(F;G). Then g o f ∈(E;G) and

$$\|g \, o \, f\| \leq \|g\| \cdot \|f\|$$

2E *Exponential Function.*   If $f \in L(E;F)$, we define $f^n = f \circ f^{n-1}$
inductively for each natural number n, and f^o is identified with the
identity map 1_E on E.   If E is a Banach space, then

$$\sum_{n=o}^{\infty} \frac{f^n}{n!}$$

converges absolutely.   Its sum is denoted by Exp (f).

2F *Linear Functional.*   A linear functional $f: E \to \mathbb{K}$ is continuous
if and only if the kernel $f^{-1}(0)$ is a closed subspace of E.

2G *Algebraic Dimension of a Banach Space.*

(a) If E is a normed space, then the only subspace of E having non-
     empty interior is E itself.

(b) The vector space dimension of a Banach space is either finite
     or uncountable.

2H *Norm.*   Let E be a normed space and $x \in E$.   Then

$$\|x^{\wedge}\| = \sup \{ |f(x)| : f \in E^*, \|f\| \le 1 \} = \|x\|$$

where $x^{\wedge} \in E^{**}$ is such that $x^{\wedge}(f) = f(x)$.

2I *Complementary Subspaces.*

(a) Let F and G be closed subspaces of E such that F + G = E and
     $F \cap G = \{o\}$.   If E is a Banach space, then the map $(x,y) \in F \times G$
     $\to x + y \in E$ is a topological isomorphism; i.e., a vector space
     isomorphism which is also a homeomorphism.   In this case, we
     say that F and G are *complementary subspaces* of E.

(b) Let E be a Banach space a finite dimensional subspace.
     Then there exists a closed subspace G such that F + G = E and
     $F \cap G = \{o\}$.   (Hint: Use the Hahn-Banach Theorem.)

2J *Closed Graph Theorem.*   Let E and F be Banach spaces and let A be
a linear map from E to F.   Then A is continuous if and only if the
graph $\{(x,A(x)): x \in E\}$ of A is closed as a subset of E × F endowed
with the product topology.   (Apply the open mapping theorem.)

2K *Banach-Steinhaus Theorem.*   Let E be a Banach space and let F be
a normed space.   If (A_n) is a sequence in L(E;F) converging to A
pointwisely, then $A \in L(E;F)$.

2L   *Dual of* ℓ^p.   If $1 < p$, $q < \infty$ and $1/p + 1/q = 1$, then there is an isometric isomorphism between $(\ell^p)^*$ and ℓ^q in which corresponding vectors y^* and $y = (y_j)$ are related by the identity

$$y^*(x) = \sum_{j=1}^{\infty} x_j y_j$$

where $x = (x_j) \in \ell^p$. Hence, we can write $(\ell^p)^* = \ell^q$.

2M   *Dual of* ℓ^1.   The dual of ℓ^1 can be identified with ℓ^∞ by the identity in 2L with x in ℓ^1 and y in ℓ^∞; i.e., $(\ell^1)^* = \ell^\infty$.

2N   *Dual of* c_o.   Let c_o be the subspace of ℓ^∞ consisting of all sequences converging to 0. Then c_o is a Banach space. The identity in 2L gives the general form of linear functional on c_o provided that $y = (y_j) \in \ell^1$; that is, $(c_o)^* = \ell^1$.

2O   If the dual E^* of a normed space E is separable, so is E.

2P   *Reflexive Spaces*.   A Banach space E is said to be *reflexive* if the natural embedding $\Phi: E \to E^{**}$ (see 2.13) is onto. The ℓ^p spaces are reflexive by 2L; but c_o is not reflexive.

(a)   A closed subspace of a reflexive Banach space is reflexive.

(b)   A Banach space E is reflexive if and only if E^* is reflexive.

2Q   *Weak Convergence*.   A sequence (x_n) in a normed space is said to be *weakly convergent* if there is an x in E with $A(x) = \lim A(x_n)$ for every A in E^*. If the sequence $(A(x_n))$ is a Cauchy sequence for every A in E^*, the (x_n) is called a *weak Cauchy sequence*. The space E is said to be *weakly complete* if every weak Cauchy sequence converges weakly. A set K in E is said to be weakly sequentially compact if every sequence (x_n) in K contains a subsequence which converges weakly to a point in E.

(a)   In a reflexive normed space every bounded set is weakly sequentially compact.

(b)   A reflexive normed space is weakly complete.   (See Dunford and Schwartz (1964), pp. 68-69.)

3

Multilinear Maps

3.1. Multilinear maps play a major role in our study of differentiable functions and holomorphic mappings. In this chapter we present the basic facts about multilinear maps, and give several necessary and sufficient conditions for a multilinear map to be continuous. We also introduce the normed space of all continuous multilinear maps.

CONTINUOUS MULTILINEAR MAPS

3.2 *Product of Normed Spaces.* Let $E_1, \ldots E_m$ be a finite family of normed spaces over the same field $\mathbb{K}$, whose norms we shall uniformly denote by $\|\cdot\|$, and let E be their Cartesian product. Then E is a vector space over $\mathbb{K}$ with coordinatewise addition and multiplication. Then the norms

$$\| (x_1, \ldots, x_m) \|_\infty = \sup \|x_i\|$$

$$\| (x_1, \ldots, x_m) \|_p = \left[\Sigma \|x_i\|^p \right]^{1/p} \qquad (1 \leq p < \infty)$$

are equivalent and each of these induces the product topology on E. Any of these norms will be considered as a natural norm on E. If each E_i is a Banach space, so is the product E.

3.3 *Multilinear Maps.* If $E_1, \ldots, E_m$ and F are normed spaces over the same field $\mathbb{K}$, a map

$$A: \ E_1 \times \cdots \times E_m \rightarrow F$$

is said to be *multilinear* if it is linear in each variable separately.
Of course, it does not mean that A is linear on the product vector
space. More precisely, A is multilinear if each partial map

$$x_i \to A(x_1,\ldots,x_i,\ldots,x_m)$$

is linear map of E_i into F. If $m = 2$, a multilinear map is called
bilinear. In general, a multilinear map on the product of m normed
spaces is called an *m-linear map*.

Many results which are true for linear maps can be extended
easily to multilinear maps.

3.4 THEOREM. *Let* A: $E_1 \times \cdots \times E_m \to F$ *be a multilinear map. Then the
following are equivalent:*

(a) A is continuous;

(b) A is continuous at the origin $(0,\ldots,0)$;

(c) There exists M > 0 such that for every $(x_1,\ldots,x_m) \in E_1 \times \cdots \times E_m$,

$$\|A(x_1,\ldots,x_m)\| \leq M\|x_1\|\cdots\|x_m\|$$

Proof. In this proof we use the norm $\|x\| = \Sigma \|x_i\|$ on the product
where $x = (x_1,\ldots,x_m)$. It is obvious that (a) implies (b).
(b) $\to$ (c): Suppose that A is continuous at the origin. Then there
exists a number r such that $\|x\| \leq r$ implies that $\|A(x)\| \leq 1$. The
inequality in (c) trivially holds if some $x_i = 0$. For any vector $x =
(x_1,\ldots,x_m)$, if each $x_i \neq 0$, set $z_i = rx_i/\|x_i\|$.
Then we have

$$\|A(z_1,\ldots,z_m)\| \leq m^m$$

But

$$\|A(z_1,\ldots z_m)\| = r^m A(x_1,\ldots,x_m)/\|x_1\|\cdots\|x_m\|$$

and hence

$$\|A(x_1,\ldots,x_m)\| \leq M\|x_1\|\cdots\|x_m\|$$

where $M = m^m/r^m$.

(c) $\rightarrow$ (a): To show the continuity of A at any point $a = (a_1, \ldots, a_m)$, we consider

$$A(x) - A(a) = A(x_1 - a_1, x_2, \ldots, x_m) + A(a_1, x_2 - a_2, x_3, \ldots, x_m)$$

$$+ \cdots + A(a_1, \ldots, a_{m-1}, x_m - a_m)$$

Then if $\|x - a\| \leq 1$,

$$\|A(x) - A(a)\| \leq M(\|x_1 - a_1\|\|x_2\|\cdots\|x_m\| + \|a_1\|\|x_2 - a_2\|\|x_3\|\cdots\|x_m\|$$

$$+ \cdots + \|a_1\|\cdots\|a_{m-1}\|\|x_m - a_m\|) \leq MK^{m-1}\|x-a\|$$

with $K = \|a\| + 1$; this proves that A is continuous. $\square$

3.5   THEOREM. *Let* $E_1, \ldots, E_m$ *be Banach spaces and let F be a normed space. Then a multilinear map* A: $E_1 \times \ldots \times E_m \rightarrow F$ *is continuous if and only if A is continuous with respect to each variable separately.*

Proof. We use the Banach-Steinhaus Theorem 2.16. For simplicity we take $m = 2$ since the general case can be proved by induction.

Let A: $E_1 \times E_2 \rightarrow F$ be a separately continuous bilinear map. For each y in E_2, the mapping $x \in E_1 \rightarrow A(x,y) \in F$ is a continuous linear map, and hence there exists a constant $M(y)$ such that

$$\|A(x,y)\| \leq M(y)\|x\|$$

By the symmetry, we also have

$$\|A(x,y)\| \leq M(x)\|y\|$$

In particular, if $\|y\| \leq 1$,

$$\|A(x,y)\| \leq M(x)$$

Therefore, the linear maps $x \in E_1 \rightarrow A(x,y) \in F$ for all y, $\|y\| \leq 1$,

satisfies the condition (b) of the uniform boundedness principle
2.16.   Thus there exists M > 0 such that

$$\sup \|A(x,y)\| \leq M$$

for $\|x\| \leq 1$ and $\|y\| \leq 1$; that is $\|A\| \leq M$.   □

3.6   COROLLARY.   *Every multilinear map of a product of finite dimen-*
sional normed spaces into a normed space is continuous.

Proof.   Since every linear map on a finite dimensional space is continuous, the corollary follows from the preceding theorem.   □

3.7   THEOREM.   *A multilinear map* A: E$_1$×...× E$_m$ → F *is continuous if*
and only if for every f ∈ F*, f o A *is continuous.*

Proof.   Suppose that A is not continuous.   Then there exists a se-
quence (a$_k$) in E$_1$ ×...× E$_m$ satisfying $\|a_k\|$ → 0 and $\|A(a_k)\|$ > k for
all k (Why?).   If f o A is continuous for all f ∈ F*, then the
sequence (f[A(a$_k$)]) is bounded for every f in F*, so the sequence
(A(a$_k$)) must be bounded by Theorem 2.17.   This contradicts the fact
that $\|A(a_k)\|$ > k; thus A is continuous.   □

THE NORMED SPACE L(E$_1$,...,E$_m$;F)

3.8   We denote by L(E$_1$,...E$_m$;F) the set of all continuous multilinear
maps of E$_1$×...× E$_m$ into F.   Then L(E$_1$,...E$_m$;F) becomes a vector space
under the pointwise addition and scalar multiplication.

For any A ∈ L(E$_1$,...,E$_m$;F), define

$$\|A\| = \sup \{\|A(x)\|: \|x\|_\infty \leq 1\}$$

Then by 3.4(c),

$$\|A(x)\| \leq \|A\|\|x_1\|...\|x_m\|$$

and $\|A\|$ is the smallest M > 0 satisfying the inequality 3.4(c).   It

is a routine exercise to verify that $\|A\|$ is a norm on the vector
space $L(E_1,\ldots,E_m;F)$.

If we are not concerned about continuity of multilinear maps, we
write the vector space of all (algebraic) multilinear mappings from
$E_1\times\ldots\times E_m$ to F by $L_a(E_1,\ldots,E_m;F)$, where the subscript 'a' indicates
that the mappings in the space are algebraic and not necessarily
continuous.

When $E = E_1 = \ldots = E_m$, we write

$$L(^mE;F) = L(E_1,\ldots,E_m;F)$$

$$L_a(^mE;F) = L_a(E_1,\ldots,E_m;F)$$

3.9 THEOREM. *If F is a Banach space, then* $L(E_1,\ldots,E_m;F)$ *is a*
Banach space for any normed spaces $E_1,\ldots,E_m$.

Proof. A proof modeled on that of Theorem 2.5 shows this result. □

3.10 *Examples of Bilinear Maps.*

(a) The inner product of $\mathbb{R}^p$, the multiplication of complex numbers,
and the cross product in $\mathbb{R}^3$ are all continuous bilinear maps.

(b) Let E, F and G be normed spaces. Then the map

$$\Phi: L(E;F) \times L(F;G) \to L(E;G)$$

defined by $\Phi(f,g) = g \circ f$ is bilinear, and furthermore, it is contin-
uous since $\|g \circ f\| \leq \|g\|\|f\|$ (see Exercise 2D). If $\|f\| \leq 1$ and $\|g\|$
≤ 1, then $\|g \circ f\| \leq 1$, hence $\|\Phi\| \leq 1$.

(c) If $A = (a_{ij})$ is an n×n matrix of numbers in $\mathbb{K}$, then A gives
rise to a continuous bilinear map $f_A: \mathbb{K}^n \times \mathbb{K}^n \to \mathbb{K}$ by the formula

$$f_A(x,y) = {}^txAy$$

where x,y are column vectors, and ${}^tx = (x_1,\ldots x_n)$ is the row vector,
the transpose of x. Conversely for any bilinear map $f: \mathbb{K}^n \times \mathbb{K}^n \to \mathbb{K}$
there exists an n×n matrix A such that $f = f_A$. [For a given n×n

matrix A, it is clear that f_A is bilinear. Conversely, let $f: \mathbb{K}^n \times \mathbb{K}^n \to \mathbb{K}$ be a bilinear map, and let $e_1,\ldots,e_n$ be the standard basis for $\mathbb{K}^n$. Then

$$f(x,y) = \Sigma\ x_i y_j f(e_i,e_j)$$

where Σ is taken for $i,j = 1,\ldots,n$. Let $a_{ij} = f(e_i,e_j)$ and $A = (a_{ij})$. Then

$$f(x,y) = \Sigma\ a_{ij}x_i y_j = {}^t xAy$$

where ${}^t x = (x_1,\ldots,x_n).]$

(d) Let $p_i: \mathbb{K}^n \to \mathbb{K}$ be the coordinate map defined by $p_i(x) = x_i$. Define $p_i \otimes p_j: \mathbb{K}^n \times \mathbb{K}^n \to \mathbb{K}$ by

$$p_i \otimes p_j(x,y) = x_i y_j$$

Then a bilinear map $f: \mathbb{K}^n \times \mathbb{K}^n \to \mathbb{K}$ can be written uniquely in the form

$$f = \Sigma\ a_{ij}p_i \otimes p_j$$

with some a_{ij} in $\mathbb{K}$. Therefore, the bilinear maps $p_i \otimes p_j$, $i,j = 1,\ldots,n$, form a basis for the space $L(^2 E;\mathbb{K})$ where $E = \mathbb{K}^n$.

3.11 *Natural Isometry.* $L(E_1,\ldots,E_m;F) \cong L(E_1;L(E_2,\ldots,E_m;F))$. In the differential calculus and other applications, we often need to identify $L(E_1,\ldots,E_m;F)$ with $L(E_1;L(E_2,\ldots,E_m;F))$ under an isometric isomorphism. For simplicity, we discuss here such an isometry between $L(E,F;G)$ and $L(E;L(F;G))$ since the general case can be studied exactly as in the bilinear case.

THEOREM. *Let* $\Phi: L(E,F;G) \to L(E;L(F;G))$ *be defined by*

$$\Phi(A)(x)(y) = A(x,y)$$

where $A \in L(E,F;G)$, $x \in E$, and $y \in F$. Then Φ is an isomorphic isometry; i.e., Φ preserves the norm.

Proof.   Clearly Φ is linear, and $\|\Phi(A)(x)(y)\| \leq \|A\|\|x\|\|y\|$. Hence for each x in E, $\Phi(A)(x) \in L(F;G)$ and $\|\Phi(A)(x)\| \leq \|A\|\|x\|$; thus $\Phi(A) \in L(E;L(F;G))$ and $\|\Phi(A)\| \leq \|A\|$. Therefore, Φ is a continuous linear map with $\|\Phi\| \leq 1$.

On the other hand, let $\Psi\colon L(E;L(F;G)) \to L(E,F;G)$ be such that

$$\Psi(B)(x,y) = B(x)(y)$$

Then Ψ is linear and $\Psi(B)$ is bilinear for each B in $L(E;L(F;G))$. Since $\|\Psi(B)(x,y)\| \leq \|B\|\|x\|\|y\|$, $\Psi(B) \in L(E,F;G)$ and $\|\Psi(B)\| \leq \|B\|$; hence $\|\Psi\| \leq 1$ and Ψ is continuous. It is clear that Ψ and Φ are inverses of each other; thus

$$1 = \|\Psi \circ \Phi\| \leq \|\Psi\|\|\Phi\|$$

and we conclude that $\|\Psi\| = \|\Phi\| = 1$ since $\|\Psi\| \leq 1$ and $\|\Phi\| \leq 1$. This shows that Φ is an isomorphic isometry.   □

3.12   THEOREM.   *There exists an isomorphic isometry*

$$L(E_1,\ldots,E_m;F) \simeq L(E_1;L(E_2;L(E_3;\ldots;L(E_m;F)\ldots)))$$

from the space of continuous multilinear maps to the space of m times repeatedly continuous linear maps.

EXERCISES

3A   *Norms on* $L(E_1,\ldots,E_m;F)$.   Let $A \in L(E_1,\ldots,E_m;F)$.   Then

$$\|A\| = \sup \{\|A(x)\|\colon \|x\| = \sup \|x_i\| \leq 1\}$$

$$= \inf \{M\colon \|A(x)\| \leq M\|x_1\|\ldots\|x_m\|,\ x = (x_1,\ldots,x_m)\}$$

and $\|\cdot\|$ is a norm on $L(E_1,\ldots,E_m;F)$.

3B *The m-Linear Maps on* $\mathbb{K}^p$. Let $E = \mathbb{K}^p$ and $A \in L(^mE;F)$. Then

$$A(x^1,\ldots,x^m) = \Sigma \; x^1_{j_1} \ldots x^m_{j_m} A(e_{j_1},\ldots,e_{j_m})$$

where $j = 1,\ldots,p$, $k = 1,\ldots,m$, $x^k = (x^k_1,\ldots,x^k_p)$, and $e_1 = (1,0,\ldots,0)$, $\ldots$, $e_p = (0,\ldots,0,1)$.

3C (a) If $A \in L(^mE;F)$ and $\Phi \in L(F;G)$, then $\Phi \circ A \in L(^mE;G)$ and $\|\Phi \circ A\| \leq \|\Phi\|\|A\|$

(b) For each $\Phi \in L(F;G)$, let $\Psi: L(^mE;F) \to L(^mE;G)$ be defined by $\Psi(A) = \Phi \circ A$. Then Ψ is a continuous linear map and $\|\Psi\| \leq \|\Phi\|$.

3D *Product of Multilinear Maps.* Let $A: E^m \to F$ and $B: E^n \to G$ be multilinear maps and let $\Phi: F \times G \to H$ be a bilinear map. We define the product AB of A and B with respect to Φ by

$$AB(x_1,\ldots,x_m,y_1,\ldots,y_n) = \Phi(A(x_1,\ldots,x_m),B(y_1,\ldots,y_n))$$

Then $AB: E^{m+n} \to H$ is an $(m+n)$-linear map. If A,B,Φ are continuous, then AB is also continuous.

3E. Let $A: E^2 \to F$ be a bilinear map, and assume that there exists a map M defined on a neighborhood of $(0,0) \in E^2$ with values in F such that $M(x,y) = 0$ as $(x,y) \to (0,0)$, and that

$$\|A(x,y)\| \leq \|M(x,y)\|\|x\|\|y\|$$

Then A is the zero map.

4
Polynomials

4.1. Polynomials are the simplest holomorphic functions which are used to approximate all other holomorphic mappings; thus their properties should be investigated first. We define a homogeneous polynomial as the restriction of a symmetric multilinear map to the diagonal. In this chapter, the multinomial and polarization formulae are developed, several necessary and sufficient conditions for a polynomial to be continuous are studied, and the Banach-Steinhaus theorem is extended to the space of m-homogeneous polynomials.

SYMMETRIC MULTILINEAR MAPS

4.2 *Symmetric Multilinear Maps.* If the normed spaces $E_1, \ldots, E_m$ are all equal to a normed space E, we recall that

$$L(^mE;F) = L(E_1, \ldots, E_m;F)$$

Notice that $L(^mE;F) \neq L(E^m;F)$.

An m-linear map $A: E^m \to F$ is said to be *symmetric* if

$$A(x_1, \ldots, x_m) = A(x_{\sigma(1)}, \ldots, x_{\sigma(m)})$$

for any permutation σ of $(1, \ldots, m)$.

We shall denote by $L_{as}(^mE;F)$ the vector space of all algebraic symmetric m-linear maps of E^m into F and let

$$L_s(^mE;F) = L(^mE;F) \cap L_{as}(^mE;F)$$

Then it is easy to see that $L_s(^mE;F)$ is a closed subspace of $L(^mE;F)$; hence $L_s(^mE;F)$ is a Banach space if F is a Banach space. For m = 0, we agree to write

$$L(^0E;F) = L_s(^0E;F) = F$$

For simplicity, when $F = \mathbb{K}$ we write

$$L(^mE) = L(^mE;\mathbb{K}) \qquad L_s(^mE) = L_s(^mE;\mathbb{K}) \qquad \ldots$$

4.3 *Symmetrization.* For each m-linear map A: $E^m \to F$ we define an m-linear map A_s: $E^m \to F$ by

$$A_s(x_1,\ldots,x_m) = \frac{1}{m!} \sum_\sigma A(x_{\sigma(1)},\ldots,x_{\sigma(m)})$$

where the summation is over the m! permutations σ of $(1,\ldots,m)$.

It is easy to see that A_s: $E^m \to F$ is a symmetric m-linear map, which we call the *symmetrization* of A. If $A \in L_s(^mE;F)$, then it is clear that $A = A_s$. We also note that the map

$$A \to A_s$$

is a continuous linear map of $L(^mE;F)$ onto $L_s(^mE;F)$ since

$$\|A_s\| \leq \|A\|$$

4.4 *Multinomial Formula.* We now generalize the multinomial formula from elementary algebra. Let $A \in L_a(E_1,\ldots,E_m;F)$, $0 \leq n \leq m$, and $(x_1,\ldots,x_n) \in E_1 \times \ldots \times E_n$. We define $A(x_1,\ldots,x_n)$ by the following formulas:

(1) If n = m, $A(x_1,\ldots,x_n) = A(x_1,\ldots,x_m)$.

(2) If n < m,

$$A(x_1,\ldots,x_n): E_{n+1} \times \ldots \times E_m \to F$$

is a mapping defined by

$$(x_{n+1}, \ldots, x_m) \to A(x_1, \ldots, x_n, x_{n+1}, \ldots, x_m)$$

Then $A(x_1, \ldots, x_n)$ is an $(m-n)$-linear map of $E_{n+1} \times \ldots \times E_m$ into F.

If $A \in L_{as}(^m E; F)$ and $x = x_1 = \ldots = x_n$, we write

$$Ax^n = A(x_1, \ldots, x_n)$$

For convenience, we define

$$Ax^o = A$$

This shows that

$$Ax^n \in L_a(^{m-n} E; F)$$

for all n, $0 \le n \le m$.

If $A \in L_a(^m E; F)$, $0 \le k \le m$, $x_1, \ldots, x_k \in E$, and $n_1, \ldots, n_k \in \mathbb{N}$ with $n_1 + \ldots + n_k = n \le m$, we define an $(m-n)$-linear map

$$Ax_1^{n_1} \ldots x_k^{n_k} = A(x_1, \ldots, x_1, x_2, \ldots, x_2, \ldots, x_k, \ldots, x_k)$$

where x_1 appears n_1 times, x_2 appears n_2 times, and so on.

MULTINOMIAL FORMULA.

$$A(x_1 + \ldots + x_k)^n = \Sigma \frac{n!}{n_1! \ldots n_k!} Ax_1^{n_1} \ldots x_k^{n_k}$$

where the summation is over all k-tuples $(n_1, \ldots, n_k)$ *satisfying* $n_1 + \ldots + n_k = n$.

The proof of this formula is by induction on n, and the procedure is very elementary. As a special case, when $k = 2$, we have the familiar binomial formula.

BINOMIAL FORMULA.

$$A(x + y)^n = \sum_{k=o}^{n} \binom{n}{k} Ax^k y^{n-k}$$

POLYNOMIALS

4.5 *Homogeneous Polynomials.* Let E and F be two vector spaces over the same field $\mathbb{K}$. A mapping P: E $\to$ F is said to be an m-*homogeneous polynomial* or a *homogeneous polynomial of degree* m if there exists an m-linear map A: $E^m \to F$ such that

$$P(x) = Ax^m$$

for all x $\in$ E.

For m = 1, an m-homogeneous polynomial is simply a linear map of E into F. We agree that a homogeneous polynomial of degree 0 is a constant map.

If P: E $\to$ F is an m-homogeneous polynomial, then

$$P(rx) = r^m P(x)$$

for any r $\in \mathbb{K}$.

We shall denote by $P_a(^mE;F)$ the vector space of all m-homogeneous polynomials of E into F. The index 'a' is to indicate that the m-homogeneous polynomials in this space are not necessarily continuous. We agree to write $F = P_a(^oE;F)$ for convenience.

4.6 *Polarization Formula.* The following formula relates m-homogeneous polynomials and symmetric m-linear maps.

THEOREM. *Let* A: $E^m \to F$ *be a symmetric m-linear map. Then*

$$A(x_1,\ldots,x_m) = \frac{1}{m!2^m} \Sigma\, e_1\ldots e_m A(e_1 x_1 + \ldots + e_m x_m)^m \qquad (1)$$

where the summation is taken for all $e_i \in \{1,-1\}$, i = 1,\ldots,m.

Proof. By the multinomial formula, we have

$$A(e_1 x_1 + \ldots + e_m x_m)^m = \Sigma\, \frac{m!}{n_1!\ldots n_m!}\, A(e_1 x_1)^{n_1}\ldots(e_m x_m)^{n_m}$$

$$= \Sigma\, \frac{m!}{n_1!\ldots n_m!}\, e_1^{n_1}\ldots e_m^{n_m} A x_1^{n_1}\ldots x_m^{n_m}$$

If we denote the right-hand side of (1) by B, then

$$B = \frac{1}{2^m} \Sigma \frac{1}{n_1! \ldots n_m!} a(n_1, \ldots, n_m) A x_1^{n_1} \ldots x_m^{n_m}$$

with

$$a(n_1, \ldots, n_m) = \Sigma e_1^{n_1+1} \ldots e_m^{n_m+1}$$

where the summation is taken over all $e_i \in \{1, -1\}$, $1 \leq i \leq m$. Note that the first three summations were over all m-tuples $(n_1, \ldots, n_m)$ with $n_1 + \ldots + n_m = m$.

If some $n_i = 0$, it is clear that $a(n_1, \ldots, n_m) = 0$; hence

$$a(n_1, \ldots, n_m) \neq 0$$

if and only if $n_1 = \ldots = n_m = 1$. Since $a(1, \ldots, 1) = 2^m$, we obtain

$$B = A(x_1, \ldots, x_m) \qquad\qquad \square$$

4.7   THEOREM.   *If* $P \in P_a(^mE;F)$, *then there exists a unique symmetric m-linear map* $A \in L_{as}(^mE;F)$ *such that*

$$P(x) = A x^m$$

for all $x \in E$.

Proof.   By definition, $P(x) = A x^m$ for some $A \in L_a(^mE;F)$. Then the symmetrization A_s of A satisfies

$$P(x) = A_s x^m$$

The uniqueness of such a symmetric m-linear map is a consequence of the polarization formula. $\square$

4.8.   To emphasize the unique correspondence between the m-homogeneous polynomial P and the symmetric m-linear map A for which $P(x) = A x^m$, we shall write

$$P = \hat{A}$$

The following theorem is self-evident.

THEOREM.   *The mapping*

$$A \in L_{as}(^mE;F) \to \hat{A} \in P_a(^mE;F)$$

is a vector space isomorphism.

4.9   EXAMPLE. If $E = \mathbb{K}$, any m-linear map $A: E^m \to F$ is of the form

$$A(x_1,\ldots,x_m) = ax_1 \ldots x_m \qquad a \in F$$

where $ax_1 \ldots x_m$ denotes $(x_1 \ldots x_m)a$. Therefore, an m-homogeneous polynomial $P: \mathbb{K} \to F$ is of the form

$$P(x) = ax^m \qquad a \in F$$

In this way, if $F = \mathbb{K}$, we regain the classical m-homogeneous poly-nomial of $\mathbb{K}$ into $\mathbb{K}$. This example motivates and justifies our definition of m-homogeneous polynomials.

4.10   *Polynomials.* A mapping $P: E \to F$ is said to be a *polynomial* if there exist m and $P_k \in P_a(^kE;F)$, $k = 0,1,\ldots,m$ such that

$$P = P_o + P_1 \ldots + P_m$$

The addition in this representation is pointwise. If $P_m \neq 0$, then we say that P is a *polynomial of degree* m.

We shall denote by $P_a(E;F)$ the vector space of all polynomials from E to F with respect to the pointwise vector operations.

4.11   THEOREM.   *Let* $P: E \to F$ *be a non-zero polynomial of degree* m. *Then the representation*

$$P = P_o + P_1 + \ldots + P_m$$
is unique.

Proof. It suffices to show that

$$P = P_o + P_1 + \ldots + P_m = 0$$

implies

$$P_o = P_1 = \ldots = P_m = 0$$

In fact, if $P(x) = 0$ for all $x \in E$, then we have for any $r \in \mathbb{K}$

$$P(rx) = P_o(x) + rP_1(x) + \ldots + r^m P_m(x) = 0$$

Dividing through by r^m if $r \neq 0$ and $r \to \infty$, we obtain $P_m = 0$. We then get by induction $P_o = P_1 = \ldots P_{m-1} = 0$.   □

CONTINUOUS POLYNOMIALS

4.12   *The Space* $P(^m E;F)$. We shall denote by $P(^m E;F)$ the vector space of all continuous m-homogeneous polynomials from E to F with respect to the pointwise vector operations. We write $F = P(^o E;F)$ as before.

The following theorem provides criteria for the continuity of an m-homogeneous polynomial.

THEOREM.   *For* $P \in P_a(^m E;F)$, *let* $A \in L_{as}(^m E;F)$ *be such that* $P = \hat{A}$. *The following statements are equivalent:*

   (a) $A \in L_s(^m E;F)$

   (b) $P \in P(^m E;F)$

   (c)   P *is continuous at the origin.*

   (d)   *There exists a constant* M > 0 *such that*

$$\|P(x)\| \leq M \|x\|^m$$

Proof. The implications (a) $\to$ (b) $\to$ (c) $\to$ (d) are evident. We shall prove that (a) follows from (d). If $\|P(x)\| \leq M$ for $\|x\| \leq 1$, it follows from the polarization formula that

$$\|A(x_1, \ldots, x_m)\| \leq M/m!$$

for $\|x_i\| \le 1/m$, $i = 1,\ldots,m$, and hence $A \in L_s(^mE;F)$. □

It is easy to show that if $P \in P(^mE;F)$, then

$$\sup \{\|P(x)\|/\|x\|^m: x \ne 0\} = \sup \{\|P(x)\|: \|x\| \le 1\}$$

$$= \inf \{M \ge 0: \|P(x)\| \le M\|x\|^m\}$$

Let $\|P\|$ be the common value of these relations. Then $\|P\|$ is a norm on $P(^mE;F)$. This norm will be considered throughout the book. It is easy to see that

$$\|P(x)\| \le \|P\|\|x\|^m$$

for all $x \in E$.

4.13 THEOREM (Martin, 1932). *The mapping*

$$A \in L_s(^mE;F) \to \hat{A} \in P(^mE;F)$$

is a vector space isomorphism and a homeomorphism of the first onto the second space. Moreover,

$$\|\hat{A}\| \le \|A\| \le \frac{m^m}{m!} \|\hat{A}\| \tag{1}$$

Proof. Since the assertion is trivially true for $m = 0,1$, we assume that $m > 1$. It is clear that the mapping is a vector space isomorphism, and it remains to show that the inequality (1) holds.

Since $\|\hat{A}(x)\| \le \|A\|\|x\|^m$, we have $\|\hat{A}\| \le \|A\|$. To obtain the other inequality, we use the polarization formula to get

$$\|A(x_1,\ldots,x_m)\| \le \frac{1}{m!2^m} \Sigma \|\hat{A}(\varepsilon_1 x_1 + \ldots + \varepsilon_m x_m)\|$$

$$\le \frac{1}{m!2^m} \Sigma \|\hat{A}\|\|\varepsilon_1 x_1 + \ldots + \varepsilon_m x_m\|^m$$

$$\le \frac{1}{m!2^m} \Sigma \|\hat{A}\| (\|x_1\| + \ldots + \|x_m\|)^m$$

(Notice that there are 2^m terms in this summation.)

$$\leq \frac{1}{m!} \|\hat{A}\| (\|x_1\|+\ldots+\|x_m\|)^m$$

Hence, if $\|x_i\| \leq 1$, $i = 1,\ldots,m$, we obtain

$$\|A\| \leq \frac{m^m}{m!} \|\hat{A}\| \qquad\qquad\qquad\qquad \square$$

COROLLARY. *If F is a Banach space, then* $P(^mE;F)$ *is a Banach space.*

4.14 *The Universal Constant* $m^m/m!$. The map $a \in L_s(^mE;F) \to \hat{A} \in P(^mE;F)$ considered in 4.13 is not in general an isometry; also notice that the coefficient $m^m/m!$ in 4.13(1) does not depend on the normed spaces E and F, but it depends only on the integer m. Therefore, we can consider $m^m/m!$ as a universal constant relating the continuous m-homogeneous polynomials and the corresponding symmetric m-linear maps. The following example shows that we cannot replace the constant $m^m/m!$ with a smaller one. In fact, in this example, we show that

$$\|A\| = \frac{m^m}{m!} \|\hat{A}\|$$

EXAMPLE (Nachbin, 1968). Consider the Banach space $E = \ell^1$ of all sequences $x = (x_n)$ of numbers in $\mathbb{K}$ such that $\|x\| = \Sigma |x_n| < \infty$ (see 1.6). For $m > 0$, let $A: E^m \to K$ be the m-linear map defined by

$$A(x^1,\ldots,x^m) = x_1^1 x_2^2 \ldots x_m^m$$

the product of the diagonal of

$$x^1 = (x_1^1,x_2^1,\ldots,x_m^1,\ldots)$$
$$x^2 = (x_1^2,x_2^2,\ldots,x_m^2,\ldots)$$
$$\ldots\ldots\ldots\ldots\ldots\ldots\ldots$$
$$\ldots\ldots\ldots\ldots\ldots\ldots\ldots$$
$$\ldots\ldots\ldots\ldots\ldots\ldots\ldots$$

$$x^m = (x_1^m, x_2^m, \ldots, x_{\underline{m}}^m, \ldots)$$

. . . .

. . . .

Then A is continuous. The symmetrization A_s of A is now given by

$$A_s(x^1, \ldots, x^m) = \frac{1}{m!} \Sigma \, x_1^{\sigma(1)} \ldots x_m^{\sigma(m)}$$

where the summation is over all permutations σ of $\{1, 2, \ldots, m\}$.

We claim that $\|A_s\| = 1/m!$. In fact

$$\|A_s(x^1, \ldots, x^m)\| \le \frac{1}{m!} \Sigma \, |x_1^{\sigma(1)}| \ldots |x_m^{\sigma(m)}| \le \frac{1}{m!} \|x^1\| \ldots \|x^m\|$$

Hence $\|A_s\| \le 1/m!$. On the other hand,

$$A_s(e^1, \ldots, e^m) = \frac{1}{m!}$$

where

$$e^i = (0, \ldots, 0, 1, 0, \ldots) \qquad (1 \text{ at the ith coordinate})$$

Therefore, $\|A_s\| = 1/m!$.

Let $\hat{A}_s(x) = A_s(x, \ldots, x)$. Then $\hat{A}_s \in P(^mE; \mathbb{K})$ and

$$\hat{A}_s(x) = x_1 \ldots x_m$$

where $x = (x_1, \ldots, x_m, \ldots)$. Since the geometric mean of positive numbers is always less than or equal to the arithmetic mean, we have

$$\|\hat{A}_s\| = |x_1| \ldots |x_m| \le \frac{1}{m^m}(|x_1| + \ldots + |x_m|)^m$$

Thus $\|\hat{A}_s\| \le 1/m^m$. If we take

$$x = (1/m, \ldots, 1/m, 0, \ldots)$$

(where 1/m appears in the first m terms of x), we obtain

$$\left\| \hat{A}_s(x) \right\| = 1/m^m$$

This shows that $\left\| \hat{A}_s \right\| = 1/m^m$; and hence

$$\left\| A_s \right\| = \frac{m^m}{m!} \left\| \hat{A}_s \right\| \qquad\qquad\qquad \square$$

REMARK. If E is a real Hilbert space (e.g., ℓ^2 or $L^2(I)$) and F is a Banach space, then the mapping $A \in L_s(^mE;F) \rightarrow A \in P(^mE;F)$ becomes an isometry. This was recognized by many authors including Banach (1938). An interesting generalization of the universal constant for L^p spaces can be found in Harris (1973).

We give a proof here that $A \rightarrow \hat{A}$ is an isometry when A is a continuous symmetric bilinear map from a Hilbert space into a Banach space. The main property we use here is the parallelogram law on such a space; i.e.,

$$2(\|x\| + \|y\|)^2 = \|x + y\|^2 + \|x - y\|^2$$

Let $A: E^2 \rightarrow F$ be a continuous symmetric bilinear map. It follows from the polarization formula 4.6 that

$$\|A(x,y)\| \leq \frac{1}{8} \Sigma \left\| \hat{A}(\varepsilon_1 x + \varepsilon_2 y) \right\| \qquad (\varepsilon = 1 \text{ or } -1)$$

$$\leq \frac{1}{8} \left\| \hat{A} \right\| (2\|x + y\|^2 + 2\|x - y\|^2)$$

$$= \frac{1}{2} \left\| \hat{A} \right\| (\|x\|^2 + \|y\|^2)$$

$$\leq \left\| \hat{A} \right\| (\|x\|^2 + \|y\|^2)$$

Thus $\|A\| \leq \|\hat{A}\|$. But $\|\hat{A}\| \leq \|A\|$ is true always; and hence we have the equality $\|A\| = \|\hat{A}\|$. $\square$

4.15 *The Space* $P(E;F)$. We showed in 4.11 that for every polynomial $P: E \rightarrow F$ of degree m there corresponds a unique set of homogeneous polynomials P_k of degree k, $k = 1,\ldots,m$, such that

$$P = P_o + P_1 + \ldots + P_m$$

We are now concerned here with continuous polynomials of a normed space E into another normed space F.

We shall denote by $P(E;F)$ the vector space of all continuous polynomials of E into F. The following theorem provides criteria for the continuity of the polynomial P.

4.16 THEOREM. *Let* P: $E \rightarrow F$ *be a polynomial of degree m such that*

$$P = P_o + P_1 + \ldots + P_m$$

Then the following are equivalent:

(a) $P_o, P_1, \ldots, P_m$ *are continuous;*

(b) P *is continuous;*

(c) P *is continuous at the origin;*

(d) P *is bounded on the unit ball.*

Proof. The implications (a) $\rightarrow$ (b) $\rightarrow$ (c) $\rightarrow$ (d) are trivial. It remains to show (d) $\rightarrow$ (a). This will be shown by induction. We first recall the polarization formula 4.6 in the following form.

POLARIZATION FORMULA. *For* f: $E \rightarrow F$ *and* $x_1, \ldots, x_m \in E$, *let*

$$\Phi_m(f) = \frac{1}{m! \, 2^m} \Sigma \, \varepsilon_1 \ldots \varepsilon_m \, f(\varepsilon_1 x_1 + \ldots + \varepsilon_m x_m)$$

where the summation is over all $\varepsilon_k = 1$ *or* -1, $k = 1, \ldots, m$. *If* $A_k: E^k$
$\rightarrow F$ *is a symmetric k-linear map and* $P_k = \hat{A}_k$, *then*

$$\Phi_m(P_k) = \begin{cases} A_m(x_1, \ldots, x_m) & \text{if } k = m \\ \\ 0 & \text{if } k < m \end{cases}$$

For future reference we shall call $\Phi_m(f)$ the *polarization of* f *with respect to* $x_1, \ldots, x_m$. We have shown the formula for the case $k = m$ in 4.6. If $k < m$, a straightforward but lengthy computation shows that $\Phi_m(P_k) = 0$ (see Proof of 4.6).

To show the implication (d) $\rightarrow$ (a), we note that

$$\Phi_m(P) = \Phi_m(P_o) + \ldots + \Phi_m(P_m) = \Phi_m(P_m)$$

or

$$\Phi_m(P) = A_m(x_1, \ldots, x_m)$$

where $\hat{A}_m = P_m$. Since P is bounded on the unit ball, $\|P(x)\| \leq M$ on the unit ball $\|x\| \leq 1$ for some $M > 0$. Now

$$\|A_m(x_1, \ldots, x_m)\| = \|\Phi_m(P)\| \leq \frac{1}{m!} \|P(\varepsilon_1 x_1 + \ldots + \varepsilon_m x_m)\|$$

Thus if $\|x_1\| + \ldots + \|x_m\| \leq 1$,

$$\|A_m(x_1, \ldots, x_m)\| \leq M/m!$$

which shows that A_m is continuous and hence $P_m = \hat{A}_m$ is continuous.

Since $P - P_m$ is also bounded on the unit ball, repeating the same argument above, we can show that P_{m-1} is continuous. Inductively, therefore, we conclude that all $P_o, P_1, \ldots, P_m$ are continuous.   □

COROLLARY. *Every polynomial on a finite dimensional space is continuous.*

Proof. It is obvious since every m-linear map on a finite dimensional space is continuous.   □

4.17. We have the following Banach-Steinhaus type theorem for homogeneous polynomials which is due to Mazur and Orlicz (1934).

THEOREM (Banach-Steinhaus). *Let E be a Banach space and let* (P_n) *be a sequence in* $P(^mE;F)$. *If P is a pointwise limit of the sequence* (P_n), *then* $P \in P(^mE;F)$.

Proof. Let $A_n \in L_s(^mE;F)$ be such that $\hat{A}_n = P_n$. By the polarization formula 4.6, $\lim A_n(x_1,\ldots,x_m)$ exists at each point $(x_1,\ldots,x_m)$ of E^m. Let $A(x_1,\ldots,x_m)$ be this limit. Then $A \in L_{as}(^mE;F)$. Since

$$L(^mE;F) \approx L(E;L(^{m-1}E;F))$$

if we consider (A_n) as a sequence in $L(E;L(^{m-1}E;F))$, then by the Banach-Steinhaus Theorem (Ex.2K), we obtain $A \in L(E;L(^{m-1}E;F))$. Hence $A \in L_s(^mE;F)$. We also have

$$P(x) = \lim_n P_n(x) = \lim_n \hat{A}_n(x) = \lim_n A_n(x,\ldots,x)$$

$$= A(x,\ldots,x) = \hat{A}(x)$$

Hence, $P = \hat{A}$; this completes the proof. □

EXERCISES

4A *The m-Homogeneous Polynomials on* $\mathbb{K}^p$. If $P: \mathbb{K}^p \to F$ is a m-homogeneous polynomial, then for any $(m_1,\ldots,m_p)$ with $m_1+\ldots+m_p = m$, there exists $a(m_1,\ldots,m_p) \in F$ such that

$$P(x_1,\ldots,x_p) = \Sigma\, a(m_1,\ldots,m_p)x_1^{m_1}\ldots x_p^{m_p}$$

where the summation is over all $(m_1,\ldots,m_p)$ with $m_1+\ldots+m_p = m$.

4B *Composite Polynomials.*

(a) If $P \in P_a(E;F)$ and $Q \in P_a(F;G)$, then $Q \circ P \in P_a(E;G)$.

(b) If $P \in P_a(E;F)$ and $\xi \in E$ is a fixed point, then the map $x \in E \to P(x+\xi) \in F$ is also a polynomial.

4C *Weak Continuity for Polynomials.*

(a) Let $P \in P_a(E;F)$. Then P is continuous if and only if the composite map $u \circ P$ is continuous for every continuous linear functional $u \in F^*$.

(b) $P \in P(^mE;F)$ if and only if $u \circ P \in P(^mE;\mathbb{K})$ for all $u \in F^*$.

4D *Successive Differences.* Let E and F be real Banach spaces and let $f: E \to F$ be a function and $h \in E$. Define a function $\Delta_h f: E \to F$ by

$$\Delta_h f(x) = f(x+h) - f(x)$$

If x_1, $x_2 \in E$, then $\Delta_{x_1}(\Delta_{x_2} f)$ is denoted by $\Delta_{x_1} \Delta_{x_2} f$ and is called the second defference of f with respect to x_1 and x_2. Inductively, the nth difference with respect to $x_1, \ldots, x_n$ is defined.

(a)   The nth difference of f is the sum of 2^n functions of the form

$$x \rightarrow (-1)^{n-p} f(x+x_{i_1} + \ldots + x_{i_p})$$

where $i_1, \ldots, i_p \in \{1, \ldots, n\}$; $i_1 < \ldots < i_p$.

(b)   The nth difference of f is a symmetric function of $x_1, \ldots, x_n$.

(c)   If P is a polynomial of degree m, then $\Delta_h P: E \rightarrow F$ is a polynomial of degree m-1.

(d)   *Polarization Formula.* If $A \in L_{as}(^m E; F)$, and $\hat{A} = P$, then

$$A(x_1, \ldots, x_m) = \frac{1}{m!} \Delta_{x_1} \ldots \Delta_{x_m} P$$

4E   *Finitely Continuous Polynomials* (Bochnak and Siciak, 1971).

(a)   If $P: \mathbb{K}^p \rightarrow F$ is a polynomial with respect to each variable separately, then P is a polynomial.

(b)   If $P: E \rightarrow F$ is a polynomial on every affine line in E and dim $E < \infty$, then for every subspace V of E, the restriction map $P|_V: V \rightarrow F$
is a polynomial.

(c)   Let $P: E \rightarrow F$ be a polynomial on each affine line contained in E. Then there exists a sequence of homogeneous polynomials $P_m \in P_a(^m E; F)$, $m \in \mathbb{N}$ such that
$P(x) = \Sigma P_m$
where for every fixed x in E, $P_m(x) = 0$ for all but finitely many m.

(d)   If $P: E \rightarrow F$ satisfies the assumption of (c), and moreover P = 0 on a non-empty open set of E, then P = 0.

(e)   Let P: E → F be continuous on a Banach space E.   If P is a polynomial on every finite dimensional subspace of E, then P ∈ P(E;F).

(f)   If dim E = ∞, there exists a function f: E → F such that its restriction to any affine line in E is a polynomial but f is not a polynomial.

5

Differentiable Maps

5.1. The fundamental notions of abstract differentials, polynomials and power series were introduced by Maurice Fréchet around 1909.

The crux of the theory of functions between normed spaces is the question of differentiability. In this general situation the differentials of Fréchet and Gâteaux appear to be the most appropriate concepts. In the present chapter we shall develop Fréchet differentials, which are more appropriate for normed spaces than Gâteaux's.

DIFFERENTIAL MAPS

5.2 *Tangency*. Let E and F be normed spaces and U a non-empty open subset of E. For f: U → F, g: U → F, and ξ ∈ U, we say that f and g are *tangent to each other* at ξ if

$$\lim_{x \to \xi} \frac{\|f(x) - g(x)\|}{\|x - \xi\|} = 0$$

i.e., for any ε > 0 there exists a δ > 0 such that if $0 < \|x-\xi\| < \delta$, x ∈ U, then $\|f(x) - g(x)\| \leq \varepsilon \|x - \xi\|$.

If f and g are tangent to each other at ξ, then f - g is continuous at ξ and f(ξ) = g(ξ); thus if one of these functions is continuous at ξ the other must be continuous at ξ. Moreover, f and g are tangent at ξ if and only if f - g is tangent to 0 at ξ. Now it is easy to verify that tangency at ξ is an equivalence relation on the vector space of all mappings from U to F which are continuous at ξ. In fact, if f and g are tangent at ξ, and g and h are tangent at

ξ, then f and h are tangent at ξ since

$$\left\| f(x) - h(x) \right\| \leq \left\| f(x) - g(x) \right\| + \left\| g(x) - h(x) \right\|$$

The notion of tangency depends only on the topologies of E and F, and not on the norms used to define these topologies. For if we replace the norms $\left\| \cdot \right\|$ on E and F by equivalent norms $\left\| \cdot \right\|_1$, then

$$\frac{1}{\left\| x-\xi \right\|_1} \leq \frac{a}{\left\| x-\xi \right\|} \quad ; \quad \left\| f(x) \right\|_1 \leq b \left\| f(x) \right\|$$

for some constants a and b. Hence

$$\frac{\left\| f(x) \right\|_1}{\left\| x-\xi \right\|_1} \leq ab \frac{\left\| f(x) \right\|}{\left\| x-\xi \right\|}$$

Since by assumption the right-hand side approaches 0 if $x \to \xi$, so does the left-hand side.

5.3 EXAMPLE. The fundamental idea involved in differential calculus is the local approximation (or tangency) of a mapping by an affine linear map (a function f: E $\to$ F is *affine linear* if f(x) = a + A(x) where a $\in$ F and A $\in$ L(E;F). Unfortunately, to understand this concept it is necessary to "unlearn" the interpretation of the derivative of a real valued function of a real variable.

If f: $\mathbb{K}$ $\to$ $\mathbb{K}$ is differentiable at $\xi \in \mathbb{K}$ and

$$f'(\xi) = \lim_{x \to \xi} \frac{f(x) - f(\xi)}{x - \xi}$$

then f and its tangent line T at ξ,

$$T(x) = f(\xi) + f'(\xi)(x-\xi)$$

are tangent to each other at ξ. Notice that we can write the affine linear map

$$T(x) = f(\xi) + A(x-\xi)$$

where A: $x \in \mathbb{K} \to f'(\xi)x \in \mathbb{K}$ is a linear map.

This motivates the following definition of a derivative.

5.4 *Differentiability.* A mapping f: U $\to$ F is said to be *differentiable* at $\xi \in U$ if there exists a linear map $A_\xi \in L(E;F)$ such that f and the continuous affine linear map $x \in E \to f(\xi) + A_\xi(x-\xi) \in F$ are tangent at ξ; i.e.,

$$\lim_{x \to \xi} \frac{\left\| f(x) - f(\xi) - A_\xi(x-\xi) \right\|}{\left\| x-\xi \right\|} = 0 \tag{1}$$

If there is such a linear map A_ξ satisfying (1), then it is unique as shown in the lemma below. We call A_ξ the *Fréchet differential* or the *differential* of f at ξ and A_ξ will be denoted by $Df(\xi)$. Therefore

$$Df(\xi) \in L(E;F)$$

We should notice that the differentiability depends only on the topologies of E and F, and not on the particular norms used to define these topologies (see 5.2).

In this book we prefer the word "differential" to the word "derivative" since in classical multivariable differential calculus we have the concept of differential but not of derivative unless we allow partial derivatives. We reserve the term "derivative" for the case where $E = \mathbb{K}$ (see 5.5).

We shall now establish the uniqueness of the differential at a point.

LEMMA. *A mapping f has at most one differential at a point.*

Proof. Suppose A_1 and A_2 are continuous linear maps satisfying (1). Let $A = A_1 - A_2$. Then

$$\lim_{x \to \xi} \frac{\|A(x-\xi)\|}{\|x-\xi\|} = \lim_{x \to 0} \frac{\|A(x)\|}{\|x\|} = 0$$

Hence for any $\varepsilon > 0$ there exists $\delta > 0$ such that

$$\|A(x)\| \leq \varepsilon \|x\|$$

for $\|x\| < \delta$. If $x \neq 0$, replacing x by $x/\|x\|$ we obtain $\|A\| \leq \varepsilon$. Since ε is arbitrary, we conclude that $A = 0$. □

If $f: U \to F$ is differentiable at each point of U, then f is said to be *differentiable on* U. In this case, the mapping

$$x \in U \to Df(x) \in L(E;F)$$

is called the *differential* of f *on* U and is denoted by Df. If Df is continuous on U, f is said to be *of class* C^1 on U.

We should notice that the differential Df is always vector valued and not scalar valued even though we may be interested in the differentiation of scalar-valued functions $f: U \to \mathbb{K}$ (i.e., when $F = \mathbb{K}$). In fact, we have $Df: U \to L(E;\mathbb{K}) = E^*$.

5.5 *Derivative*. If U is a non-empty subset of $\mathbb{K}$ and $f: U \to F$, then the differential $Df(\xi)$ of f at $\xi \in U$ is a linear map in $L(\mathbb{K};F)$. However, if $A \in L(\mathbb{K};F)$, then for any $x \in \mathbb{K}$, we have

$$A(x) = A(x \cdot 1) = xA(1)$$

Hence A is a multiplication (on the right) by the vector $A(1)$ in F, and we usually may identify A with this vector. It is easy to see that this identification is a natural isometry between $L(\mathbb{K};F)$ and F. Let $f'(\xi)$ denote the identification of $Df(\xi)$ and we call it the *derivative* of f at ξ. Notice that $Df(\xi)$ and $f'(\xi)$ are two distinct objects. We also use the Leibnitz notation $\frac{df}{d\lambda}(\xi)$ for $f'(\xi)$. Hence if $f: U \to F$ is differentiable at ξ, then

$$\lim_{x \to \xi} \frac{f(x) - f(\xi)}{x - \xi} = f'(\xi)$$

This conforms with the classical notation of a derivative.

PROPERTIES OF THE DIFFERENTIAL

5.6 *Linearity of the Differential.* Let E, F be normed spaces and
U a non-empty open subset of E. For f: U → F and g: U → F, the sum
f + g of f and g and the product λf of f by a scalar $\lambda \in \mathbb{K}$ are
respectively defined by

$(f + g)(x) = f(x) + g(x)$

$(\lambda f)(x) = \lambda f(x)$

THEOREM. *(a) If f and g are differentiable at $\xi \in U$, then f + g
is differentiable at ξ and*

$D(f + g)(\xi) = Df(\xi) + Dg(\xi)$

*(b) If f is differentiable at ξ, then λf is differentiable at ξ for
all $\lambda \in \mathbb{K}$ and*

$D(\lambda f)(\xi) = \lambda Df(\xi)$

Proof. This is obvious from the definition.   □

As a consequence of the theorem, we have the following vector
subspaces of the space of all mappings of U into F:
   (i)   the vector space of all mappings of U into F which are
         differentiable at ξ;
   (ii) the vector space of all mappings of U into F which are
         differentiable on U;
   (iii) the vector space of all mappings of U into F which are
         of class C^1 on U.

5.7 *Local Differentiability.* Let E, F be normed spaces of U a non-empty open subset of E. Then the notion of differentiability is a local property in the following sense:

THEOREM. *(a) If f: U → F is differentiable on U, then f is differentiable on any open subset V of U, and*

$$D(f\big|_V) = (Df)\big|_V$$

(b) If U = $\bigcup_{i \in I} V_i$, where each V_i is open in U, then f is differentiable on U if and only if f is differentiable on each V_i.

Proof. It is an easy exercise for the reader. □

5.8 THEOREM (Lipschitzian Property). *Let E and F be normed spaces and U a non-empty open subset of E. If f: U → F is differentiable at $\xi \in U$, then there exist C > 0 and δ > 0 such that*

$$\|f(x) - f(\xi)\| \leq C\|x - \xi\|$$

for $x \in U$, $\|x - \xi\| < \delta$. In particular, it follows that f is continuous at ξ.

Proof. Let $A = Df(\xi)$ and $\Delta(x) = f(x) - f(\xi) - A(x-\xi)$ for $x \in U$. Then $\|f(x) - f(\xi)\| \leq \|A\| \cdot \|x - \xi\| + \|\Delta(x)\|$. Since $\lim_{x \to \xi} \Delta(x)/\|x-\xi\| = 0$ and $\Delta(\xi) = 0$, there exists δ > 0 such that if $\|x-\xi\| \leq \delta$, $x \in U$ then $\|\Delta(x)\| \leq \|x-\xi\|$. Hence

$$\|f(x) - f(\xi)\| \leq (\|A\| + 1)\|x-\xi\|$$

for $x \in U$, $\|x-\xi\| \leq \delta$. □

5.9 THEOREM (Chain Rule). *Let E, F, G be normed spaces, U an open subset of E, and V an open subset of F. Let f: U → V and g: V → G. If f is differentiable at $\xi \in U$ and g is differentiable at $f(\xi) \in V$,* *then* g o f: U → G *is differentiable at* ξ *and*

$$D(g \circ f)(\xi) = Dg(f(\xi)) \circ Df(\xi)$$

Before giving the proof, we make explicit the meaning of the chain rule by the following arrow diagram.

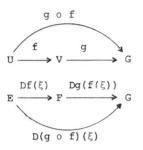

Note that $Df(\xi) \in L(E;F)$, $Dg(f(\xi)) \in L(F;G)$ and $D(g \circ f)(\xi) \in L(E;G)$.

Proof. Let $\eta = f(\xi)$ and $h = g \circ f$. By assumption, given ε, $0 < \varepsilon < 1$, there is $\delta > 0$ such that $x \in U$, $\|x-\xi\| \leq \delta$ and $y \in V$, $\|y-\eta\| \leq \delta$, so that we can write

$$A(x) = f(x) - f(\xi) - Df(\xi)(x-\xi)$$
$$B(y) = g(y) - g(\eta) - Dg(\eta)(y-\eta)$$

with $\|A(x)\| \leq \varepsilon\|x-\xi\|$ and $\|B(y)\| \leq \varepsilon\|y-\eta\|$. Let $a = \|Df(\xi)\|$ and $b = \|Dg(\eta)\|$. Then for $x \in U$, $\|x-\xi\| \leq \delta$ we have

$$\|f(x) - f(\xi)\| = \|A(x) + Df(\xi)(x-\xi)\| \leq (a+1)\|x-\xi\|$$

Therefore, if $x \in U$, $\|x-\xi\| \leq \delta/(a+1)$,

$$\|B(f(x))\| \leq \varepsilon\|f(x) - f(\xi)\|$$

Since f is differentiable at ξ, by 5.8 we have

$$\|f(x) - f(\xi)\| \leq c\|x-\xi\|$$

for some c > 0 if δ is small enough. Thus if x ∈ U, $\|x-\xi\| \leq \delta/(a+1)$,

$$\|B(f(x))\| \leq \varepsilon c \|x-\xi\|$$

If we write

$$C(x) = h(x) - h(\xi) - (Dg(\eta) \circ Df(\xi))(x-\xi)$$

then

$$C(x) = B(f(x)) + Dg(f(\xi))(A(x))$$

Hence

$$\|C(x)\| \leq \|B(f(x))\| + \|Dg(f(\xi))\|\|A(x)\| \leq \varepsilon(b+c)\|x-\xi\|$$

for x ∈ U, $\|x-\xi\| \leq \delta/(a+1)$, which proves the theorem.   □

EXAMPLES OF DIFFERENTIALS

5.10 *Constant Maps and Linear Maps*. Let E and F be normed spaces.
Then a *constant map* is the map x ∈ F ↦ b ∈ F for a fixed point b in
F.

THEOREM.   *(a) If* f: E → F *is a constant map, then* Df = 0; *i.e.,*
Df(x) = 0 *for all* x ∈ E.
(b) If A ∈ L(E;F), *then* DA *is the constant map satisfying* DA(x) = A
for all x ∈ E.

Proof. These are obvious from the definition of differentiability
and are left to the reader.   □

For simplicity, we shall write the constant map DA: x ∈ E ↦ A ∈
L(E;F) by DA = A. However, the rule DA = A should not be confused
with the formula $\frac{d}{dx} e^x = e^x$ in elementary calculus. The rule DA = A
should be compared with $\frac{d}{dx}(cx) = c$ by noting that $\mathbb{K}^* = \mathbb{K}$.

5.11   THEOREM. *Let* f: U → F *be differentiable at* ξ ∈ U *and let*

A ∈ L(F;G). *Then*

$$D(A \circ F)(\xi) = A \circ Df(\xi)$$

Proof. This follows from Theorem 5.10 and the chain rule. Of course, we can prove this directly by observing that

$$\|A[f(x) - f(\xi) - Df(\xi)(x-\xi)]\| \le \|A\| \|f(x) - f(\xi) - Df(\xi)(x-\xi)\| \qquad \square$$

5.12 *Mappings into a Product Space.* Let F_n be normed spaces for n = 1,...,m and let $F = F_1 \times \ldots \times F_m$. Then F is a normed space and the topology induced by a norm on F is the product topology (see 3.2). Let

$$\pi_n \colon F \to F_n$$

be the projection onto the nth factor F_n and let

$$u_n \colon F_n \to F$$

be the natural embedding map defined by

$$u_n(x_n) = (0,\ldots,0,x_n,0,\ldots,0)$$

Then both π_n and u_n are continuous linear maps and

$$\pi_n \circ u_n = 1_{F_n} \qquad \text{(the identity map on } F_n \text{)}$$

$$\Sigma \, u_n \circ \pi_n = 1_F \qquad \text{(the identity map on F)}$$

Let U be an open subset of E and let f: U → F and let $f_n = \pi_n \circ f \colon U \to F_n$ be the nth coordinate map. Then

$$f = \Sigma \, u_n \circ \pi_n \circ f = \Sigma \, u_n \circ f_n = (f_1,\ldots,f_m)$$

If f is differentiable at ξ, then $u_n \circ \pi_n \circ f$ is differentiable at ξ by

5.11 and we obtain

$$Df(\xi) = \Sigma \ u_n \circ Df_n(\xi) = (Df_1(\xi), \ldots, Df_m(\xi))$$

Conversely, if each f_n is differentiable at ξ, then f is clearly differentiable at ξ. This proves the following theorem.

THEOREM. *A mapping* f: U $\to$ F$_1$ $\times \ldots \times$ F$_m$ *is differentiable at* ξ *if and only if each coordinate map* $f_n = \pi_n$of *is differentiable at* ξ. *In this case*

$$Df(\xi) = (Df_1(\xi), \ldots, Df_m(\xi))$$

5.13 *Multilinear Maps.* Let U be a non-empty open set in $\mathbb{K}$. The product rule in elementary calculus $(fg)' = f'g + fg'$ can be derived from the following theorem by observing that the map $x \in U \to f(x)g(x)$ $\in \mathbb{K}$ is a composition of the following two maps:

$$x \in U \to (f(x), g(x)) \in \mathbb{K}^2$$

$$(y_1, y_2) \in \mathbb{K}^2 \to y_1 y_2 \in \mathbb{K}$$

the first map taking values in the product space $\mathbb{K}^2$ and the second being a continuous bilinear map as mentioned in 3.9(a).

THEOREM. *Let* E$_1, \ldots,$E$_m$ *and F be normed spaces. If* A $\in$ L(E$_1, \ldots,$E$_m$;F), *then* A *is differentiable at each* $(\xi_1, \ldots, \xi_m)$. *Furthermore,*

$$DA(\xi_1, \ldots, \xi_m)(x_1, \ldots, x_m) = \sum_{i=1}^{m} A(\xi_1, \ldots, \xi_{i-1}, x_i, \xi_{i+1}, \ldots, \xi_m)$$

$$DA: E_1 \times \ldots \times E_m \to L(E_1, \ldots, E_m; F)$$

Proof. For m = 1, this is just 5.10(b). If m = 2, we want to show

$$DA(\xi_1, \xi_2)(x_1, x_2) = A(x_1, \xi_2) + A(\xi_1, x_2)$$

Since

$$A(x_1, x_2) = A(\xi_1, \xi_2) + A(h_1, \xi_2) + A(\xi_1, h_2) + A(h_1, h_2)$$

where $h_1 = x_1 - \xi_1$, $h_2 = x_2 - \xi_2$, we get

$$\|A(x_1, x_2) - A(\xi_1, \xi_2) - A(x_1 - \xi_1, \xi_2) - A(\xi_1, x_2 - \xi_2)\|$$

$$\leq \|A\| \|x_1 - \xi_1\| \|x_2 - \xi_2\| \leq \|A\| \|x - \xi\|^2$$

where we used the norm $\|x\| = \sup \{\|x_1\|, \|x_2\|\}$ for $x = (x_1, x_2) \in E_1 \times E_2$. If we write

$$A_\xi(x_1, x_2) = A(x_1, \xi_2) + A(\xi_1, x_2)$$

then $A_\xi \in L(E_1 \times E_2; F)$ and

$$\|A(x) - A(\xi) - A_\xi(x - \xi)\| \leq \|A\| \|x - \xi\|^2$$

hence $DA(\xi) = A_\xi$, as we wanted.

The proof for the general case can be completed by induction and is left as an exercise.   □

5.14 *Polynomials.* Let $P \in P(^m E; F)$ and let $A \in L_s(^m E; F)$ be such that $P = \hat{A}$; i.e.,

$$P(x) = A(x, \ldots, x) = Ax^m$$

THEOREM. *If* $P \in P(^m E; F)$ *and* $P = \hat{A}$, *then*

$$DP(x) = mAx^{m-1}$$

$$DP: E \to L(E; F) = P(^1 E; F)$$

Proof. We use the multinomial formula 4.4. In particular, we have the binomial formula

$$A(x+y)^m = A(x+y,\ldots,x+y) = \sum_{i=0}^{m} \binom{m}{i} Ax^{m-i}y^i$$

where $x,y \in E$. Hence

$$\lim_{y \to 0} \frac{\|A(x+y)^m - Ax^m - mAx^{m-1}y\|}{\|y\|} = 0$$

which proves that $DP(x) = mAx^{m-1}$.   □

5.15 *Inverse Maps*. Let U and V be open subsets of E and F respectively. If a mapping f: U → V is a homeomorphism and is differentiable on U, it does not necessarily follow that f is a diffeomorphism, i.e., the inverse f^{-1} of f is differentiable on V. Consider, for example, the function f: $\mathbb{R} \to \mathbb{R}$ defined by

$$f(x) = x^3$$

Then f is a homeomorphism of class C^1, but the inverse $g = f^{-1}$ is not differentiable at the origin.

THEOREM. *Let U and V be open subsets of E and F respectively, and let f: U → V be a homeomorphism. Assume that f is differentiable at a point P ∈ U. Then the inverse $g = f^{-1}$ is differentiable at the point q = f(p) if and only if Df(p) ∈ L(E;F) is a homeomorphism of E onto F ; i.e., Df(p) is a topological isomorphism. In this case,*

$$Dg(q) = (Df(p))^{-1}$$

Proof. If g is differentiable at q = f(p) then by the chain rule and 5.10, we have gof = 1_U, fog = 1_V and

$$Dg(q) \circ Df(p) = 1_E \qquad\qquad Df(p) \circ Dg(q) = 1_F$$

which proves that Df(p) is a topological isomorphism of E onto F.

Conversely, suppose that Df(p) is a topological isomorphism of E onto F. We want to show that g is differentiable at q = f(p). We

will first prove that g has the Lipschitzian property at q. For
simplicity, let $A = Df(p)$ and

$$\Delta(x) = f(x) - f(p) - A(x-p) \tag{1}$$

for $x \in U$. Since $A^{-1} \in L(F;E)$, by applying A^{-1} to both sides of (1)
we obtain

$$x-p = A^{-1}(f(x) - f(p) - \Delta(x)) \tag{2}$$

$$\|x-p\| \leq a(\|f(x) - f(p)\| + \|\Delta(x)\|) \tag{3}$$

with $a = \|A^{-1}\|$. Since f is differentiable at p, for $\varepsilon = 1/2a$ there
is $r > 0$ such that $B_r(p) \subset U$ and $\|\Delta(x)\| \leq \|x-p\|/2a$ for $x \in B_r(p)$.
Hence from (3) we get

$$\|x-p\| \leq 2a\|f(x) - f(p)\| \tag{4}$$

for $x \in B_r(p)$. Since g is also continuous, for some $s > 0$, $g(B_s(q))$
$\subset B_r(p)$. If we set $x = g(y)$ in (4), we have

$$\|g(y) - g(q)\| \leq 2a\|y-q\| \tag{5}$$

for $y \in B_s(q)$, which shows that g is Lipschitzian at q.

Now we show that g is differentiable. From the relation (2), we
have

$$\|x - p - A^{-1}(f(x) - f(p))\| \leq a\|\Delta(x)\| \tag{6}$$

Since f is differentiable at p, for any $\varepsilon > 0$ there exists $r > 0$ such
that $B_r(p) \subset U$ and if $x \in B_r(p)$, $\|\Delta(x)\| \leq \varepsilon\|x-p\|$. Hence (6) becomes

$$\|x - p - A^1(f(x) - f(p))\| \leq a\varepsilon\|x-p\|$$

for $x \in B_r(p)$. Now choose $\delta > 0$ such that $B_\delta(q) \subset V$ and $g(B_\delta(q)) \subset$
$B_r(p)$. Then if $y \in B_\delta(q)$ and $x = g(y)$, using (5) we obtain

$$\|g(y) - g(q) - A^{-1}(y-q)\| \le a\epsilon\|g(y) - g(q)\| \le 2a^2\epsilon\|y-q\|$$

This completes the proof. □

5.16 *Real and Complex Cases.* We have developed the differentiability of a function between two normed spaces over a field $\mathbb{K}$ without specifying whether $\mathbb{K} = \mathbb{R}$ or $\mathbb{K} = \mathbb{C}$. If E and F are two complex normed spaces, they can be considered as real normed spaces since $\mathbb{R}$ is a subfield of $\mathbb{C}$. If we denote these real normed spaces by $E_{\mathbb{R}}$ and $F_{\mathbb{R}}$, then $L(E;F)$ becomes a subspace of $L(E_{\mathbb{R}};F_{\mathbb{R}})$ since a complex linear map is *a fortiori* a real linear map.

Let U be an open subset of E, $f\colon U \to F$, and let $\xi \in U$. We say that f is *complex* (resp., *real*) *differentiable at* ξ if f is differentiable at ξ for the vector space structure over $\mathbb{C}$ (resp., $\mathbb{R}$). Since $L(E;F) \subseteq L(E_{\mathbb{R}};F_{\mathbb{R}})$, it is clear that *if f is complex differentiable at ξ then it is real differentiable at ξ and its differential in the real sense is equal to its differential* $Df(\xi)$ *in the complex sense.*

However, the converse is not true as shown by the following example. If we let $E = F = \mathbb{C}$, then $E_{\mathbb{R}} = F_{\mathbb{R}} = \mathbb{R}^2$, and the function $f\colon z \in \mathbb{C} \to \bar{z} \in \mathbb{C}$ is real differentiable, but not complex differentiable. In general, if f is real differentiable at ξ and $Df(\xi) \in L(E_{\mathbb{R}};F_{\mathbb{R}})$, then *for f to be complex differentiable at ξ it is necessary and sufficient that* $Df(\xi)$ *should belong to the subspace $L(E;F)$ of* $L(E_{\mathbb{R}};F_{\mathbb{R}})$.

A function which is complex differentiable on an open set is said to be *holomorphic*. We will study such functions in Part II.

EXERCISES

5A *Differentiability.* Let E and F be normed spaces, U a non-empty open subset of E, and $f\colon U \to F$. Then f is differentiable at a point $\xi \in U$ if and only if there exist $r > 0$, $A \in L(E;F)$, and $q\colon U \to F$, q being continuous and satisfying $q(\xi) = 0$, such that for all $x \in B_r(\xi)$,

$$f(x) = f(\xi) + A(x-\xi) + \|x-\xi\|q(x)$$

5B *Vector Operations.* If E is a normed space, then the mappings

$(x,y) \in E \times E \to x+y \in E$

$(\lambda,x) \in \mathbb{K} \times E \to \lambda x \in E$

are differentiable.

5C. Let $U \subset E$ be an open set containing 0. If f: $U \to F$ is differentiable such that $f(tx) = tf(x)$, then $f(x) = Df(0)(x)$; i.e., $f \in$
$L(E;F)$.

5D *Product Rule.* Let E, F_1, F_2, G be normed spaces and let $\phi \in$
$L(F_1,F_2;G)$. For an open subset U of E, let f: $U \to F_1$ and g: $U \to F_2$.
Then the product fg: $U \to G$ with respect to ϕ is defined by

$$fg(x) = \phi(f(x),g(x))$$

If both f and g are differentiable at $p \in U$, then fg is differentiable at p and

$$D(fg)(p)(x) = \phi(Df(p)(x),g(p)) + \phi(f(p),Dg(p)(x))$$

or

$$D(fg)(p) = Df(p)g(p) + f(p)Dg(p)$$

In particular, if $E = F_1 = F_2 = G = \mathbb{K}$, then we have the usual
product rule for differentiation.

5E *Quotient Rule.* Let E be a Banach space.

(a) If $A \in L(E;E)$ satisfies $\|A\| < 1$, then $I - A$ has an inverse
$B \in L(E;E)$ such that $A \circ B = B \circ A$ where I is the identity map. *Hint:*
Let $B = \Sigma A^n$, where $A^n = A \circ A^{n-1}$.

(b) The set GL(E;E) of all $A \in L(E;E)$ which have an inverse $A^{-1} \in$
$L(E;E)$ is open in $L(E;E)$.

(c) Let ϕ: $A \in GL(E;E) \to A^{-1} \in GL(E;E)$. Then ϕ is differentiable
on GL(E;E), and

$$D\phi(A)(B) = -A^{-1} \circ B \circ A^{-1}.$$

for all A $\in$ GL(E;E) and B $\in$ L(E;E).

(d)   If E = $\mathbb{K}$, then (c) is the classical quotient rule $\left(\dfrac{1}{f}\right)' = -\dfrac{f'}{f^2}.$

6

Mean Value Theorem

6.1. We now turn to the problem of obtaining a generalization of the classical mean value theorem for differentiable mappings on E to F and its applications.

It might be expected that if f: E → F is differentiable on E and a,b are members of E, then there exists a point c lying between a and b such that

$$f(b) - f(a) = Df(c)(b-a) \tag{1}$$

This conclusion fails even for functions defined on $\mathbb{R}$ to $\mathbb{R}^2$ or on $\mathbb{C}$ to $\mathbb{C}$. However, the formula (1) holds for a real-valued differentiable map defined on a real normed space (see 6.3)

Although the most natural extension of the mean value theorem does not hold for functions of a normed space to another, there are some extensions which are based on an inequality rather than an equality.

6.2 Line Segments. Let $a,b \in E$ and

$$[a,b] = \{a + \lambda(b-a): 0 \leq \lambda \leq 1\}$$

[a,b] is called the *closed line segment* joining a and b. The line segment [a,b] can be ordered in the most natural way. If $x_1, x_2 \in$ [a,b] with $x_1 = a + \lambda_1(b-a)$ and $x_2 = a + \lambda_2(b-a)$, then we write $x_1 < x_2$ if $\lambda_1 < \lambda_2$. It is clear that [a,b] is a linearly ordered set

under this ordering. Moreover, every non-empty subset of [a,b] has a supremum in [a,b]. Indeed, for A ⊂ [a,b] non-empty, let

$$A_{\mathbb{R}} = \{\lambda: a + \lambda(b-a) \in A\}$$

Then the supremum $s = \sup A_{\mathbb{R}}$ exists and the corresponding vector $a + s(b-a)$ is then the supremum of A. We also define an *open line segment* (a,b) similarly.

MEAN VALUE THEOREM FOR REAL NORMED SPACES

6.3 Let E and F be *real* normed spaces and let U be an open subset of E.

THEOREM. *Suppose that* U *contains the line segment* [a,b]. *If* f: U → $\mathbb{R}$ *is a differentiable map, then there exists* c ∈ (a,b) *such that*

$$f(b) - f(a) = Df(c)(b-a)$$

Proof. Let $\phi: \mathbb{R} \to E$ be defined by

$$\phi(\lambda) = a + \lambda(b-a)$$

so that $\phi(0) = a$, $\phi(1) = b$, and $\phi([0,1]) = [a,b]$. Since U is open and ϕ is continuous, there is a number $\delta > 0$ such that $\phi((-\delta,1+\delta)) \subset$ U. By the chain rule it follows that

$$(f\circ\phi)'(\lambda) = Df(a + \lambda(b-a))(b-a)$$

If we apply the classical mean value theorem to f∘ϕ, we infer that there exists $\lambda_0 \in (0,1)$ such that

$$f(b) - f(a) = Df(a + \lambda_0(b-a))(b-a) \qquad\qquad \square$$

6.4 THEOREM. *Suppose that* U *contains the line segment* [a,b]. *If* f: U → F *is a differentiable map, then there exists* c ∈ [a,b] *such that*

$$\|f(b) - f(a)\| \qquad \|Df(c)(b-a)\|$$

Proof. Let y = f(b) - f(a). If y is the zero vector in F, then the result is trivial. If y ≠ 0, by the Hahn-Banach Theorem 2.10 there exists φ ∈ F* such that φ(y) = $\|y\|$ and $\|φ\|$ = 1. Let A: U → ℝ be given by A = φof. Clearly A is differentiable on U and

$$A(b) - A(a) = φ(y) = \|f(b) - f(a)\|$$

$$DA(x) = φoDf(x)$$

It follows from Theorem 6.3 above that there is c ∈ [a,b] such that

$$A(b) - A(a) = (φoDf(c))(b-a)$$

hence

$$\|f(b) - f(a)\| \leq \|φ\| \cdot \|Df(c)(b-a)\| = \|Df(c)(b-a)\| \qquad\qquad \square$$

MEAN VALUE THEOREM FOR GENERAL SPACES

6.5 We now extend Theorem 6.4 to arbitrary normed spaces. Since the proof of Theorem 6.4 depended on the classical mean value theorem, we cannot use the same proof for complex normed spaces. However, the following proof is independent of the classical mean value theorem.

THEOREM. *Let E and F be normed spaces over ℝ and let U be an open subset of E such that U contains the line segment [a,b]. If f: U → F is a differentiable map, then n*

$$\|f(b) - f(a)\| \leq \|b-a\| \sup \{\|Df(x)\|: x \in [a,b]\}$$

Proof. If f(a) = f(b) or the mapping x → $\|Df(x)\|$ is unbounded on [a,b], then the theorem is evident. We now assume that M = sup $\{\|Df(x)\|: x \in [a,b]\} < ∞$. Let ε > 0 and let

$$X = \{x \in [a,b]: \|f(t) - f(a)\| \le (M + \varepsilon)\|t-a\|, \ t \in [a,x]\}$$

We notice that $a \in X$, and if $x \in X$, then $[a,b] \subset X$. Denote $c = \sup X$ (as discussed in 6.2), and we claim that $c = b$. If $c < b$, then by the differentiability of f at c, there exists $\delta > 0$ such that $B_\delta(c) \subset U$ and

$$\|f(t)-f(c)-Df(c)(t-c)\| \le \varepsilon\|t-c\| \tag{1}$$

for $t \in B_\delta(c)$. Hence if $t,t' \in B_\delta(c) \cap [a,b]$, $a < c$, $t < c < t' < b$, then $t \in X$ and from (1) we have

$$\|f(t')-f(a)\| \le \|f(t')-f(c)\| + \|f(c)-f(t)\| + \|f(t)-f(a)\|$$

$$\le (\|Df(c)\|+\varepsilon)\|t'-t\| + (M+\varepsilon)\|t-a\| \le (M+\varepsilon)(\|t'-t\| + \|t-a\|)$$

$$= (M + \varepsilon)\|t'-a\|$$

(The above argument holds trivially for $c = a$.) This shows that $t' \in X$ and also $c \in X$, contradicting the fact that $c = \sup X$. Hence $c = b$, and

$$\|f(b)-f(a)\| \le (M+\varepsilon)\|b - a\|$$

Since the left-hand side of this inequality is independent of ε, we conclude

$$\|f(b)-f(a)\| \le M\|b-a\| \qquad \qquad \square$$

COROLLARY. *Suppose the hypotheses of the preceding theorem are satisfied and that there exists $M > 0$ such that $\|Df(x)\| \le M$ for all $x \in [a,b]$. Then*

$$\|f(b) - f(a)\| \le M\|b - a\|$$

6.6 THEOREM. *Let E and F be normed spaces and let U be an open*

subset of E such that U contains the line segment [a,b]. If $f: U \to F$
is a differentiable map, then

$$\left\| f(b) - f(a) - Df(a)(b-a) \right\| \leq \left\| b-a \right\| \sup_{a \leq x \leq b} \left\| Df(x) - Df(a) \right\|$$

Proof.   If the map $g: U \to F$ is given by

$$g(x) = f(x) - Df(a)(x)$$

then

$$Dg(x) = Df(x) - Df(a)$$

If we apply the Mean Value Theorem 6.5 to g we infer that

$$\left\| f(b)-f(a)-Df(a)(b-a) \right\| = \left\| g(b)-g(a) \right\| \leq \left\| b-a \right\| \sup_{a \leq x \leq b} \left\| Df(x)-Df(a) \right\| \quad \square$$

The next lemma is useful later.

6.7   APPROXIMATION LEMMA.   *Let U be an open subset of E and let* $f: U$
$\to F$ *be of class* C^1 *on U. If* $a \in U$, *then for any* $\varepsilon > 0$ *there*
exists $\delta > 0$ *such that* $B_\delta(a) \subseteq U$ *and if* $x_1, x_2 \in B_\delta(a)$ *then*

$$\left\| f(x_1) - f(x_2) - Df(a)(x_1-x_2) \right\| \leq \varepsilon \left\| x_1-x_2 \right\|$$

Proof.   Since $Df: U \to L(E;F)$ is continuous on U, given $\varepsilon > 0$ there
exists $\delta > 0$ such that if $\left\| x-a \right\| < \delta$ then $x \in U$ and $\left\| Df(x)-Df(a) \right\| < \varepsilon$.
Now let $x_1, x_2 \in B_\delta(a)$, so $[x_1,x_2] \subseteq B_\delta(a)$, and then apply Theorem
6.6 to obtain the desired conclusion. $\square$

SOME APPLICATIONS OF THE MEAN VALUE THEOREM

6.8   *Locally Constant Maps.*   Let X and Y be topological spaces.   A
function $f: X \to Y$ is said to be *locally constant* if f is a constant
function on a neighborhood of each point of X.   We first prove the
following lemma.

LEMMA. *Let* X *be a non-empty connected topological space and let* Y *be a Hausdorff space.* If* f: X → Y *is a locally constant continuous function, then* f *is constant on* X.

Proof. For a fixed point a ∈ f(X), let A = f^{-1}(a). Then A is a non-empty closed subset of X. Since f is locally constant, A must be open. Hence A = X since X is connected. This proves that f(x) = a for all x ∈ X. □

6.9 THEOREM. *Let* E *and* F *be normed spaces,* U *an open connected subset of* E, *and* f: U → F *a differentiable map. If* Df(x) = 0 *for all* x ∈ U, *then* f *is a constant map of* U.

Proof. On the basis of the preceding lemma, it is sufficient to show that f is locally constant on U. If a ∈ U, we have $B_r(a)$ ⊂ U for some r > 0. By the Mean Value Theorem, $\|f(x) - f(a)\|$ = 0 for any x ∈ $B_r(a)$; hence f is constant on $B_r(a)$. This shows that f is locally constant. □

6.10 *Convergent Sequences of Differentiable Maps.* We use the Mean Value Theorem to obtain the following result which explicitly states a sufficient condition for the limit of a sequence of differentiable maps to be differentiable.

THEOREM. *Let* U *be an open connected subset in a normed space* E *and let* (f_n) *be a sequence of differentiable maps of* U *into a Banach space* F. *Assume the following:*
(a) There exists a point ξ ∈ U *such that the sequence* $(f_n(\xi))$ *converges in* F;
(b) For each point x ∈ U *there is a ball* $B_r(x)$ ⊂ U *such that the sequence* (Df_n) *converges uniformly on* $B_r(x)$ *to a map* g: U → L(E;F).
Then for each x ∈ U, *the sequence* (f_n) *converges uniformly on some open ball centered at* x. *Furthermore, if for each* x ∈ U *we let* f(x) = lim $f_n(x)$, *then* f *is differentiable on* U *and* Df = g; *i.e.,*

$$Df(x) = \lim Df_n(x)$$

---

* A topological space is called *Hausdorff* if each pair of distinct points has a pair of disjoint neighborhoods of these points.

Proof.   To show that $(f_n(x))$ converges at each point x in U, let D
be the set of all points x in U such that $(f_n(x))$ converges.   Then
$D \neq \emptyset$ by condition (a).   We now prove that D is a closed and open
subset of the connected set U, and hence D = U.   In fact, if $p \in D$,
by condition (b) there exists r > 0 such that (Df_n) converges to g
uniformly on $B_r(p)$.   Let ε > 0.   There exists n_o > 0 such that if
$n,m > n_o$,

$$\|Df_n(y) - Df_m(y)\| < \varepsilon/r \tag{1}$$

for all $y \in B_r(p)$; moreover,

$$\|f_n(p) - f_m(p)\| < \varepsilon \tag{2}$$

Now by the Mean Value Theorem 6.5 applied to $f_n - f_m$, we have, for
$x \in B_r(p)$,

$$\|f_n(x) - f_m(x) - [f_n(p) - f_m(p)]\| \leq \|x-p\| \sup_{y \in B_r(p)} \|Df_n(y) - Df_m(y)\| \tag{3}$$

or

$$\|f_n(x) - f_m(x)\| \leq \|f_n(p) - f_m(p)\| + \|x-p\| \sup_{y \in B_r(p)} \|Df_n(y) - Df_m(y)\| \tag{4}$$

hence, $\|f_n(x) - f_m(x)\| < 2\varepsilon$ if $n,m > n_o$.   Thus $(f_n(x))$ is a Cauchy
sequence in the Banach space F, which shows that $(f_n(x))$ converges.
Therefore, $B_r(p) \subset D$, and hence D is open.   On the other hand, if
$x \in \overline{D}$, then there exists r > 0 such that (Df_n) converges to g uniformly on $B_r(x)$.   Let $p \in B_r(x) \cap D$.   Then again by (4) we infer
that $(f_n(x))$ converges; hence $\overline{D} \subset D$.   This completes the proof that
D = U.

Consequently, let f: U → F be defined by

$$f(x) = \lim f_n(x)$$

for each $x \in U$.

It remains to be shown that f is differentiable and that $Df(x) = g(x)$ for all $x \in U$. Consider $p \in U$. We shall prove that f is differentiable at p and $Df(p) = g(p)$. To show this, we estimate

$$\left\| f(x) - f(p) - g(p)(x-p) \right\| \leq \left\| f(x) - f(p) - [f_n(x) - f_n(p)] \right\|$$

$$+ \left\| f_n(x) - f_n(p) - Df_n(p)(x-p) \right\| + \left\| Df_n(p)(x-p) - g(p)(x \cdot p) \right\| \qquad (5)$$

Let $\varepsilon > 0$. Since (Df_n) converges uniformly on $B_r(p)$ there exists $n_0 > 0$ such that for $m > n_0$, $n > n_0$, we have $\left\| Df_n(y) - Df_m(y) \right\| < \varepsilon$ for every $y \in B_r(p)$, and moreover,

$$\left\| g(p) - Df_n(p) \right\| < \varepsilon \qquad (6)$$

If we let $m \to \infty$ in (3), we obtain

$$\left\| f(x) - f(p) - [f_n(x) - f_n(p)] \right\| \leq \varepsilon \left\| x-p \right\| \qquad (7)$$

Since $Df_n(p)$ exists, there exists $r_1 \leq r$ such that

$$\left\| f_n(x) - f_n(p) - Df_n(p)(x-p) \right\| \leq \varepsilon \left\| x-p \right\| \qquad (8)$$

whenever $\left\| x-p \right\| \leq r_1$ and $x \in U$.

Therefore, we can have the following estimation from the inequalities (5) − (8):

$$\left\| f(x) - f(p) - g(p)(x-p) \right\| \leq 3\varepsilon \left\| x - p \right\|$$

whenever $\left\| x-p \right\| < r_1$, $x \in U$, which proves that $Df(p)$ exists and is equal to $g(p)$.

To show that for each $p \in U$, (f_n) converges to f uniformly on some open neighborhood of p, estimate

$$\left\| f_n(x) - f(x) - [f_n(p) - f(p)] \right\|$$

We will leave the detail to the reader. □

COROLLARY. *Let* U *be an open connected subset of a normed space* E *and let* (f_n) *be a sequence of differentiable maps of* U *into a Banach space* F. *Suppose that*

(a)   There exists a point $\xi \in$ U *such that the series* $\Sigma \, f_n(\xi)$ *converges;*

(b)   For each point $x \in$ U *there is a ball* $B_r(x) \subseteq$ U *such that the series* $\Sigma \, Df_n$ *converges uniformly on* $B_r(x)$ *to a map* $g: U \to L(E;F)$. *Then for each* $x \in$ U *the series* $\Sigma \, f_n$ *converges uniformly on some open ball centered at* x. *Furthermore, if* $f(x) = \Sigma \, f_n(x)$ *for all* $x \in$ U, *then* f *is differentiable on* U *and* $Df = g$.

The assumption that F is a Banach space was used in the preceding theorem only to guarantee that the sequence $(f_n(x))$ is convergent for all x in U. Therefore, if we assume in the theorem that $(f_n(x))$ converges for all x in U, we may dispense with the assumption that F is a Banach space.

EXERCISES

6A   *On the Classical Mean Value Theorem.*   Let $f: \mathbb{R} \to \mathbb{C}$ be defined by $f(x) = e^{ix}$. Then for any real numbers a and b, $a \neq b$, there is no real number c such that

$$f(b) - f(a) = f'(c)(b-a)$$

6B   *Lipschitz Property.*   If U is a convex open subset of a normed space E and $f: U \to F$ is differentiable on U, and $\|Df(x)\| \leq M$ for all x in U, then for any $x, y \in$ U,

$$\|f(x) - f(y)\| \leq M \|x-y\|$$

6C   *Affine Linear Maps.*   Let U be a convex open subset of E and let $f: U \to F$ be differentiable. If $Df: U \to L(E;F)$ is constant, then f is a restriction of an affine linear map.

6D   *Maps of Class* C^1.   Let $f: U \to F$ be a differentiable map and let $p \in$ U.

(a)   The derivative Df is continuous at p if and only if for any $\varepsilon > 0$ there exists $\delta > 0$ such that the relations $\|x\| < \delta$, $\|y\| < \delta$ imply

$$\|f(p+x) - f(p+y) - Df(p)(x-y)\| \lesssim \varepsilon\|x-y\|$$

(b)   The derivative Df is uniformly continuous on U if and only if for any $\varepsilon > 0$ there exists $\delta > 0$ such that the relations $\|y\| < \delta$, $x \in A$, $x + \lambda y \in U$ for $0 < \lambda < 1$ imply

$$\|f(x+y) - f(x) - Df(x)y\| \lesssim \varepsilon\|y\|$$

6E *Strongly Differentiable Maps.* Let U be an open subset of E. A mapping $f: U \to F$ is said to be *strongly differentiable* at a point $p \in U$ if (i) f is differentiable at p and (ii) for any $\varepsilon > 0$ there exists $r > 0$ such that the mapping

$$g(x) = f(x) - f(p) - Df(p)(x-p)$$

has the ε-Lipschitzian property on $B_r(p)$; i.e., if $x,y \in B_r(p)$, then $\|g(x) - g(y)\| \lesssim \varepsilon\|x-y\|$.

THEOREM. *If $f: U \to F$ is differentiable on U and the derivative $Df: U \to L(E;F)$ is continuous at $p \in U$, then f is strongly differentiable at the point p.*

6F *Sequences of Differentiable Maps.* Let E be a normed space, F a Banach space, and U a non-empty connected open subset of E. Let (f_n) be a sequence of differentiable maps of U into F and assume that

(a)   There exits a point $p \in U$ such that the sequence $(f_n(p))$ converges in F;

(b)   The sequence (Df_n) is uniformly convergent on every compact subset of U.

Then if we let $f(x) = \lim f_n(x)$ for all $x \in U$, the mapping $f: U \to F$ is differentiable on U and $Df(x) = \lim Df_n(x)$ for all $x \in U$. (When E is finite dimensional and therefore locally compact, this result coincides with Theorem 6.10.)

7

Higher Differentials

7.1. We now introduce higher order differentials and prove the Schwarz Symmetric Theorem, which states that all higher differentials can be considered as symmetric multilinear maps. We also introduce the Gâteaux differential and study its relationship with the Fréchet differential.

SECOND DIFFERENTIALS

7.2 Let E,F be normed spaces, U an open subset of E, and f: U → F a differentiable map on U. Then

   Df: U → L(E;F)

Since L(E;F) is a normed space, we are in a position to define the differential of Df, the second differential of f.

If Df is differentiable at a point $\xi \in$ U, we say that f is *twice differentiable at* ξ and the differential of Df at ξ is called the *second differential* of f at ξ, and written $D^2f(\xi)$. Then we have

$$D^2f(\xi) \in L(E;L(E;F))$$

Without assuming that f is differentiable on the entire subset U, we can say more generally that f is *twice differentiable* at the point $\xi \in$ U if:

   (1)   f is differentiable on a neighborhood V of ξ;

   (2)   the map Df: V → L(E;F) is differentiable at the point ξ.

If Df: $U \to L(E;F)$ is differentiable on U, we say that f is *twice differentiable on* U. In this case, we have the *second differential*

$$D^2f: U \to L(E;L(E;F))$$

7.3 *Second Differentials and Bilinear Maps*. Although the space $L(E;L(E;F))$ arises naturally in defining the second differential, it is very complicated to elucidate the meaning of the mapping $D^2f(\xi): E \to L(E;F)$. To avoid this complication, we identify $L(E;L(E;F))$ with $L(^2E;F)$, the space of continuous bilinear maps of $E \times E$ into F, as we studied in 3.10. We recall that this is done by identifying $A \in L(E;L(E;F))$ with the bilinear map A*: $(x,y) \in E \times E \to A(x)(y) \in F$. This correspondence is a natural isometric isomorphism.

Thus for the second differential

$$D^2f(\xi) \in L(E;L(E;F))$$

we can denote by

$$d^2f(\xi) \in L(^2E;F)$$

the corresponding bilinear map in $L(^2E;F)$ under the natural isometric isomorphism. Then

$$d^2f(\xi)(x,y) = D^2f(\xi)(x)(y)$$

for any $x,y \in E$. Hence we have

$$D^2f: U \to L(E;L(E;F))$$

$$d^2f: U \to L(^2E;F)$$

Both notations D^2f and d^2f will be used interchangeably and will be called the second differential, but keep in mind that they represent two different objects.

7.4 *The Schwarz Symmetric Theorem.* For an open set $U \subset \mathbb{R}^2$, let
f: $U \rightarrow \mathbb{R}$ and $\xi \in U$. The following theorem from the elementary
calculus is well known.
If the partial derivatives $\dfrac{\partial f}{\partial x}, \dfrac{\partial f}{\partial y}$: $U \rightarrow \mathbb{R}$ *are differentiable on* U,
then

$$\frac{\partial^2 f}{\partial x \partial y} = \frac{\partial^2 f}{\partial y \partial x}$$

This is the so-called Schwarz Symmetric Theorem. We will derive this
result in 7.13 from the following generalization.

THEOREM. *Let* f: $U \rightarrow F$ *be twice differentiable at* $\xi \in U$. *Then* $d^2 f(\xi)$
is a symmetric bilinear map; i.e., for all $h, k \in E$,

$$d^2 f(\xi)(h,k) = d^2 f(\xi)(k,h)$$

Proof. Since f is twice differentiable at ξ, there is some r > 0
such that $B_r(\xi) \subset U$ and f is differentiable on $B_r(\xi)$. Consider the
function

$$\Delta(h,k) = f(\xi+h+k) - f(\xi+h) - f(\xi+k) + f(\xi)$$

where $h, k \in E$ are such that $\xi+h+k$, $\xi+h$, $\xi+k$ belong to U. More pre-
cisely, we may assume $\|h\| \leq r/2$ and $\|k\| \leq r/2$. Then

$$\Delta(h,k) = \Delta(k,h)$$

We wish to approximate $d^2 f(\xi)(h,k)$ with $\Delta(h,k)$ to show $d^2 f(\xi)$ is
symmetric. Fix h and define g: $B_{r/2}(\xi) \rightarrow F$ by

$$g(x) = f(x+h) - f(x)$$

Then

$$Dg(x) = Df(x+h) - Df(x) \quad ;$$

$$\Delta(h,k) = g(\xi+k) - g(\xi)$$

Hence

$$\left\| \Delta(h,k) - d^2 f(\xi)(h,k) \right\| \leq \left\| g(\xi+k) - g(\xi) - Dg(\xi)(k) \right\|$$

$$+ \left\| Dg(\xi)(k) - D^2 f(\xi)(h)(k) \right\| \leq \left\| g(\xi+k) - g(\xi) - Dg(\xi)(k) \right\|$$

$$+ \left\| Dg(\xi) - D^2 f(\xi)(h) \right\| \left\| k \right\| \tag{1}$$

On the other hand, by Theorem 6.6 we have

$$\left\| g(\xi+k) - g(\xi) - Dg(\xi)(k) \right\| \leq \left\| k \right\| \sup \left\| Dg(\xi+x) - Dg(\xi) \right\|$$

$$\leq \left\| k \right\| \sup \left(\left\| Dg(\xi+x) - D^2 f(\xi)(h) \right\| + \left\| D^2 f(\xi)(h) - Dg(\xi) \right\| \right)$$

$$\leq 2 \left\| k \right\| \sup \left\| Dg(\xi+x) - D^2 f(\xi)(h) \right\| \tag{2}$$

where the suprema are taken over $\|x\| \leq \|k\|$.

The inequality (1) now becomes

$$\left\| \Delta(h,k) - d^2 f(\xi)(h,k) \right\| \leq 3 \left\| k \right\| \sup \left\| Dg(\xi+x) - D^2 f(\xi)(h) \right\| \tag{3}$$

for $\|x\| \leq \|k\|$.

We first estimate $\left\| Dg(\xi+x) - D^2 f(\xi)(h) \right\|$. Notice

$$Dg(\xi+x) - D^2 f(\xi)(h) = [Df(\xi+x+h) - Df(\xi) - D^2 f(\xi)(x+h)]$$

$$- [Df(\xi+x) - Df(\xi) - D^2 f(\xi)(x)] \tag{4}$$

Since Df is differentiable at ξ, for any $\varepsilon > 0$ there is s, $0 < 4s < r$ such that

$$\left\| Df(\xi+t) - Df(\xi) - D^2 f(\xi)(t) \right\| \leq \varepsilon \|t\| \tag{5}$$

for $\|t\| \leq 2s$.   Then, from (4) and (5) we have

$$\|Dg(\xi+x)-D^2f(\xi)(h)\| \leq \varepsilon(\|x+h\| + \|x\|) \leq \varepsilon(\|h\| + 2\|x\|) \tag{6}$$

if $\|x\| \leq s$ and $\|h\| \leq s$.

   The inequality (3) can be rewritten as

$$\|\Delta(h,k)-d^2f(\xi)(h,k)\| \leq 3\varepsilon(\|h\|\|k\| + 2\|k\|^2) \tag{7}$$

provided that $\|h\| \leq s$ and $\|k\| \leq s$.   Interchanging h and k in (7), we obtain

$$\|\Delta(h,k)-d^2f(\xi)(k,h)\| \leq 3\varepsilon(\|h\|\|k\| + 2\|h\|^2) \tag{8}$$

Thus it follows from (7) and (8) that

$$\|d^2f(\xi)(h,k)-d^2f(\xi)(k,h)\| \leq 6\varepsilon(\|h\| + \|k\|)^2 \tag{9}$$

provided that $\|h\| \leq s$ and $\|k\| \leq s$.   We now claim that the inequality (9) is true for any h and k in E.   In fact, for any h and k in E choose $\lambda > 0$ such that $\|\lambda h\| \leq s$ and $\|\lambda k\| \leq s$.   If we replace h and k by λh and λk, we obtain the inequality (9) multiplied by λ^2 on both sides; hence the result is true for any h and k in E.   Since $\varepsilon > 0$ was arbitrary in (9), we conclude that

$$d^2f(\xi)(h,k) = d^2f(\xi)(k,h)$$

for any h and k in E, which completes the proof.   □

HIGHER DIFFERENTIALS

7.5   *The mth Differential.*   The third differential is the differential of the second, the fourth is the differential of the third, and so on.   The general definition is best expressed inductively.

   Let $f: U \to F$ be twice differentiable.   Then we have the second differential

$$D^2 f: U \to L(E;L(E;F))$$

$$d^2 f: U \to L(^2 E;F)$$

By repeating the procedure inductively we can define the mth differential of f. We first simplify some notation. For m = 1, we write

$$L^1(E;F) = L(E;F)$$

and for m = 2,

$$L^2(E;F) = L(E;L(E;F))$$

Inductively we write

$$L^m(E;F) = L(E;L^{m-1}(E;F))$$

For example, if f: U → F is twice differentiable on U, we have

$$Df: U \to L^1(E;F) = L(E;F)$$

$$D^2 f: U \to L^2(E;F)$$

After defining

$$D^{m-1} f: U \to L^{m-1}(E;F)$$

if $D^{m-1}f$ is differentiable at the point $\xi \in U$, we let

$$D^m f(\xi) = D(D^{m-1}f)(\xi) \in L^m(E;F)$$

which will be called the mth *differential of f* at ξ. Of course, we also have

$$D^m f: U \to L^m(E;F)$$

if $D^m f(\xi)$ exists at each point ξ of U. The identity

$$D^p(D^q f) = D^m f$$

is trivially true if $p + q = m$ and $D^m f$ exists. Also, the mth differential D^m is linear in the sense that

$$D^m(f + g) = D^m f + D^m g \quad \text{and} \quad D^m(\lambda f) = \lambda D^m f$$

where $\lambda \in \mathbb{K}$. For convenience, we identify

$$D^0 f = f$$

For $\xi \in U$ and $x_1, \ldots, x_m \in E$, we have

$$D^m f(\xi)(x_1)(x_2) \cdots (x_k) \in L^{m-k}(E;F)$$

for $1 \leq k \leq m$. To avoid the cumbersome parentheses, we often write

$$D^m f(\xi)(x_1, \ldots, x_k) = D^m f(\xi)(x_1) \cdots (x_k)$$

If $D^m f : U \to L^m(E;F)$ is continuous, we say that f is of *class* C^m.
If f is of class C^m for every m, f is said to be of *class* C^∞ ; in
this case f is called *infinitely many times differentiable*.

7.6 *The Differentials* $D^m f$ *and* $d^m f$. As in 7.3 we will identify the
mth differential $D^m f(\xi)$ with the corresponding m-linear map $d^m f(\xi)$.
More explicitly, let

$$\phi : L(^m E:F) \to L^m(E;F)$$

be the natural isometric isomorphism defined by

$$\phi(A)(x_1) \cdots (x_m) = A(x_1, \ldots, x_m)$$

For $D^m f : U \to L^m(E;F)$, we define

$$d^m f : U \to L(^m E;F)$$

to be such that

$$d^m f = \phi^{-1} \circ D^m f$$

We call $d^m f$ also the m*th differential of* f on U. Then it is evident that

$$d^m f(\xi)(x_1, \ldots, x_m) = D^m f(\xi)(x_1, \ldots, x_m)$$

Notice that if m = 1,

$$df = Df$$

Again we have to keep in mind that the mth differentials $D^m f$ and $d^m f$ represent two different objects if m ≠ 1. We will prefer the mth differential $d^m f$ to $D^m f$ in this book because the former is much simpler.

We should notice that although for m = p + q we have $D^m f(\xi)$ = $D^p(D^q f)(\xi)$, we cannot expect to have $d^m f(\xi) = d^p(d^q f)(\xi)$ since the last two objects are elements of two different spaces. (Why?)

7.7   THEOREM. *If* A ∈ $L_s(^m E; F)$ *and* P = $\hat{A}$ ∈ P($^m E; F$), *then*

$$d^k P(x) = \begin{cases} m(m-1) \cdots (m-k+1) \ Ax^{m-k} & 0 \leq k \leq m \\ 0 & k > m \end{cases}$$

$$d^k P: E \to L(^k E; F)$$

Proof. We have shown this result for k = 1 in 5.14. The rest can be proved by induction and is left to the reader.   □

7.8 *The mth Differential in Relation to Linear Maps.* For the higher differentials, we have similar statements to those obtained with the first differential in relation to linear maps (see 5.11). Notice that if A ∈ $L(^m E; F)$ and ϕ ∈ L(F; G), then the composite map

ϕoA $\in$ L(^{m}E;G). Let ϕ*: L(^{m}E;F) $\to$ L(^{m}E;G) be given by

$$\phi*(A) = \phi oA$$

Then it is immediate that ϕ* is linear. Since

$$\|\phi oA(x_1,\ldots,x_m)\| \leq \|\phi\|\|A\|\|x_1\| \cdot\cdot\|x_m\|$$

ϕ* is also continuous.

THEOREM. *For an open subset* U *of* E, *let* f: U $\to$ F *be* m *times differ-*
entiable and let ϕ $\in$ L(F;G). *Then for any* ξ $\in$ U *we have*

$$d^m(\phi of)(\xi) = \phi od^m f(\xi)$$

Proof. Proceed by induction on m. Consider the map x $\in$ U $\to$
$d^{m-1}(\phi of)(x)$. By induction we get

$$d^{m-1}(\phi of)(x) = \phi od^{m-1} f(x)$$

Now the composite map of

$$d^{m-1}f: U \to L(^{m-1}E;F)$$

$$\phi*: L(^{m-1}E;F) \to L(^{m-1}E;G)$$

is differentiable by (5.11) and

$$d(\phi*od^{m-1}f) = \phi*od^m f$$

Therefore, for any ξ,

$$d^m(\phi of)(\xi) = \phi od^m f(\xi) \qquad\qquad \square$$

7.9 *The Schwarz Symmetric Theorem*. We have seen in 7.4 that the
second differential $d^2 f(\xi)$ is a symmetric bilinear map. This result

can be extended to the mth differential $d^m f(\xi)$ by induction. First we need the following lemma.

LEMMA. *Let* $f: U \to F$ *be* (m-1) *differentiable on* U *and* m-*differentiable at* $\xi \in U$. *For* $m \geq 2$ *and for any fixed* $x_2, \ldots, x_m \in E$, *if* $g: U \to F$ *is defined by*

$$g(x) = d^{m-1} f(x) (x_2, \ldots, x_m)$$

then g *is differentiable at* ξ *and*

$$dg(\xi)(x) = d^m f(\xi)(x, x_2, \ldots, x_m)$$

for all $x \in E$.

Proof. The map $g: U \to F$ satisfies the composition

$$g = \lambda \circ \phi^{-1} \circ D^{m-1} f$$

where $\phi: L(^{m-1}E;F) \to L^{m-1}(E;F)$ is the natural isometric isomorphism (see 3.10) and $\lambda: L(^{m-1}E;F) \to F$ is given by $\lambda(A) = A(x_2, \ldots, x_m)$ for all $A \in L(^{m-1}E;F)$. λ is clearly continuous and linear. Now

$$Dg = D(\lambda \circ \phi^{-1} \circ D^{m-1} f) = \lambda \circ \phi^{-1} \circ D(D^{m-1} f) = \lambda \circ \phi^{-1} \circ D^m f$$

Hence, for all $x \in E$, we have

$$dg(\xi)(x) = Dg(\xi)(x) = (\lambda \circ \phi^{-1} \circ D^m f(\xi))(x) = \lambda(\phi^{-1}(D^m f(\xi)(x)))$$

$$= \lambda(d^m f(\xi)(x)) = d^m f(\xi)(x, x_2, \ldots, x_m)$$

which is what we set out to prove. $\square$

THEOREM (Schwarz). *Let* $f: U \to F$ *be* m-*differentiable at* $\xi \in U$. *Then* $d^m f(\xi)$ *is a symmetric* m-*linear map; i.e.,* $d^m f(\xi) \in L_s(^m E;F)$.

Proof. For $m = 2$, this was proved in Theorem 7.4. We now proceed by induction on m for $m \geq 3$. Assume that the theorem has been proved

for m-1. Then $d^m f(\xi)$ is the differential of the mapping $d^{m-1}f$,
which by our assumption exists in a neighborhood V of ξ. By the
induction hypothesis, we can assume that

$$d^{m-1}f: V \to L_s(^{m-1}E;F)$$

By the lemma, for fixed $x_3, \ldots, x_m \in E$, the mapping g: $V \to F$ defined
by

$$g(x) = d^{m-2}f(x)(x_3, \ldots, x_m)$$

is differentiable at ξ and its differential at ξ is given by

$$dg(x)(y) = d^{m-1}f(x)(y, x_3, \ldots, x_m)$$

for all $x \in V$ and for all $y \in E$. Similarly, applying the lemma to
the map $x \mapsto d^{m-1}f(x)(x_2, \ldots, x_m)$, where $x_2, \ldots, x_m$ are fixed, we
obtain

$$d^2 g(\xi)(x_2, x_1) = d^m f(\xi)(x_1, x_2, \ldots, x_m) \tag{1}$$

for all $x_1 \in E$. From 7.4 we have

$$d^2 g(\xi)(x_1, x_2) = d^2 g(\xi)(x_2, x_1)$$

and hence

$$d^m f(\xi)(x_2, x_1, x_3, \ldots, x_m) = d^m f(\xi)(x_1, \ldots, x_m) \tag{2}$$

But $d^{m-1}f(\xi)$ is symmetric and $x_2, \ldots, x_m$ were fixed but arbitrary,
so from (1), for $x_1 \in E$, the map

$$d^m f(\xi)(x_1): (x_2, \ldots, x_m) \to d^m f(\xi)(x_1, \ldots, x_m) \tag{3}$$

is symmetric. This shows that the multilinear map $d^m f(\xi): E^m \to F$
is symmetric by (2) and (3), so the proof is complete.   □

DIRECTIONAL DERIVATIVES

7.10 *Gâteaux Differential.* Let f: U → F, where U is an open subset
of E. If given $\xi \in$ U and h $\in$ E, and the limit

$$\frac{\partial f}{\partial h}(\xi) = \lim_{\lambda \to 0} \frac{f(\xi+\lambda h) - f(\xi)}{\lambda}$$

exists (here $\lambda \in$ $\mathbb{K}$), then f is said to be *Gâteaux differentiable at*
ξ *in the direction* h; $\frac{\partial f}{\partial h}(\xi)$ is called the *Gâteaux derivative of* f
at ξ *in the direction* h. If f is Gâteaux differentiable at ξ in
any direction h, we say that f is *Gâteaux differentiable at* ξ; *the*
mapping

$$\partial f(\xi): x \in E \to \frac{\partial f}{\partial x}(\xi) \in F$$

is called the *Gâteaux differential* or *derivative.*

If f is Gâteaux differentiable at each point of U in the direction
h, the mapping

$$\frac{\partial f}{\partial h}: x \in U \to \frac{\partial f}{\partial h}(x) \in F$$

is called the *derivative of f on* U *in the direction* h.

7.11 *EXAMPLES.* (a) Consider f: $\mathbb{K}^n \to \mathbb{K}$, and the basis $e_i =$
$(0,\ldots,0,1,0,\ldots,0)$, i=1,$\ldots$,n. Then

$$x = (x_1,\ldots,x_n) = \Sigma\ x_i e_i$$

and

$$\frac{\partial f}{\partial e_i}(x) = \frac{\partial f}{\partial x_i}(x)$$

where $\frac{\partial f}{\partial x_i}$ is the partial derivative of f with respect to the ith coordinate. Thus the partial derivative is the derivative in the direction e_i.

(b) Let $f: \mathbb{K}^2 \to \mathbb{K}$ be defined by

$$f(x) = \begin{cases} \dfrac{x_1 x_2^2}{x_1^2 + x_2^2} & (x \neq 0) \\[2mm] 0 & (x = 0) \end{cases}$$

Then $\dfrac{\partial f}{\partial h}(0) = f(h)$, which shows that the Gâteaux derivative is not, in general, a linear map.

(c) Let $f: \mathbb{K}^2 \to \mathbb{K}$ be defined by

$$f(x) = \begin{cases} x_1^3 / x_2 & (x \neq 0) \\[2mm] 0 & (x = 0) \end{cases}$$

Then $\dfrac{\partial f}{\partial h}(0) = 0$ for all $h \in \mathbb{K}^2$. Hence f is Gâteaux differentiable at 0, but f is not continuous.

7.12 *Gâteaux vs. Fréchet Differentials.* Unlike the Fréchet differential, the Gâteaux differential may not be linear, and Gâteaux differentiability does not necessarily imply the continuity of the function. We now show that Fréchet differentiability is stronger than Gâteaux differentiability.

THEOREM. *If f: U $\to$ F is Fréchet differentiable at the point $\xi \in U$, then f is Gâteaux differentiable at ξ and*

$$\frac{\partial f}{\partial h}(\xi) = Df(\xi)h$$

$$\partial f(\xi) = Df(\xi)$$

Proof. We have

$$\lim_{x \to 0} \frac{\|f(\xi+x) - f(\xi) - Df(\xi)x\|}{\|x\|} = 0$$

Replace x with λh to get the desired result. ◻

COROLLARY. *If* f: U → F *is differentiable at the point* $\xi \in U$, *then*

$$\frac{df(\xi+\lambda h)}{d\lambda} \bigg|_{\lambda = 0} = \frac{\partial f}{\partial h}(\xi)$$

Proof. Let g: V → F be defined by

$$g(\lambda) = f(\xi+\lambda h)$$

where V is a neighborhood of O in $\mathbb{K}$. Then by the chain rule, we obtain the derivative (see 5.5)

$$g'(0) = Df(\xi)h$$

Hence

$$g'(0) = \frac{\partial f}{\partial h}(\xi)$$ ◻

7.13 *The Schwarz Theorem Again.* As promised in 7.4, we are going to derive the classical form of the Schwarz theorem. If $\frac{\partial f}{\partial h}$: U → F has a derivative on U in the direction h, we write

$$\frac{\partial^2 f}{\partial h \partial k} = \frac{\partial}{\partial h} \left(\frac{\partial f}{\partial k} \right)$$

THEOREM. *Let* f: U → F *be twice differentiable on* U. *Then for any* h,k $\in$ E, $\frac{\partial f}{\partial k}$: U → F *is differentiable on* U *such that for any* x $\in$ U,

$$\frac{\partial^2 f}{\partial h \partial k}(x) = d^2 f(x)(h,k)$$

Hence,

$$\frac{\partial^2 f}{\partial h \partial k} = \frac{\partial^2 f}{\partial k \partial h}$$

Proof. Since $\frac{\partial f}{\partial k}(x) = Df(x)k$, $\frac{\partial f}{\partial k}$ is a composition of the differentiable mappings

$$x \in U \to Df(x) \in L(E;F)$$

$$A \in L(E;F) \to A(x) \in F$$

Thus $\frac{\partial f}{\partial k}$ is differentiable and its differential at x is given by

$$D(\frac{\partial f}{\partial k}(x))(h) = \frac{\partial^2 f}{\partial h \partial k}(x)$$

We have then

$$d^2 f(x)(h,k) = \frac{\partial^2 f}{\partial h \partial k}(x)$$

Since $d^2 f(x)$ is symmetric by Theorem 7.4, this establishes the commutativity of directional derivatives.   □

INVERSE MAPPING THEOREM

7.14. The inverse mapping theorem and the implicit mapping theorem are two major results in differential calculus and main pillars of nonlinear analysis. They play important roles in differential equations and in the study of manifolds. In this book these results are not used later and may be omitted until the reader needs them. To state the inverse mapping theorem we need the following concept.

      A map $f\colon U \subseteq E \to V \subseteq F$ (where U and V are open) is said to be a C^m-*diffeomorphism* if f is of class C^m, f is a bijection, and f^{-1} is also of class C^m.

      All of the normed spaces in this section are assumed to be Banach spaces since we use the following fixed point theorem.

BANACH FIXED POINT THEOREM. Let X be a complete metric space and let f: X → X be a contraction map; i.e., for some C, $0 < C < 1$,

$$d(f(x),f(y)) \leq Cd(x,y)$$

for all x,y ∈ X. Then f has a unique fixed point in X.

We prove the inverse mapping theorem 7.18 by assembling first a few standard theorems.

7.15 *Banach Algebra* L(E;E). On the Banach space L(E;E) if we define the multiplication AB of two members A,B in L(E;E) by the composition A ∘ B, then

$$\|AB\| \leq \|A\|\|B\|$$

This makes L(E;E) a Banach algebra with the multiplicative identity I, the identity map.

THEOREM. *If* A ∈ L(E;E) *satisfies* $\|A\| < 1$, *then* I − A *has a multiplicative inverse* B *in* L(E;E) *such that* AB = BA.

Proof. Let $B = \sum_{n=0}^{\infty} A^n$ where $A^0 = I$. Then B is clearly linear and

$$\|B\| \leq \sum_{n=0}^{\infty} \|A\|^n = \frac{1}{1-\|A\|}$$

Hence B ∈ L(E;E) by Theorem 1.18. It is easy to see that B is the inverse of I − A and AB = BA. □

7.16 THEOREM. *Let* GL(E;F) *denote the set of linear isomorphisms in* L(E;F). *Then* GL(E;F) *is open in* L(E;F).

Proof. We may assume that E = F. Indeed, if A ∈ GL(E;F), then the map

$B \in L(E;F) \rightarrow A^{-1}B \in L(E;E)$

is continuous and $GL(E;F)$ is the inverse image of $GL(E;E)$.

For $A \in GL(E;E)$, we claim that if $B \in L(E;E)$ and $\|A - B\| < \|A^{-1}\|^{-1}$, then $B \in GL(E;E)$, which proves that $GL(E;E)$ is open in $L(E;E)$.

By the preceding theorem, $I - A^{-1}(A - B)$ is invertible since $\|A^{-1}(A - B)\| < 1$. Hence AB is invertible since $AB = A[I - A^{-1}(A - B)]$. From this we can see that B is invertible.   □

7.17 THEOREM (Quotient Rule). *Let* $\phi: GL(E;F) \rightarrow GL(F;E)$ *be defined by* $\phi(A) = A^{-1}$. *Then* ϕ *is of class* C^{∞} *and*

$$D\phi(A)(B) = -A^{-1}BA^{-1}$$

Proof. We may assume that $GL(E;F)$ is nonempty. It is clear that ϕ is of class C^{∞} if the formula is true. To verify the formula, since $B \rightarrow -A^{-1}BA^{-1}$ is linear on $L(E;F)$, it suffices to show

$$\lim_{B \rightarrow A} \frac{\left\| B^{-1} - A^{-1} + A^{-1}(B - A)A^{-1} \right\|}{\|B - A\|} = 0$$

We now estimate the numerator of the above limit:

$$\left\| B^{-1} - A^{-1} + A^{-1}(B - A)A^{-1} \right\| = \left\| B^{-1}(B - A)A^{-1}(B - A)A^{-1} \right\|$$

$$< \|B^{-1}\| \|B - A\|^2 \|A^{-1}\|^2$$

From this inequality we can see that the above limit is clearly zero.   □

7.18 THEOREM (Inverse Mapping Theorem). *Let* $f: U \rightarrow F$ *be of class* C^m, $m \geq 1$, $\xi \in U$, *and suppose that* $Df(\xi)$ *is a linear isomorphism; i.e.*, $Df(\xi) \in GL(E;F)$. *Then* f *is a* C^m-*diffeomorphism of some neighborhood of* ξ *onto some neighborhood of* $f(\xi)$.

Proof. For simplicity we may assume that $\xi = 0$, $E = F$, $f(0) = 0$, and $Df(0) = I$. In fact, we can replace f by

$$g(x) = Df(\xi)^{-1}[f(x+\xi) - f(\xi)]$$

Let $h(x) = x - f(x)$. Then $Dh(0) = 0$. Since Dh is continuous, we can find $r > 0$ such that $\|Dh(x)\| < 1/2$ for $\|x\| < r$. From the mean value theorem 6.5, it follows that $\|h(x)\| < r/2$ if $\|x\| < r$.

Let $h_y(x) = y + h(x)$. Then, by the mean value theorem again, if $y \in \bar{B}_{r/2}(0)$ and $x_1, x_2 \in \bar{B}_r(0)$, then

$$\|h_y(x)\| \le \|y\| + \|h(x)\| < r \tag{1}$$

$$\|h_y(x_1) - h_y(x_2)\| \le \tfrac{1}{2}\|x_1 - x_2\| \tag{2}$$

Therefore h_y is a continuous mapping on $\bar{B}_r(0)$; hence h_y has a unique fixed point in $\bar{B}_r(0)$ by the Banach fixed point theorem. Let x be the fixed point of h_y. Then $f(x) = y$ or $x = f^{-1}(y)$. Thus f^{-1} is now well-defined and

$$f^{-1}: B_{r/2}(0) \to f^{-1}(B_{r/2}(0)) \subset B_r(0)$$

From (2) above,

$$\|f^{-1}(y_1) - f^{-2}(y_2)\| \le 2\|y_1 - y_2\| \tag{3}$$

so f^{-1} is continuous. Thus f is a homeomorphism on a neighborhood of 0.

It remains to show that f^{-1} is differentiable. Since Df is continuous, $Df(0) = I \in GL(E;E)$, and $GL(E;E)$ is open in $L(E;E)$, we can find $s > 0$ such that $Df(x) \in GL(E;E)$ for $\|x\| < s$. Let $t = \min\{r,s\}$. Then f^{-1} is a homeomorphism on $B_{t/2}(0)$ and $Df(x) \in GL(E;E)$ for $\|x\| < r$. Hence by Theorem 5.15, f^{-1} is differentiable on $B_{t/2}(0)$ and $Df^{-1}(y) = [Df(f^{-1}(y))]^{-1}$.

It follows from 7.17 that f^{-1} is of class C^1, and from the formula for $Df^{-1}(y)$ we can see that f^{-1} is of class C^2. The general case now follows by induction. $\qquad\square$

IMPLICIT MAPPING THEOREM

7.19 *Partial Differentials.* In order to state the implicit mapping theorem, we need to introduce partial differentials. Consider a product $E = E_1 \times E_2$ of normed spaces and let U_1 and U_2 be open in E_1 and E_2 respectively. For a map

$$f: U_1 \times U_2 \to F$$

consider the partial map

$$x \in U_1 \to f(x,b) \in F$$

for a fixed b. If this mapping is differentiable at $a \in U_1$, we call its differential *the partial differential* of f at (a,b) and denote it by

$$D_1 f(a,b)$$

Similarly, we define $D_2 f(a,b)$.

It is clear that if $f: U_1 \times U_2 \to F$ is differentiable at $(a,b) \in U_1 \times U_2$, then $D_1 f(a,b)$ and $D_2 f(a,b)$ exist and

$$Df(a,b)(x,y) = D_1 f(a,b)(x) + D_2 f(a,b)(y) = \begin{bmatrix} D_1 f(a,b) & D_2 f(a,b) \end{bmatrix} \begin{pmatrix} x \\ y \end{pmatrix}$$

(The last expression is in matrices.)

If $f: U_1 \times U_2 \to F_1 \times F_2$ is differentiable at (a,b), then we can represent the differential Df(a,b) in the following form

$$Df(a,b)(x,y) = \begin{pmatrix} D_1 f_1(a,b) & D_2 f_1(a,b) \\ \\ D_1 f_2(a,b) & D_2 f_2(a,b) \end{pmatrix} \begin{pmatrix} x \\ \\ y \end{pmatrix}$$

where $f(x,y) = (f_1(x,y), f_2(x,y))$ by 5.18.

7.20 THEOREM (Implicit Mapping Theorem). *Let U and V be open sub-sets of Banach spaces E and F respectively, and let* $f: U \times V \to G$ *be of class* C^m, $m \geq 1$ *where G is also a Banach space. Assume that for some* $(a,b) \in U \times V$, $D_2 f(a,b) \in GL(F;G)$. *Then there exist neighborhoods* U_o *of a and* W_o *of f(a,b) and a unique* C^m-*map*

$$g: U_o \times W_o \to V$$

such that for all $(x,w) \in U_o \times W_o$,

$$f(x,g(x,w)) = w$$

Proof. Consider the map $\phi: U \times V \to E \times G$ defined by

$$\phi(x,y) = (x, f(x,y))$$

Then ϕ is differentiable and $D\phi(a,b)$ is given by

$$D\phi(a,b)(x,y) = \begin{pmatrix} I & O \\ D_1 f(a,b) & D_2 f(a,b) \end{pmatrix} \begin{pmatrix} x \\ y \end{pmatrix}$$

Since $D_2 f(a,b)$ is an isomorphism, we see that $D\phi(a,b)$ is invertible and an isomorphism of $E \times F$ onto $E \times G$. Therefore ϕ has a unique local inverse

$$\phi^{-1}: U_o \times W_o \to U \times V$$

by the inverse mapping theorem 7.18 for some neighborhoods U_o and W_o of a and f(a,b). Let g such that

$$\phi^{-1}(x,w) = (x, g(x,w))$$

Then clearly g is of class C^m. $\qquad\qquad\square$

7.21 *Complementary Subspaces.* Let E be a normed space (or more
generally, a topological vector space) and F a closed subspace of E.
Then F is said to be a *complementary subspace* of E if there is a
closed subspace G of E such that $F \cap G = \{0\}$, $E = F + G$, and the map

$$(x,y) \in F \times G \to x + y \in E$$

is a topological isomorphism.

It is an exercise to show that F is a complementary subspace of
E if and only if F is closed and there exists $T \in L(E;E)$, $T \neq I$, such
that

$$T^2 = T$$

on F.

7.22 THEOREM. *Let U be an open subset of E and let* f: U → F *be of
class* C^m, $m \geq 1$. *If* Df(a) *is surjective and* ker Df(a) =
$\{x \in E: Df(a)(x) = 0\}$ *is complementary, then* f(U) *contains a neighborhood of* f(a).

Proof. Let E_1 = Ker Df(a) and E_2 a complementary subspace of E
corresponding to E_1. Then

$$D_2 f(a): E_2 \to F$$

is an isomorphism. Thus the hypotheses of the implicit mapping
theorem are satisfied and so f(U) contains W_o provided by that
theorem.   □

SURJECTIVE MAPPING THEOREM

7.23 In this section we eliminate the condition in Theorem 7.22
that ker Df(a) is complementary. This can be achieved by using the
Banach open mapping theorem 2.15, a result due to Graves, 1935.
First we need the following important theorem.

THEOREM. *Let E and F be Banach spaces. The subspace of* $L(E;F)$ *consisting of all surjective maps is open in* $L(E;F)$.

Proof. Let $A:E \to F$ be a continuous linear map and assume that A is surjective. By the open mapping theorem 2.15, A is an open map. By the openness and linearity of A, (*) there exists $M > 0$ such that for any $y \in F$, $\|y\| < 1$, if $A(x) = y$, then $\|x\| \leq M\|y\|$ (why?). We may assume $M = 1$, if necessary, by changing the norm on F to an equivalent norm.

Let $0 < \varepsilon < 1$ and $B \in L(E;F)$ be such that $\|A - B\| < \varepsilon$. We prove that B is surjective; then the subspace in question is open in $L(E;F)$. It is sufficient to show that B maps the ball of radius $1/(1-\varepsilon)$ in E onto the unit ball of F.

Let $y_1 \in F$ be such that $\|y_1\| \leq 1$. We want to find x in E with $\|x\| \leq 1/(1-\varepsilon)$ such that $y_1 = B(x)$. Since $\|y_1\| \leq 1$, it follows from the remark (*) above that there exists x_1 in E such that $A(x_1) = y_1$ and $\|x_1\| \leq \|y_1\| \leq 1$. Let

$$y_2 = A(x_1) - B(x_1)$$

Then $\|y_2\| \leq \varepsilon$. Again by the same remark, we can find x_2 in E such that $A(x_2) = y_2$ and $\|x_2\| \leq \varepsilon$.

Now let

$$y_3 = A(x_2) - B(x_2)$$

Then $\|y_3\| \leq \varepsilon^2$. Inductively, we continue to find sequences (x_n) in E and (y_n) in F such that $y_{n+1} = A(x_n) - B(x_n)$, $\|y_{n+1}\| \leq \varepsilon^n$, and $\|x_{n+1}\| \leq \varepsilon^n$. If we set $x = \Sigma \ x_n$, then $x \in E$ and $\|x\| \leq 1/(1-\varepsilon)$. Since

$$y_1 = A(x_1) = B(x_1) + y_2 = \ldots$$

$$= B(x_1) + B(x_2) + \ldots + B(x_n) + y_{n+1}$$

we see that $B(x) = y_1$, which completes the theorem. $\qquad\square$

7.24 THEOREM *(Surjective* Mapping Theorem). *Let* U *be an open set in a Banach space* E, *and let* f: U $\rightarrow$ F *be a* C^1 *map into a Banach space* F. *If* Df(ξ) *is surjective for a point* ξ *in* U, *then there exists an open set* V *with* $\xi \in V \subset U$ *having the following property: If* x $\in$ V *and* $B_r(x) \subset V$, *then* f($B_r(x)$) *contains an open neighborhood of* f(x).

Proof. For simplicity, we assume $\xi = 0$ and f(0) = 0. Since the subspace of L(E;F) consisting of surjections is open and f is of class C^1 on U, it suffices to prove that if W is an open ball centered at 0 in E, then f(W) contains a neighborhood of 0 in F.

Let A = Df(0). Since A is surjective, A is an open mapping. Hence as in the proof of 7.23, (*) there exists M > 0 such that for any y in F with $\|y\| \leq 1$, if A(x) = y, then $\|x\| \leq M\|y\|$. We assume M = 1 again. Let $0 < \varepsilon < 1$. Since f is of class C^1 on U, by the approximation lemma 6.7, there exists $\delta > 0$ with $B_\delta(0) \subset U$ and if x,z $\in B_\delta(0)$, then

$$\|f(x) - f(z) - A(x - z)\| \leq \varepsilon\|x - z\| \qquad (**)$$

We claim that

$$f(B_\delta(0)) \supset A(B_r(0))$$

where r = $\delta(1-\varepsilon)$. In fact, let $\|x_1\| < r$ and $y_1 = A(x_1)$. Set $y_2 = A(x_1) - f(x_1)$. Then it follows from (**) that

$$\|y_2\| = \|A(x_1) - f(x_1)\| \leq \varepsilon\|x_1\| < \varepsilon r$$

By the remark (*) above, there is x_2 in E such that $A(x_2) = y_2$ and $\|x_2\| \leq \varepsilon r$. We then have

$$\|x_1 + x_2\| \leq (1+\varepsilon)r$$

Since $(1+\varepsilon)r < \delta$, it follows from (**) that

$$\|A(x_1) - f(x_1+x_2)\| = \|f(x_1) - f(x_1+x_2) + A(x_2)\| \leq \varepsilon\|x_2\| < \varepsilon^2 r$$

Let $y_3 = A(x_1) - f(x_1+x_2)$. There exists x_3 in E with $\|x_3\| \le \|y_3\| \le \varepsilon^2 r$ such that $A(x_3) = y_3$. Hence

$$\|x_1 + x_2 + x_3\| \le (1 + \varepsilon + \varepsilon^2)r$$

Since $(1+\varepsilon+\varepsilon^2)r < \delta$, we have from (**)

$$\|y_1 - f(x_1+x_2+x_3)\| = \|A(x_1) - f(x_1+x_2) + f(x_1+x_2) - f(x_1+x_2+x_3)\|$$

$$= \|A(x_3) + f(x_1+x_2) - f(x_1+x_2+x_3)\| \le \varepsilon\|x_3\| < \varepsilon^3 r$$

In general, by induction, we find sequences (x_n) and (y_n) such that

$$\|x_n\| \le \varepsilon^{n-1} r$$

$$y_n = y_1 - f(x_1+\ldots+x_{n-1}) = A(x_n)$$

$$\|y_1 - f(x_1+\ldots+x_n)\| \le \varepsilon^n r$$

Set $x = \Sigma \, x_n$. Then $x \in E$ and $\|x\| \le r/(1-\varepsilon) = \delta$; we see that $f(x) = y_1$. This shows that

$$f(B_\delta(0)) \supset A(B_r(0)) \qquad\qquad \square$$

EXERCISES

7A *Composition of Differentiable Maps.* Let $U \subset E$ and $V \subset F$ be two open sets, and f: $U \to V$ and g: $V \to G$.

(1) If f is m times differentiable at $\xi \in U$ and g is m times differentiable at $f(\xi)$, then gof is m times differentiable at ξ.

(2) If f and g are of class C^m, so is gof.

7B *Product of Differentiable Maps.* With the notations appearing in Ex. 5D, if f and g are mappings of class C^m from an open set U into F_1 and F_2 respectively, then the product $f \cdot g$ with respect to a bilinear map $\phi \in L(F_1,F_2;G)$ is of class C^m and

$$D^m f \cdot g(\xi)(x_1, \ldots, x_m) = \sum_\sigma \phi_\sigma (\xi; x_1, \ldots, x_m)$$

where ϕ_σ is defined as follows: if $\sigma = \{i_1, \ldots, i_p\} \subset \{1, 2, \ldots, m\}$
and $\{k_1, \ldots, k_{m-p}\} = \{1, 2, \ldots, m\} \backslash \sigma$, then

$$\phi_\sigma(\xi; x_1, \ldots, x_m) = \phi(D^p(\xi)(x_{i_1}, \ldots, x_{i_p}), D^{m-p} f(\xi)(x_{k_1}, \ldots, x_{k_{m-p}}))$$

7C *Exponential Function.* (see 2E). Let E be a Banach space. Then

$$\text{Exp:} \ f \in L(E;E) \to \text{Exp}(f) \in L(E;E)$$

is of class C^∞.

7D *Commutativity of Directional Derivatives.* Let $f: U \to F$ be m-differentiable on U. Then for any $x_1, \ldots, x_m \in E$,

$$\frac{\partial^m f}{\partial x_1 \ldots \partial x_m} = \frac{\partial^m f}{\partial x_{i_1} \ldots \partial x_{i_m}}$$

for any permutation $(i_1, \ldots, i_m)$ of $(1, 2, \ldots, m)$.

Finite Expansions and Taylor's Formula

8.1 In this chapter we prove Taylor's theorem for mappings between normed spaces. This theorem says that if a function f is m-differentiable, then f may be approximated locally by a polynomial of degree m. In fact, f admits a finite expansion of order m, which is introduced in the following section.

FINITE EXPANSIONS

8.2 *Tangency of Higher Order.* Let E and F be normed spaces and U a non-empty open subset of E. Two functions f,g: U → F are said to be m-*tangent* at a point ξ ∈ U if

$$\lim_{x \to \xi} \frac{\|f(x) - g(x)\|}{\|x-\xi\|^m} = 0$$

i.e., for ε > 0 there exists δ > 0 such that if $\|x-\xi\| < \delta$,

$$\|f(x) - g(x)\| \leq \varepsilon\|x-\xi\|^m$$

For m = 1, we regain the definition of tangency at ξ which was introduced in (5.2). As in 5.2, we can easily show the following properties:

If f and g are m-tangent at ξ, then

(a) f - g is continuous at ξ and f(ξ) = g(ξ);

(b) f and g are (m-1) tangent at ξ;

(c) f - g is m-tangent to 0 at ξ.

(d) m-tangency is an equivalence relation on the vector space of
all mappings from U to F which are continuous at ξ.

(e) The notion of m-tangency depends only on the topologies of E
and F, not on the norms used to induce the topologies.

8.3 *Finite Expansion.* If f: U → F is differentiable at a point
$\xi \in U$, f is 1-tangent to an affine linear map (i.e., a polynomial of
degree ≤ 1) at ξ. Then f is "almost" a polynomial of degree ≤ 1 on
some neighborhood of ξ. More precisely, we have the following
approximation:

For every $\varepsilon > 0$ there exists $\delta > 0$ such that if $\|x-\xi\| < \delta$,

$$\|f(x) - f(\xi) - Df(\xi)(x-\xi)\| \leq \varepsilon\|x-\xi\|$$

In general, we are interested in approximating a function f: U → F
with a polynomial on some neighborhood of ξ with an "infinitely
small" error. This leads us to the following concept.

A polynomial $P \in P(E;F)$ of degree $\leq m$ is said to be an m-*expansion*
of a function f: U → F at a point $\xi \in U$ if P is m-tangent to f at ξ;
i.e.,

$$\lim_{x \to \xi} \frac{\|f(x) - P(x)\|}{\|x-\xi\|^m} = 0$$

The above expresses an asymptotic property between f and P as x
approaches ξ.

It is clear that if P is a polynomial of degree $\leq m$, then P is
an m-expansion of itself at any point.

8.4 *Uniqueness of Finite Expansion.* If f: U → F is given, it is not
clear whether f admits an m-expansion at a point $\xi \in U$. However,
we will prove that every function admits at most one m-expansion at
a point. To show this fact, we need two lemmas.

First we recall the polarization formula presented in (4.16).
Let U be an open subset of E containing an open ball $B_\delta(0)$. If

f: $U \to F$ and $x_1, \ldots, x_m \in U$ are such that $\|x_1\| + \ldots + \|x_m\| < \delta$, then the polarization $\phi_m(f)$ of f with respect to $x_1, \ldots, x_m$ is defined by

$$\phi_m(f) = \frac{1}{m! \, 2^m} \Sigma \, \varepsilon_1 \cdots \varepsilon_m f(\varepsilon_1 x_1 + \ldots + \varepsilon_m x_m) \tag{1}$$

where the summation is over all $\varepsilon_k = \pm 1$, $k=1, \ldots, m$. We have shown in 4.16 that if $P = P_0 + \ldots + P_m \in P(E;F)$ with $P_m = \hat{A}_m$, where $A_m \in L_s(^mE;F)$ then

$$\phi_m(P) = A_m(x_1, \ldots, x_m) \tag{2}$$

LEMMA 1. *Let U be an open subset of E which contains the origin. If f: $U \to F$ is m-tangent to 0 at the origin, then for any $\varepsilon > 0$ there exists $\delta > 0$ with $B_\delta(0) \subset U$ such that for any m elements $x_1, \ldots, x_m$ in U with $\|x_1\| + \ldots + \|x_m\| < \delta$, the polarization $\phi_m(f)$ of f with respect to $x_1, \ldots, x_m$ satisfies the inequality*

$$\|\phi_m(f)\| \leq \varepsilon(\|x_1\| + \ldots + \|x_m\|)^m$$

Proof. By the definition of $\phi_m(f)$, it suffices to show

$$\|f(\varepsilon_1 x_1 + \ldots + \varepsilon_m x_m)\| \leq \varepsilon(\|x_1\| + \ldots + \|x_m\|)^m$$

for all $\varepsilon_k = \pm 1$, $k=1, \ldots, m$. But this is obvious since f is m-tangent to 0 at the origin.   □

LEMMA 2. *If $P \in P(E;F)$ is a polynomial of degree $\leq m$ and if P is m-tangent to 0 at a point $\xi \in E$, then $P = 0$.*

Proof. If P is m-tangent to 0 at the point ξ; i.e.,

$$\lim_{x \to \xi} \|P(x)\| / \|x - \xi\|^m = 0$$

then the polynomial P_ξ defined by $P_\xi(x) = P(x + \xi)$ is m-tangent to 0

at the origin. Therefore, it is sufficient to consider the case
where ξ is the origin.

Let $P = P_0 + P_1 + \ldots + P_m$. Then for $A_m \in L_s(^m E; F)$, $P_m = \hat{A}_m$, we
have from Lemma 1 and the relation (2) above that for any $\varepsilon > 0$
there exists $\delta > 0$ such that

$$\|A_m(x_1, \ldots, x_m)\| \leq \varepsilon(\|x_1\| + \ldots + \|x_m\|)^m \tag{3}$$

if $\|x_1\| + \ldots + \|x_m\| < \delta$. The inequality (3) holds for any $x_1, \ldots, x_m$
since A_m is an m-linear map. In fact, we have

$$\|A_m(\lambda x_1, \ldots, \lambda x_m)\| = |\lambda|^m \|A_m(x_1, \ldots, x_m)\|$$

Since the inequality (3) is valid for any $\varepsilon > 0$ and for any $x_1, \ldots, x_m$,
we conclude that

$$A_m(x_1, \ldots, x_m) = 0$$

for all $x_1, \ldots, x_m$. Thus $P_m = 0$. Consequently, the degree of P is
less than m. Since P is m-tangent to 0 at the origin, P is also
(m-1)-tangent to 0 at the origin. Repeating the above argument,
we have $P_{m-1} = 0$. Thus by induction, we obtain

$$P_0 = P_1 = \ldots = P_m = 0 \qquad \qquad \square$$

THEOREM. *Let* f: U $\to$ F *and* $\xi \in U$. *If* P_1 *and* P_2 *are two m-expansions*
of f *at* ξ, *then* $P_1 = P_2$.

Proof. Let $P = P_1 - P_2$. We claim that $P = 0$. Since both P_1 and
P_2 are m-expansions of f, P is m-tangent to 0 at the point ξ. By
Lemma 2, we conclude that $P = 0$. $\qquad \qquad \square$

TAYLOR'S FORMULA

8.5 *The Differential* $\hat{d}^m f$. Let E and F be normed spaces, and f: U $\to$
F, where U is an open subset of E. If f is m-differentiable at a

point $\xi \in U$, then it was shown that

$$d^k f(\xi) \in L_s(^k E; F)$$

for $k = 0, 1, \ldots, m$. Hence we can associate $d^k f(\xi)$ with the unique k-homogeneous polynomial which we will denote by $\hat{d}^k f(\xi)$ (see 4.8). Thus

$$\hat{d}^k f(\xi) \in P(^k E; F)$$

and

$$\hat{d}^k f(\xi)(x) = d^k f(\xi) x^k$$

Since $\hat{d}^k f(\xi)$ is a k-homogeneous polynomial, the mapping

$$x \in E \to \hat{d}^k f(\xi)(x-\xi) \in F$$

is a polynomial of degree k.

8.6 *Taylor Polynomials.* By a Taylor polynomial $T_{m,f,\xi}$ of order m of f at ξ we mean the polynomial defined by

$$T_{m,f,\xi}(x) = \sum_{k=0}^{m} \frac{1}{k!} \hat{d}^k f(\xi)(x-\xi)$$

i.e.,

$$T_{m,f,\xi}(x) = \sum_{k=0}^{m} \frac{1}{k!} d^k f(\xi)(x-\xi)^k$$

EXAMPLE. If $P \in P(E; F)$ is of degree $\leq m$, then

$$P = T_{m,P,\xi}$$

for every $\xi \in E$.

To show this, we may assume without loss of generality that P is a homogeneous polynomial, since the general case follows from the

homogeneous case by adding a finite number of homogeneous polynomials.
Let $P = \hat{A}$, where $A \in L_s(^mE;F)$. Then we have by the binomial formula
and Theorem 7.7

$$P(x) = Ax^m = A(\xi + (x-\xi))^m = \sum_{k=0}^{m} \binom{m}{k} A\xi^{m-k}(x-\xi)^k$$

$$= \sum_{k=0}^{m} \frac{1}{k!} d^k f(\xi)(x-\xi)^k$$

which shows that $P = T_{m,P,\xi}$ for any $\xi \in E$.

8.7 *Differentiation of Taylor Polynomials.* If $f: U \to F$ is m-differentiable at a point $\xi \in U$, then

$$df(\xi) \in L(E;F)$$

and

$$d^{m-1}(df)(\xi) \in L(^{m-1}E;L(E;F))$$

On the other hand,

$$d^m f(\xi) \in L(^mE;F)$$

Thus we do *not* have $d^m f = d^{m-1}(df)$, unlike $D^m f = D^{m-1}(Df)$. However,
the map $\phi: L(^mE;F) \to L(^{m-1}E;L(E;F))$ defined by

$$\phi(A)(x_1,\ldots,x_{m-1})(x_m) = A(x_1,\ldots,x_m)$$

is an isometric isomorphism. Under this correspondence, we can
identify $d^m f(\xi)$ with $d^{m-1}(df)(\xi)$ with respect to ϕ.

Then

$$d^{m-1}(df)(\xi)(x_1,\ldots,x_{m-1})(x_m) = d^m f(\xi)(x_1,\ldots,x_m)$$

In particular,

$$d^{m-1}(df)(\xi)(x-\xi)^{m-1}(x-\xi) = d^m f(\xi)(x-\xi)^m$$

Also we have from Theorem 5.14

$$d(d^k f(\xi)(x-\xi)^k) = kd^k f(\xi)(x-\xi)^{k-1}$$

THEOREM. *If* $f: U \to F$ *is* m-*differentiable at a point* $\xi \in U$, *then we have*

$$dT_{m,f,\xi} = T_{m-1,df,\xi}$$

Proof.

$$dT_{m,f,\xi}(x) = d(\sum_{k=0}^{m} \frac{1}{k!} d^k f(\xi)(x-\xi)^k) = \sum_{k=0}^{m} \frac{1}{k!} d(d^k f(\xi)(x-\xi)^k)$$

$$= \sum_{k=1}^{m} \frac{1}{(k-1)!} d^k f(\xi)(x-\xi)^{k-1} = \sum_{k=1}^{m} \frac{1}{(k-1)!} d^{k-1}(df)(\xi)(x-\xi)^{k-1}$$

$$= \sum_{k=0}^{m-1} \frac{1}{k!} d^k(df)(\xi)(x-\xi)^k = T_{m-1,df,\xi}(x) \qquad \square$$

8.8 *Mean Value Theorem.* If $f: U \to F$ is a given function, it is not certain that f admits an m-expansion at a point ξ in U. However, if f is m-differentiable, we will see that f admits an m-expansion. This result is the so-called Taylor's theorem. We need the following generalization of the mean value theorem 6.5 for Taylor's theorem.

LEMMA. *Let* U *be an open subset of* E *such that* U *contains the line segment* [a,b]. *If* $f: U \to F$ *and* $\phi: [0,1] \to \mathbb{R}$ *are differentiable such that*

$$\|Df(a + \lambda(b-a))\| \leq \phi'(\lambda)$$

for all $\lambda \in [0,1]$, *then*

$$\|f(b) - f(a)\| \leq \|b-a\|[\phi(1) - \phi(0)]$$

Proof.   We modify the proof of Theorem 6.5.   If $x \in [a,b]$, then we
denote by $\lambda_x \in [0,1]$ the unique real number satisfying

$$x = a + \lambda_x (b-a)$$

For $\varepsilon > 0$, let X be the set of all points $x \in [a,b]$ satisfying the
following condition:

$$\left\| f(t) - f(a) \right\| \leq \left\| b-a \right\| [\phi(\lambda_t) - \phi(0)] + \varepsilon \left\| t-a \right\| \tag{1}$$

for $a \leq t \leq x$.   Then $X \neq \emptyset$ since $a \in X$.   Notice that if $x \in X$, then
$[a,x] \subset X$.   Let $c = \sup X$ (see 6.2).   We claim that $c \in X$ and $c = b$.
If $c < b$, then by the differentiability of f at c, we can find $\delta > 0$
such that $\left\| b-c \right\| > \delta$, $B_\delta(c) \subset U$ and if $t \in B_\delta(c)$,

$$\left\| f(t) - f(c) - Df(c)(t-c) \right\| \leq \varepsilon \left\| t-c \right\| / 2$$

or

$$\left\| f(t) - f(c) \right\| \leq (\left\| Df(c) \right\| + \varepsilon/2) \left\| t-c \right\| \tag{2}$$

Since ϕ is also differentiable at $\lambda_c \in [0,1]$, there exists $\eta > 0$
with $\eta \leq \delta / \left\| b-a \right\|$ such that if $\left| \lambda_t - \lambda_c \right| < \eta$,

$$\left| \phi(\lambda_t) - \phi(\lambda_c) - \phi'(\lambda_c)(\lambda_t - \lambda_c) \right| \leq \varepsilon \left| \lambda_t - \lambda_c \right| / 2$$

or

$$\left| \phi'(\lambda_c)(\lambda_t - \lambda_c) \right| \leq \left| \phi(\lambda_t) - \phi(\lambda_c) \right| + \varepsilon \left| \lambda_t - \lambda_c \right| / 2 \tag{3}$$

Let $\gamma = \eta \left\| b-a \right\| \leq \delta$, and let $t,t' \in B_r(c) \cap [a,b]$ be such that
$t < c < t'$.   Then

$$\left\| f(t') - f(a) \right\| \le \left\| f(t')-f(c) \right\| + \left\| f(c)-f(t) \right\| + \left\| f(t)-f(a) \right\|$$

$$\le \left\| Df(c) \right\| \left\| t'-c \right\| + \left\| Df(c) \right\| \left\| c-t \right\| + \varepsilon \left\| t'-t \right\| /2$$

$$+ \left\| b-a \right\| [\phi(\lambda_t)-\phi(0)] + \varepsilon \left\| t-a \right\| \tag{4}$$

by (2) and (1). Notice that $\left\| t'-t \right\| = \left\| t'-c \right\| + \left\| c-t \right\|$, $t'-t = (\lambda_{t'}-\lambda_t) \times$ $\cdot(b-a)$, and $\left\| Df(c) \right\| \le \phi'(\lambda_c)$. Also note that ϕ is an increasing function since $\phi'(\lambda) \ge 0$. Keeping this in mind and applying (3) to the first two terms in the extreme right hand side of (4) we get after some computation that

$$\left\| f(t') - f(a) \right\| \le \left\| b-a \right\| [\phi(\lambda_{t'})-\phi(0)] + \varepsilon \left\| t'-a \right\|$$

This shows that $t' \in X$ and also $c \in X$, which contradicts the fact that $c = \sup X$. Thus $c = b$, and $X = [a,b]$. It follows that for any $\varepsilon > 0$ we have

$$\left\| f(b) - f(a) \right\| \le \left\| b-a \right\| [\phi(1)-\phi(0)] + \varepsilon \left\| b-a \right\|$$

and hence

$$\left\| f(b) - f(a) \right\| \le \left\| b-a \right\| [\phi(1)-\phi(0)] \qquad \qquad \square$$

ANOTHER PROOF OF THEOREM 6.5. Let $\phi: [0,1] \to \mathbb{R}$ be defined by $\phi(\lambda) = M\lambda$ where $M = \sup\{ \left\| Df(x) \right\| : x \in [a,b] \}$. Then we obtain the mean value theorem from the preceding lemma.

8.9 *Taylor's Formula.* We now prove the major theorem of the chapter. The following theorem is sometimes called *Taylor's formula* with *Lagrange remainder*. Theorems 8.10 and 8.11 will also be called Taylor's formulae.

THEOREM. *Let* U *be an open subset of* E *such that* U *contains the line segment* [a,b]. *If* f: U → F *is* (m+1)-*differentiable on* U, *then*

$$\left\| f(b) - T_{m,f,a}(b) \right\| \leq \frac{M \left\| b-a \right\|^{m+1}}{(m+1)!}$$

where

$$M = \sup\{ \left\| d^{m+1} f(t) \right\| : t \in [a,t] \}$$

Proof. If m = 0, we regain the mean-value theorem 6.5. We now proceed by induction on m. Suppose that the theorem is true for m-1. Then if x ∈ [a,b], by Theorem 8.7 we have

$$\left\| df(x) - T_{m-1,df,a}(x) \right\| \leq \frac{\left\| x-a \right\|^m}{m!} \sup_{t \in [a,b]} \left\| d^m (df)(t) \right\| = \frac{M \left\| x-a \right\|^m}{m!}$$

Consider $g = f - T_{m,f,a}$. Then g(a) = 0 and

$$dg = df - T_{m-1,df,a}$$

Hence for x ∈ [a,b],

$$\left\| dg(x) \right\| \leq \frac{M \left\| x-a \right\|^m}{m!}$$

Replacing x by a + λ(b-a), we obtain

$$\left\| dg(a + \lambda(b-a)) \right\| \leq \frac{M \left\| b-a \right\|^m \lambda^m}{m!}$$

We now apply Lemma 8.8 to g and the map φ: [0,1] → ℝ defined by

$$\phi(\lambda) = \frac{M \left\| b-a \right\|^m \lambda^{m+1}}{(m+1)!}$$

to conclude that

$$\left\| g(b) \right\| = \left\| g(b) - g(a) \right\| \leq \frac{M \left\| b-a \right\|^{m+1}}{(m+1)!}$$

This completes the proof.   □

COROLLARY. *If* f: U → F *is* (m+1)-*differentiable on* U, *then for any*
ξ ∈ U, f *admits the* m-*expansion* $T_{m,f,\xi}$ *at* ξ.

8.10 Taylor's formula with Lagrange remainder has been obtained by
assuming that f is (m+1)-differentiable. We can weaken this condition
and obtain the following theorem.

THEOREM. *If* f: U → F *is* m-*differentiable on* U *and* [a,b] ⊂ U, *then*
for x ∈ [a,b]

$$\left\| f(x) - T_{m,f,a}(x) \right\| \le \frac{\left\| x-a \right\|^m}{m!} M_x$$

where

$$M_x = \sup\{ \left\| d^m f(t) - d^m f(a) \right\| : t \in [a,x] \}$$

Proof. Let $g = f - T_{m,f,a}$. Then

$$d^k g = d^k f - T_{m-k,d^k f,a}$$

by Theorem 8.7, and for k=0,1,...,m

$$d^k g(a) = 0$$

Therefore, $T_{m-1,g,a} = 0$. Also notice that

$$d^m g(t) = d^m f(t) - d^m f(a)$$

It follows from Theorem 8.9 that

$$\left\| g(x) \right\| = \left\| g(x) - T_{m-1,g,a}(x) \right\| \le \frac{\left\| x-a \right\|^m}{m!} \sup_{t \in [a,x]} \left\| d^m g(t) \right\|$$

This proves the theorem.   □

COROLLARY. If f: U → F is of class C^m on U then f admits the unique m-expansion $T_{m,f,\xi}$ at every point $\xi \in U$.

Proof. This follows from 8.4 and the fact that $d^m f$ is continuous. □

8.11 Finally, we present Taylor's formula with asymptotic property.

THEOREM. If f: U → F is m-differentiable at a point $\xi \in U$, then f admits the unique m-expansion $T_{m,f,\xi}$ at ξ.

Proof. The uniqueness follows from 8.4. It remains to show that $T_{m,f,\xi}$ is an m-expansion of f at ξ. Let $g = f - T_{m,f,\xi}$. Then we know

$$d^k g = d^k f - T_{m-k, d^k f, \xi}$$

and $d^k g(\xi) = 0$ for $k = 0, 1, \ldots, m$. Thus $T_{m-2, g, \xi} = 0$. By Taylor's Theorem 8.9, for $[\xi, x] \subseteq U$

$$\|g(x)\| = \|g(x) - T_{m-2, g, \xi}(x)\| \le \frac{\|x - \xi\|^{m-1}}{(m-1)!} M_x \tag{1}$$

where

$$M_x = \sup\{\|d^{m-1} g(t)\| : t \in [\xi, x]\} \tag{2}$$

But

$$d^{m-1} g(t) = d^{m-1} f(t) - T_{1, d^{m-1} f, \xi}(t)$$

$$= d^{m-1} f(t) - d^{m-1} f(\xi) - d(d^{m-1} f)(\xi)(t - \xi)$$

Since $d^{m-1} f$ is differentiable at ξ, for $\varepsilon > 0$ there exists $\delta > 0$ such that $B_\delta(\xi) \subseteq U$ and if $x \in B_\delta(\xi)$ then

$$\left\| d^{m-1}f(x) - d^{m-1}f(\xi) - d(d^{m-1}f)(\xi)(x-\xi) \right\| \leq \varepsilon \left\| x-\xi \right\|$$

Therefore, if $\left\| x-\xi \right\| < \delta$ then $M_x \leq \varepsilon \left\| x-\xi \right\|$, and

$$\left\| g(x) \right\| \leq \frac{\varepsilon \left\| x-\xi \right\|^m}{(m-1)!}$$ $\square$

EXERCISES

8A *Identity Theorem for Polynomials.* If $P,Q \in P(E;F)$ and $P = Q$ on some non-empty open subset of E, then $P = Q$.

8B *Polynomials of Degree m.* Let U be a connected open subset of E. If $f: U \to F$ is (m+1)-differentiable on U, then $D^{m+1}f = 0$ on U if and only if there exists a polynomial $P \in P(E;F)$ of degree $\leq m$ such that $P = f$ on U.

8C *m-Homogeneous Polynomials.* Let U be a ξ-balanced* neighborhood of ξ. Assume that $f: U \to F$ is m-differentiable at ξ and that f is m-homogeneous on U; i.e., for $0 \leq \lambda \leq 1$ and $x \in U$,

$$f(\lambda x) = \lambda^m f(x)$$

Then there is a polynomial $P \in P(^mE;F)$ such that $P = f$ on U.

8D *Truncation of Polynomials.* If $P = P_0 + P_1 + \ldots + P_m$ is a polynomial of degree $\leq m$ and $k < m$, the polynomial $P_0 + P_1 + \ldots + P_k$ is called the *truncation* of P of the order k. If $f: U \to F$ admits an m-expansion P at a point ξ, then for any $k < m$, the truncation of P of order k is the k-expansion of f at ξ.

8E *m-Expansion vs. m-Differentiability.* The existence of an m-expansion of f does not imply m-differentiability. For example, the function $f(x) = x^3\sin(1/x)$ for $x = 0$, $f(0) = 0$ has a 2-expansion at the origin, but f is not twice differentiable.

———

*Let A be a subset of a vector space over $\mathbb{K}$ and $\xi \in A$. Then A is said to be ξ-balanced if $\xi + \lambda A \subseteq A$ for any $\lambda \in \mathbb{K}$, $\left| \lambda \right| \leq 1$.

PART II

HOLOMORPHIC MAPPINGS

9

Holomorphic Functions of a Complex Variable

9.1 In this chapter we shall be concerned with holomorphic mappings. We will assume that the reader has some elementary knowledge of the theory of complex variables. In particular, the student should be familiar with the two equivalent ways of defining a holomorphic mapping, the one based on complex differentiability and the other based on convergent power series.

In the study of Banach algebras, as well as in some other contexts, the concept of vector-valued holomorphic mappings arises naturally. We first enlarge the classical definition of holomorphic mappings from complex-valued ones to vector-valued ones. Of course, we will generalize the domains by going from the complex plane to normed spaces in later chapters.

The theory of vector-valued holomorphic mappings, like that of complex-valued holomorphic mappings, can be developed most efficiently through the use of curvilinear integrals. First we need a generalization of the Riemann integral to vector-valued functions.

RIEMANN INTEGRATION ON [a,b]

9.2 *Step Maps.* Let $I = [a,b]$ be a closed interval in $\mathbb{R}$, and E a Banach space over $\mathbb{K}$. By a *step map* $f: [a,b] \to E$ we mean a map for which there exists a partition

$$P: a = a_0 < a_1 < \ldots < a_n = b$$

and elements $y_1, \ldots, y_n \in E$ such that

$$f = \Sigma \; y_i \chi_{(a_i, a_{i+1})}$$

where χ_A is the characteristic function on the set A.

Let $S(I;E)$ be the vector space of all step maps from I to E.   Then $S(I;E)$ is a vector subspace of the vector space $B(I;E)$ of all bounded maps from I to E.   With the sup norm of $B(I;E)$, $B(I;E)$ becomes a Banach space.

Let $R(I;E)$ be the closure of $S(I;E)$ in $B(I;E)$.   A map in $R(I;E)$ will be called *regular* for convenience.   Since a continuous map $f \colon I \to E$ is uniformly continuous, we can conclude that

$$C(I;E) \subseteq R(I;E)$$

9.3 *Integrals.* We define the integral of a step map

$$f = \Sigma \; y_i \chi_{(a_i, a_{i+1})}$$

by

$$\int_a^b f = \Sigma \; (a_{i+1} - a_i) y_i$$

It is easy to see that the integral

$$\int_a^b \colon S(I;E) \to E$$

is linear and

$$\left\| \int_a^b f \right\| \leq (b-a) \| f \|$$

so $\int_a^b$ is continuous.   We can therefore extend $\int_a^b$ to the closure $R(I;E)$ of $S(I;E)$ by the linear extension theorem 2.7; hence if $f \in R(I;E)$, we can write

$$\int_a^b f$$

and call it the *integral* of f.

9.4 *Fundamental Theorem of Calculus.* If $a \leqslant c \leqslant d \leqslant b$, we define

$$\int_d^c f = - \int_c^d f$$

Then for any three points c,d,e in any order, lying in I, and $f \in R(I;E)$, we can prove without difficulty that

$$\int_c^d f = \int_c^e f + \int_e^d f$$

THEOREM. *Let* $f \in R(I;E)$, *and assume that* f *is continuous at a point* c *of* I. *Then the map*

$$F(t) = \int_a^t f$$

is differentiable at c *and*

$$F'(c) = f(c)$$

Proof. The standard proof in the elementary calculus works. That is,

$$F(c+h) - F(c) = \int_c^{c+h} f$$

and

$$F(c+h) - F(c) - hf(c) = \int_c^{c+h} (f - f(c))$$

Thus

$$\| F(c+h) - F(c) - hf(c) \| \leq |h| \sup \| f(x) - f(c) \|$$

where x lies between c and c+h. Since f is continuous at c, we can
see that

$$F'(c) = f(c) \qquad\qquad\qquad \square$$

9.5   THEOREM.   *Let E and F be Banach spaces and* $A \in L(E;F)$.   *If*
$f \in R(I;E)$, *then*

$$\int_a^b A \circ f = \left(\int_a^b f \right)$$

Proof.   If f is the uniform limit of a sequence (f_n) of step maps,
then each $A \circ f_n$ is a step map of I into F.   Then the sequence $(A \circ f_n)$
clearly converges to $A \circ f$; hence $A \circ f \in R(I;F)$. For a step map f_n
we have from the definition of the integral that

$$\int_a^b A \, f_n = A \left(\int_a^b f_n \right)$$

Since the integral is continuous, taking the limit proves the identity
we want. $\square$

We now introduce the curvilinear integral.

9.6   *Path*.   A *path* in the complex plane $\mathbb{C}$ is a continuous mapping

$$\gamma: I = [a,b] \to \mathbb{C}$$

of a closed interval I of $\mathbb{R}$ (not reduced to a point) into $\mathbb{C}$, which
is piecewise-continuously differentiable.   A path $\gamma: I \to \mathbb{C}$ is said
to be *contained in* an open set $U \subset \mathbb{C}$ if $\gamma(I) \subseteq U$.   The point $\gamma(a)$
is called the *initial point* of the path γ, and the point $\gamma(b)$ the
terminal point.   If $\gamma(t_1) \neq \gamma(t_2)$ unless $t_1 = t_2$ or t_1 and t_2 are
the same as a and b, γ is said to be *simple*.   If $\gamma: I \to \mathbb{C}$ is a path
and $\gamma(a) = \gamma(b)$, then γ is said to be *closed*.   If a point p lies in
the image of γ, we often say that p is *on the path* γ.

9.7 *Curvilinear Integrals.* Let γ: I → ℂ be a path and let f: γ(I) → F be continuous where F is a complex normed space. The the composed function

t ∈ I → f(γ(t))γ'(t) ∈ F

is piecewise-continuous, and hence its integral exists on I by 9.3. Then the *curvilinear integral* of f over the path γ is defined by

$$\int_\gamma f(z)\,dz = \int_a^b f(\gamma(t))\gamma'(t)\,dt$$

Note that the curvilinear integral above is an element of F.

It follows immediately from Theorem 9.5 that

$$A \quad \int_\gamma f(z)\,dz = \int_\gamma A(f(z))\,dz$$

for every A ∈ F*.

REVIEW OF COMPLEX-VALUED FUNCTIONS

9.8 *Characterization of Holomorphic Functions.* As a prerequisite for a study of vector-valued holomorphic mappings, the reader should be familiar with the following characterization of holomorphic mappings. This will be applied later to obtain the extension to vector-valued mappings.

Let γ be a closed path in ℂ and let p be a point of ℂ which is not on the path γ. Recall that the *index* of γ with respect to p, denoted by I(γ,p), is defined to be the value of the integral

$$\frac{1}{2\pi i} \int_\gamma \frac{dz}{z-p}$$

THEOREM. *Let U be a non-empty open subset of the complex plane* ℂ *and let f: U → ℂ. Then the following conditions are equivalent:*

(a)   f is differentiable on U.

(b)   f is continuous on U and if γ is a closed path in U such that I(γ;p) = 0 for every p ∉ U, then

$$\int_\gamma f(z)\,dz = 0$$

(c). f is continuous on U and if γ is a closed path in U as in (b), and if z ∈ U and I(γ;z) = 1, then

$$f(z) = \frac{1}{2\pi i} \int_\gamma \frac{f(t)\,dt}{t - z}$$

(d)   f is of class C^∞ on U and at any point p ∈ U, the Taylor series

$$f(z) = \Sigma \frac{f^{(m)}(p)}{m!} (z-p)^m$$

converges absolutely and uniformly for $|z-p| \leq r$ for every r < R, where R = d(p,∂U), the distance between p and the boundary ∂U of U.

(e)   For any point p ∈ U, there exists a sequence (a_m) in $\mathbb{C}$ and a real number R > 0 such that the power series

$$f(z) = \Sigma \, a_m (z-p)^m$$

converges absolutely and uniformly for $|z-p| \leq r$ for every r < R.

The implication (a) ⇒ (b) in the above theorem is called the *Cauchy integral theorem,* (b) ⇒ (a) the *Morera theorem,* and the integral in (c) is called the *Cauchy integral formula.*

9.9   *Cauchy and Weierstrass Viewpoints.*   A function f: U → $\mathbb{C}$ is said to be *holomorphic* on U if f is differentiable on U, and f is said to be *analytic* on U if f satisfies condition (e) of the preceding theorem.   Then the implication (a) ⇒ (e) says that a holomorphic function is analytic.   Conversely, an analytic function is holomorphic ((e) ⇒ (a)).   Thus the concepts of holomorphy and analyticity are equivalent for functions of a complex variable.

The theory of holomorphic functions based on complex differentiability is referred to as the *Cauchy viewpoint*, and the theory based on power series is called the *Weierstrass viewpoint*. These viewpoints are equivalent for functions of a complex variable as we have remarked above. However, we should note that these two viewpoints are not equivalent for functions of a real variable. In fact, a function of a real variable can be continuously differentiable without even having a second derivative, as is shown by the example

$$f(x) = x|x|$$

A function represented by a convergent power series is indefinitely differentiable within the radius of convergence regardless of whether the underlying domain is the real or complex numbers. The Weierstrass approach therefore is generally more suitable for both the real and complex cases.

9.10 *Some Properties of Holomorphic Mappings*. The consequences of complex differentiability (i.e., holomorphy) are quite involved, and we refer the reader to the standard texts on the subject appearing in the bibliography. We recall here the following properties of holomorphic mappings which will be used later in this chapter.

Cauchy Integral Formula for Derivatives. Let U be a non-empty open subset of the complex plane $\mathbb{C}$. If f: U $\to \mathbb{C}$ is holomorphic on U, and if γ is a closed path in U such that if $z \in U$ and $I(\gamma, z) = 1$, then

$$f^{(m)}(z) = \frac{m!}{2\pi i} \int_\gamma \frac{f(t)\,dt}{(t-z)^{m+1}}$$

Maximum Modulus Theorem. Let f: U $\to \mathbb{C}$ be holomorphic on U, where U is a connected open subset of $\mathbb{C}$. If at a point $p \in U$, the function $z \to |f(z)|$ attains a relative maximum, then f is constant in U.

Liouville Theorem.   If $f: \mathbb{C} \to \mathbb{C}$ is holomorphic on the entire plane $\mathbb{C}$ and bounded on $\mathbb{C}$, then f is constant.

Identity Theorem.   If $f,g: U \to \mathbb{C}$ are holomorphic on U and if there exists a sequence (a_n) in U with a cluster point in U such that $f(a_n) = g(a_n)$ for all n, then f = g.

VECTOR-VALUED HOLOMORPHIC MAPPINGS

9.11   *Holomorphy and Weak Holomorphy.*   Let U be a non-empty open subset of $\mathbb{C}$ and let F be a complex normed space. For a function $f: U \to F$ there are at least two very natural definitions of holomorphy available, a "weak" one and a "strong" one. However, they turn out to be equivalent.

(a)   A function $f: U \to F$ is said to be *holomorphic on* U if f is differentiable.

(b)   A function $f: U \to F$ is said to be *weakly holomorphic on* U if $\lambda \circ f: U \to \mathbb{C}$ is holomorphic in the ordinary sense for every $\lambda \in F^*$.

The differentiability of continuous linear functionals in F^* makes it obvious that every holomorphic mapping is weakly holomorphic. Although the converse is true, it is far from obvious. We will prove this using the Cauchy integral formula. The set of all holomorphic mappings of U into F will be denoted by $H(U;F)$.

9.12   THEOREM (Dunford, 1938).   *Let U be an open subset in $\mathbb{C}$, F a complex Banach space, and $f: U \to F$. The following conditions are equivalent:*

(a)   f is weakly holomorphic on U.

(b)   f is continuous and if γ is a closed path in U such that $I(\gamma,p)=0$ for $p \notin U$, then

$$\int_\gamma f(z)\,dz = 0$$

(c)   f is continuous and if γ is a closed path in U as described

in (b) such that if z ∈ U and I(γ,z) = 1,

$$f(z) = \frac{1}{2\pi i} \int_\gamma \frac{f(t)\,dt}{t-z}$$

(d) f is holomorphic on U.

Proof. (a) ⇒ (b): Since f is weakly holomorphic, λ∘f is continuous for any λ ∈ F*. To prove that f is continuous at each point p ∈ U, we may assume 0 ∈ U and show that f is continuous at 0; otherwise we consider the mapping x → f(x-p).

For λ ∈ F*, let g = λ∘f. Then g: U → $\mathbb{C}$ is holomorphic in the ordinary sense. Since 0 ∈ U, $\overline{B}_{2r}(0) \subset U$ for some r > 0. Then by the Cauchy integral formula (Theorem 9.8(c)), if 0 < $|z|$ < 2r we have

$$\frac{g(z) - g(0)}{z} = \frac{1}{2\pi i} \int_\gamma \frac{g(t)\,dt}{(t-z)\,t} \tag{1}$$

where $\gamma(\theta) = 2re^{i\theta}$, $0 \le \theta \le 2\pi$. Let M(λ) be the maximum of $|g|$ on the closed ball $\overline{B}_{2r}(0)$. If 0 < $z| \le r$, it follows from (1) that

$$\left| \frac{g(z) - g(0)}{z} \right| \le \frac{M(\lambda)}{r} \tag{2}$$

Let

$$Q = \left\{ \frac{f(z) - f(0)}{z} : 0 < |z| \le r \right\}$$

Then the inequality (2) implies that λ(Q) is bounded for λ ∈ F*. Since λ was arbitrary, we conclude that Q is bounded by (2.17). Thus there exists A > 0 such that

$$\left\| \frac{f(z) - f(0)}{z} \right\| \le A \qquad \text{or} \qquad \| f(z) - f(0) \| \le A|z|$$

for all z, $0 < |z| \leq r$.   Consequently, $f(z) \to f(0)$ as $z \to 0$.

It remains to show that $\int_\gamma f(z)\,dz = 0$.   Since f is continuous on U, the integral $\int_\gamma f$ exists, and hence for any $\lambda \in F*$

$$\lambda \int_\gamma f(z)\,dz = \int_\gamma (\lambda \circ f)(z)\,dz = 0$$

by 9.5 and the Cauchy integral theorem 9.8.   Therefore, we have $\int_\gamma f(z)\,dz = 0$ by 2.11.

(b) $\Rightarrow$ (c):   First notice that the integral

$$\int_\gamma \frac{f(t)\,dt}{t-z}$$

exists because f is continuous and z is not on the path γ.   Using (9.5) and the classical Cauchy integral formula 9.8, we conclude that

$$f(z) = \frac{1}{2\pi i} \int_\gamma \frac{f(t)\,dt}{t-z}$$

(c) $\Rightarrow$ (d):   As in the proof (a) $\Rightarrow$ (b), we assume $0 \in U$ and show that f is differentiable at 0.   For $\gamma(\theta) = 2re^{i\theta}$, $0 \leq \theta \leq 2\pi$, and $0 < |z| \leq 2r$, we have

$$\frac{f(z) - f(0)}{z} = \frac{1}{2\pi i} \int_\gamma \frac{f(t)\,dt}{(t-z)t} \tag{3}$$

If we substitute $t = 2re^{i\theta}$ in (3), after some computation we obtain

$$\frac{f(z) - f(0)}{z} = \frac{1}{2\pi i} \int_\gamma \frac{f(t)\,dt}{t^2} + zg(z) \tag{4}$$

where

$$g(z) = \frac{1}{2\pi i} \int_0^{2\pi} \frac{f(2re^{i\theta})d\theta}{2re^{i\theta}(2re^{i\theta}-z)} \tag{5}$$

Notice that the integrals in (4) and (5) are well-defined members in F. Since the circle $S = \gamma([0,2\pi])$ is compact, $f(S)$ is a compact set in F. Hence $f(S)$ is bounded. This shows that for some $s > 0$,

$$\|f(2re^{i\theta})\| \le s$$

for any $\theta \in [0,2\pi]$. Thus if $|z| \le r$, it follows from (5) that

$$\|g(z)\| \le s/2r^2 \tag{6}$$

Applying the estimate (6) to (4), we conclude that f is differentiable at 0 and

$$f'(0) = \frac{1}{2\pi i} \int_\gamma \frac{f(t)dt}{t^2}$$

We should recall that $f'(0)$ is the derivative associated with the differential $Df(0)$ (see 5.5). This completes the proof of Theorem 9.12. $\qquad\qquad\square$

9.13 *Operator-Valued Holomorphic Mappings.* Let E and F be complex Banach spaces, U an open subset in $\mathbb{C}$, and $f: U \to L(E;F)$. The following theorem states the necessary and sufficient condition for f to be holomorphic.

THEOREM. *A mapping* $f: U \to L(E;F)$ *is holomorphic if and only if for each* $x \in E$, *the mapping*

$$f_x: U \to F$$

defined by $f_x(z) = f(z)(x)$ *is holomorphic.*

Proof. If f is holomorphic, then f_x is obviously holomorphic since f_x is a composition of f and a linear map $A \in L(E;F) \to A(x) \in F$.

To prove the converse it is sufficient to show that f is continuous on U and $\int_\gamma f = 0$ for any closed path γ in U such that $I(\gamma,p) = 0$ for $p \notin U$. We imitate the proof (a) $\Rightarrow$ (b) in Theorem 9.12. To show that f is continuous on U, we may assume $\overline{B}_{2r}(0) \subseteq U$ for some $r \succ 0$ and show that f is continuous at 0. Since f_x is holomorphic, by the Cauchy integral formula (9.12(c)) we have

$$\frac{f_x(z) - f_x(0)}{z} = \frac{1}{2\pi i} \int_\gamma \frac{f_x(t)\,dt}{(t-z)\,t} \tag{1}$$

where γ: $|t| = 2r$. Then if we set $M(x,r) = \max\{\|f_x(t)\| : \|t\| < 2r\}$, from (1) we obtain

$$\|f_x(z) - f_x(0)\| \leq M(x,r)|z| \tag{2}$$

for $0 < |z| \leq r$. Let

$$Q = \frac{f(z) - f(0)}{z} : 0 < |z| \leq r$$

Then $Q \subset L(E;F)$ and Q is pointwisely bounded by (2), and hence Q is uniformly bounded by the Banach-Steinhaus Theorem 2.16. Thus for some M > 0,

$$\|f(z) - f(0)\| \leq M|z|$$

for all z, $0 \leq |z| \leq r$. Consequently f is continuous at 0.

We now show that $\int_\gamma f = 0$. Since f is continuous on U, $\int_\gamma f$ exists. The function $f_x: U \to F$ is holomorphic for each $x \in E$, and hence we have

$$\int_\gamma f_x(t)\,dt = \int_\gamma f(t)(x)\,dt = 0$$

for any closed path γ in U such that $I(\gamma,p) = 0$ for $p \notin U$. Since

$$\int_\gamma f \text{ exists and } A \in L(E;F) \to A(x) \in F \text{ is linear,}$$

$$\int_\gamma f(t)(x)\,dt = \int_\gamma f(t)\,dt(x) = 0$$

for all $x \in E$. Therefore $\int_\gamma f = 0$. This shows that f is holomorphic by Theorem 9.12. □

9.14 *Derivatives of Holomorphic Mappings.* In 5.5 we have identified $L(\mathbb{K};F)$ with F and $Df(\xi) \in L(\mathbb{K};F)$ with $f'(\xi) \in F$. In general, we have from 3.11 and 5.5,

$$L(^m\mathbb{K};F) \approx L(\mathbb{K};L(^{m-1}\mathbb{K};F)) \approx L(^{m-1}\mathbb{K};F)$$

Hence

$$L(^m\mathbb{K};F) \approx F$$

for any natural number $m = 0,1,\dots$.

If $f: U \to F$, where $U \subset \mathbb{K}$, is m-differentiable, then

$$D^m f: U \to L(^m\mathbb{K};F)$$

can be naturally identified with

$$f^{(m)}: U \to F$$

where $f^{(m)} = (f^{(m-1)})'$. The function $f^{(m)}$ is called the m-*th derivative* of f, and $f^{(m)}$ and $D^m f$ satisfy the identity

$$D^m f(\xi)(x_1,\dots,x_m) = f^{(m)}(\xi) \cdot x_1 \cdots x_m$$

for any $x_1,\dots,x_m \in \mathbb{K}$.

9.15 THEOREM (*Cauchy Integral Formula for* $f^{(m)}$). *Let* U *be open in*
$\mathbb{C}$ *and* $f \in H(U;F)$. *Then* f *has derivatives of all orders and if* γ *is*
a closed path in U *such that* $I(\gamma,p) = 0$ *for* $p \notin U$, *then*

$$f^{(m)}(z) = \frac{m!}{2\pi i} \int_\gamma \frac{f(t)\,dt}{(t-z)^{m+1}}$$

if $z \in U$ *and* $I(\gamma,z) = 1$.

Proof. Since f is holomorphic on U, f has the derivative $f': U \to F$,
and hence

$$(\lambda \circ f)' = \lambda \circ f' \tag{1}$$

for any $\lambda \in F^*$. Because a holomorphic mapping is weakly holomorphic
$(\lambda \circ f)': U \to \mathbb{C}$ is then holomorphic in the ordinary sense. It follows
immediately from (1) that f' is weakly holomorphic, and hence f' is
holomorphic. Repeating this argument, we conclude that $f^{(m)}$ exists
for all m. Now the integral representation of $f^{(m)}$ is automatic.
In fact, since $\lambda \in F^*$ we have

$$(\lambda \circ f)^{(m)} = \lambda \circ f^{(m)}$$

for all m, and

$$(\lambda \circ f)^{(m)}(z) = \frac{m!}{2\pi i} \int_\gamma \frac{\lambda \circ f(t)\,dt}{(t-z)^{m+1}} = \lambda\, \frac{m!}{2\pi i} \int_\gamma \frac{f(t)\,dt}{(t-z)^{m+1}}$$

we obtain

$$f^{(m)}(z) = \frac{m!}{2\pi i} \int_\gamma \frac{f(t)\,dt}{(t-z)^{m+1}} \qquad\qquad \square$$

9.16 THEOREM (*Cauchy Inequalities*). *If* $f \in H(U;F)$ *and* $\overline{B}_r(p) \subset U$,
then

$$\left\| f^{(m)}(p) \right\| \leq \frac{m!}{r^m} \sup\{\left\| f(z) \right\| : \left\| z-p \right\| = r\}$$

for any $m = 1, 2, \ldots$

Proof. In the Cauchy integral formula for $f^{(m)}$ in (9.15), let $z = p$ and $\gamma(\theta) = p + re^{i\theta}$, $\theta \in [0, 2\pi]$. Then the inequality we want to prove can be obtained by estimating $\|f^{(m)}(p)\|$ using the integral inequality appearing in 9.3. □

9.17 *Taylor Series.* Once we have the Cauchy inequalities at our disposal for vector-valued holomorphic mappings, we are ready to prove that a holomorphic map is analytic.

THEOREM. *If* $f \in H(U;F)$ *and* $p \in U$, *then* f *as the Taylor series*

$$f(z) = \sum_{m=0}^{\infty} \frac{1}{m!} f^{(m)}(p)(z-p)^m$$

which converges absolutely and uniformly on any closed ball of the form $\overline{B}_r(p)$ *which is contained in* U.

Proof. We can prove this theorem using linear functionals. Instead, we use Taylor's formula, which was presented in Chapter 8.

Let $R = d(p, \partial U)$ be the distance from p to the boundary ∂U of U. The absolute convergence of the series for $|z-p| < R$ follows from the Cauchy inequalities. It remains to show that if $0 < r < R$, the series converges to f uniformly on $\overline{B}_r(p)$. We first apply the Taylor formula 8.10 to f:

$$\left\| f(x) - T_{m,f,p}(x) \right\| \leq \frac{\|x-p\|^{m+1}}{(m+1)!} \sup \left\| f^{(m+1)}(t) \right\| \tag{1}$$

where the supremum is over $\overline{B}_r(p)$ since $[p,x] \subset \overline{B}_r(p)$. On the other hand, if $r < s < R$, then for any $t \in \overline{B}_r(p)$, it follows from the Cauchy inequalities that

$$\left\| f^{(m+1)}(t) \right\| \leq \frac{(m+1)!}{s^{m+1}} M_s \tag{2}$$

where M_s is the supremum of $\|f(y)\|$ over $\overline{B}_s(p)$. Therefore, combining (1) and (2), we obtain

$$\|f(x) - T_{m,f,p}(x)\| \le (r/s)^{m+1} M_s \tag{3}$$

for all $x \in \overline{B}_r(p)$. Since $r < s$, we infer that $T_{m,f,p}$ converges to f uniformly on $\overline{B}_r(p)$.   □

9.18 *Power Series*. The converse of the preceding theorem is also true. We state the following theorem without proof.

THEOREM. *Let* (a_n) *be a sequence in a Banach space F, and assume that*

$$f(z) = \sum_{m=0}^{\infty} a_m(z-p)^m$$

converges for $|z-p| < R$. *Then*

   (a) For each $k > 1$, *the series*

$$\sum_{m=k}^{\infty} m(m-1)\ldots(m-k+1) a_m(z-p)^{m-k} \tag{*}$$

converges for $|z-p| < R$;

   (b) The function f is infinitely differentiable on the ball $|z-p| < R$, *and furthermore,* $f^{(k)}(z)$ *is given by the series (*)* *for all* $k > 1$ *and* $|z-p| < R$;

   (c) For $m \ge 0$,

$$a_m = \frac{1}{m!} f^{(m)}(p)$$

9.19 *Entire Functions and Liouville's Theorem*. A function f defined and holomorphic on the whole complex plane is said to be *entire*. Liouville's theorem has a natural extension to vector-valued functions.

THEOREM. *If f: $\mathbb{C} \to F$ is a bounded entire function, then f is constant.*

Proof. Although it is easy to prove Liouville's theorem directly using Taylor's series and Cauchy's inequalities and imitating the classical proof, we will use linear functionals in this proof. For a bounded entire function f: $\mathbb{C} \to F$, define $g(z) = f(z) - f(0)$. Then g is bounded and entire, and $g(0) = 0$. For each $\lambda \in F^*$, $\lambda \circ g$ is a complex-valued bounded and entire function, and $\lambda \circ g(0) = 0$. Hence by the classical Liouville theorem, $\lambda \circ g = 0$, and by (2.11), $g = 0$. Thus $f(z) = f(0)$, and f is constant. □

9.20 A great deal of the standard classical theory of complex variables can be taken over intact to Banach space valued functions, as we have seen. Cauchy's integral theorem, Cauchy's integral formulae for a function and its derivatives, Taylor's theorem, Liouville's theorem, the identity theorem, and many other theorems retain their validity. Some of these are stated in Exercises for the chapter. The proofs are in general just as in the classical theory, except that norms replace absolute values and the Hahn-Banach theorem is used. Laurent's theorem also holds in this general setting. These theorems in this generality are used in *spectral theory* (see Dunford & Schwartz (1957), Taylor (1958), or Yosida (1968)). However, the maximum modulus theorem needs more special attention for vector-valued functions, and we devote the next chapter to a study of this topic.

EXERCISES

9A *Riemann Sum.* Let $f \in R(I;E)$. Then for any $\varepsilon > 0$ there exists $\delta > 0$ such that for any partition

P: $a = a_1 < \cdots < a_n = b$

with $a_{k+1} - a_k \leq \delta$, we have if $a_k \leq t_k \leq a_{k+1}$,

$$\left\| \int_a^b f - \Sigma \, f(t_k)(a_{k+1} - a_k) \right\| < \epsilon$$

9B *Limit of Zeros.* Let U be a connected open set in $\mathbb{C}$ and f: U → F a non-zero or non-constant holomorphic map. If z_1 is the limit of zeros of f, then $z_1 \notin U$.

9C *The Identity Theorem.* Let U be a connected open set in $\mathbb{C}$ and let (x_n) be a sequence of distinct points in U which converges to a point in U. If two holomorphic mappings f,g: U → F satisfy $f(x_n) = g(x_n)$ for all n, then f = g on U.

9D *Limit of Holomorphic Maps.* Let U be a nonempty open set in $\mathbb{C}$ and f_n: U → F a sequence of holomorphic maps. If (f_n) converges uniformly on each compact subset of U to a function f, then f is holomorphic.

9E Let $V \subset \mathbb{C}$ be open, g: V → F be holomorphic, and $0 < r \leq R$. If the closed annulus in $\mathbb{C}$ of center at ξ and radii r and R is contained in V, we have

$$\int_{|\lambda - \xi| = r} g(\lambda) \, d\lambda = \int_{|\lambda - \xi| = R} g(\lambda) \, d\lambda$$

10

The Strong Maximum Modulus Theorem

10.1 For a connected open set U in the complex plane, if $f: U \to \mathbb{C}$ is holomorphic on U, then the classical maximum modulus theorem states that *if* $|f(z)|$ *has a maximum at a point of* U *then* $|f(z)|$ *is constant on* U. Furthermore, if $|f(z)|$ is a constant function, it follows that $f(z)$ is constant.

This theorem has a generalization for Banach space valued functions, but, unlike complex-valued functions, we cannot infer that if $\|f(z)\|$ is constant then $f(z)$ is itself constant (see 10.3).

The maximum modulus theorem which claims that if $\|f(z)\|$ has a maximum at a point in U, then $f(z)$ is a constant on U is referred to as the *strong maximum modulus theorem*. The purpose of this chapter is to characterize those Banach spaces which satisfy the strong maximum modulus theorem.

10.2 THEOREM *(Maximum Modulus Theorem)*. *Let* U *be a connected open subset of* $\mathbb{C}$ *and* $f \in H(U;F)$. *If* $\|f(z)\|$ *has a maximum at a point in* U, *then* $\|f(z)\|$ *is a constant on* U.

Proof. We use linear functionals in this proof. Suppose that the theorem is false, and for some $p \in U$, $\|f(z)\| \leq \|f(p)\|$ for all $z \in U$. We may assume $\|f(p)\| \neq 0$, since otherwise the theorem holds trivially. By 2.10, there exists $\lambda \in F^*$ such that $\|\lambda\| = 1$ and $\lambda(f(p)) = \|f(p)\|$. Since f is holomorphic, $\lambda \circ f: U \to \mathbb{C}$ is holomorphic and for all $z \in U$, we have

$$|\lambda(f(z))| \le |\lambda(f(p))|$$

Now the classical maximum modulus theorem implies that $\lambda \circ f$ is a constant map on U, and hence $\lambda(f(z)) = \|f(p)\|$ for all $z \in U$. Since $\|\lambda\| = 1$ and $\|f(z)\|$ is not a constant map, there exists $q \in U$ such that $\|f(q)\| < \|f(p)\|$. Then

$$\|f(p)\| = \lambda(f(q)) \le \|f(q)\| < \|f(p)\|$$

a contradiction, which completes the proof.   □

10.3 EXAMPLE. If f is a complex-valued function on U such that $|f(z)|$ is constant for all $z \in U$, then f is a constant function on U. Therefore, for complex-valued functions the strong maximum modulus theorem holds. However, this is not the case in general for vector-valued functions, as the following example shows.

Let U be the open unit disc in the complex plane and let $F = \mathbb{C}^2$ be normed by

$$\|x\| = \max \{|x_1|, |x_2|\}$$

If $f: U \to F$ is defined by

$$f(z) = (1, z)$$

then f is obviously a non-constant holomorphic mapping and

$$\|f(z)\| = \max \{1, |z|\} = 1$$

for all $z \in U$. This shows that the maximum modulus theorem fails to hold for the Banach space $\mathbb{C}^2$.

EXTREME POINTS

10.4 Thorp and Whitley (1967) were first to notice that the discussion

of the maximum modulus theorem for vector-valued mappings leads to
a study of certain generalized extreme points, which they called
complex extreme points. This chapter is based on their work.

10.5 *Real Extreme Points.* Let X be a convex set in a vector space.
A point $x \in X$ is called an *extreme point* of X if whenever x_1, x_2 are
points of X such that if we can write

$$x = ax_1 + (1-a)x_2$$

with $0 < a < 1$, then $x_1 = x_2$. In other words, an extreme point of a
convex set X is a point in X that is not an interior point of any
line segment in X. If an extreme point is an interior point of a
line segment $[x_1, x_2]$, then either x_1 or x_2 must be in the complement
of X. Thus a "real" disc centered at x "sticks out of" X, no matter
in which direction it is tilted. Henceforth we call an extreme
point a *real extreme point*, and denote the set of all real extreme
points of a convex set X by $\text{Ext}_R X$.

It is easy to see that $\text{Ext}_R X$ may be empty. For example, if
$E = \mathbb{R}^2$, then the open unit ball has no extreme points. On the
other hand, every point in the boundary of the closed unit ball is
a real extreme point.

10.6 For each p, $1 \leq p < \infty$, let B^p be the closed unit ball of the
Lebesgue space $L^p(I)$, where I is an interval in $\mathbb{R}$. Using the
Hölder inequalities it can be shown that

$$\text{Ext}_R(B^1) = \emptyset$$

$$\text{Ext}_R(B^p) = \{f \in L^p(I) : \|f\|_p = 1\}, \ 1 < p < \infty$$

See, for example, Larson (1973), pp. 317-320.

10.7 *Complex Extreme Points.* A point x in a convex subset X of a
complex vector space E is said to be a *complex extreme point* of X
if there is no non-zero vector $y \in E$ with

$\{x + zy: z \in \mathbb{C}, |z| \leq 1\} \subset X$

It is easy to show that a real extreme point of X is a complex extreme point of X. In fact, if x is a real extreme point of X, then

$x = (x + y)/2 + (x - y)/2$

and hence x + y = x - y, which shows that y = 0.

We shall denote the set of all complex extreme points of X by $Ext_\mathbb{C}X$. For 1 < p < ∞, since

$Ext_R B^p = \{f \in L^1(I): \|f\|_1 = 1\}$

we can see that

$Ext_\mathbb{C} B^p = Ext_R B^p$

Though the notions of real and complex extreme points often coincide, as we have seen above for L^p-spaces (1 < p < ∞), the notions differ significantly in the case of $L^1(I)$.

10.8 THEOREM. *Let* B^1 *be the closed unit ball of* $L^1(I)$. *Then*

$Ext_\mathbb{C} B^1 = \{f \quad L^1(I): \|f\|_1 = 1\}$

We need the following lemma. For $f \in L^1(I)$, let $S(f) = \{x \in I: f(x) \neq 0\}$.

LEMMA. *For* $f,g \in L^1(I)$, $\|f + g\| = \|f\| + \|g\|$ *if and only if there exists a function h on I such that* f=hg *almost everywhere on* $S(f) \cap S(g)$ *and* h(x) > 0 *for* $x \in S(f) \cap S(g)$.

Proof. The sufficiency is clear. We show the necessity. Suppose that $\|f + g\| = \|f\| + \|g\|$. Then

$$\int_I (|f + g| - |f| - |g|) = 0$$

and hence

$$|f + g| = |f| + |g| \quad \text{a.e.}$$

An elementary calculation with complex numbers shows that if $|z + w| = |z| + |w|$, then the arguments of the numbers z, w are equal, and $z/w > 0$ if $zw \neq 0$. Therefore, we conclude that if we set $h = f/g$, then $h(x) > 0$ for $x \in S(f) \cap S(g)$. □

Proof of Theorem. For simplicity, we omit the index "1" in the L^1-norm $\|\cdot\|_1$. Suppose that f and g are in $L^1(I)$ with $\|f\| = 1$ and $\|f + zg\| \leq 1$ for all $|z| \leq 1$. We claim that $g = 0$ a.e. Suppose $g \neq 0$ a.e. If $\|f + z'g\| < 1$ for some $|z'| \leq 1$, then

$$2\|f\| \leq \|f + z'g\| + \|f - z'g\| < 2$$

a contradiction. Thus

$$\|f + zg\| = 1$$

for all $|z| \leq 1$.

Let $S(f) = \{x \in I : f(x) \neq 0\}$. Then

$$2 = 2\int_{S(f)} |f| \leq \int_{S(f)} (|f + g| + |f - g|)$$

$$\leq \int_{S(f)} (|f + g| + |f - g|) + 2\int_{I \setminus S(f)} |g|$$

$$= \int_I (|f + g| + |f - g|) = 2$$

Thus $\int_{I \setminus S(f)} |g| = 0$. Hence $S(g) \subseteq S(f) \cup N$, where N is a null set. Now write for each z, $|z| \leq 1$,

$$2f = (f + zg) + (f - zg)$$

Then

$$2\|f\| = \|f + zg\| + \|f - zg\|$$

Applying the lemma above, we can find a function h_z corresponding to z, $|z| \leq 1$, such that $h_z(x) > 0$ on $S(f + zg) \cap S(f - zg)$ and

$$f + zg = h_z(f - zg) \quad \text{a.e.}$$

on $S(f + zg) \cap S(f - zg)$. Then

$$zg = (h_z - 1)f/(h_z + 1) \quad \text{a.e.}$$

on $S(f + zg) \cap S(f - zg)$. We also have

$$zg\overline{f} = (h_z - 1)|f|/(h_z + 1) \quad \text{a.e.}$$

on $S(f + zg) \cap S(f - zg)$. This shows that for any z, $|z| \leq 1$, $zg(x)\overline{f}(x)$ is always a real number for almost all x in $S(f + zg) \cap S(f - zg)$. Therefore $g = 0$ a.e. on $S(f)$. But we know that $g = 0$ a.e. on $I \backslash S(f)$; hence $g = 0$ a.e. on I. This completes the proof. $\square$

THE STRONG MAXIMUM MODULUS THEOREM

10.9 *Strictly Convex Banach Spaces.* A Banach space is said to be *strictly convex* (or *rotund*) if the unit sphere is the set of real extreme points of the closed unit ball.

The Lebesgue space $L^p(I)$ is strictly convex for $1 < p < \infty$, but $L^1(I)$ is not strictly convex. In 10.11 we shall see that if f is a holomorphic mapping of an open subset in $\mathbb{C}$ into a strictly convex Banach space, then f satisfies the strong maximum modulus theorem.

First we introduce the following notations. Let E be a vector space over $\mathbb{K}$ and suppose $X \subseteq E$. Then the *convex hull* of X is defined as

$$co(X) = \{ \sum_{k=1}^{n} a_k x_k : a_k \geq 0 , \sum_{k=1}^{n} a_k = 1 , x_k \in X\}$$

If E is a normed space, then the *closed convex hull* of X is defined as the closure in E of co(X). It will be denoted by $\overline{co}(X)$. It is not difficult to see that co(X) and $\overline{co}(X)$ are, respectively, the smallest convex and the smallest closed convex sets that contain X.

10.10 LEMMA. *Let U be a connected open subset in the complex plane, F a complex Banach space, and $f \in H(U;F)$ with $\|f(z)\| = 1$ for all z in U. Then for each point y in $\overline{co}(f(U))$ we have $\|y\| = 1$.*

Proof. Let $y \in co(f(U))$. Then there are $a_k \geq 0$, $z_k \in U$ such that $\sum_{k=1}^{n} a_k = 1$ and $y = \sum_{k=1}^{n} a_k f(z_k)$. Since $\|f(z_1)\| = 1$, by the Hahn-Banach theorem there is a continuous linear functional $\lambda \in F^*$ with $\lambda(f(z_1)) = 1$ and $\|\lambda\| = 1$. Since $|\lambda(f(z))| \leq \|\lambda\| \cdot \|f(z)\| = 1$ for all $z \in U$, the holomorphic map λ of attains its maximum modulus at z_1. From the classical maximum modulus theorem, $\lambda(f(z)) = 1$ on U. Thus $\lambda(y) = \sum a_k \lambda(f(z_k)) = 1$. Since $\|y\| = \sup\{|\phi(y)| : \phi \in F^*, \|\phi\| \leq 1\}$ (see 2.11), $\|y\| \geq 1$. On the other hand, $\|y\| \leq \sum a_k \|f(z_k)\| = 1$, and hence $\|y\| = 1$. It follows immediately that every point of $\overline{co}(f(U))$ has norm one, which completes the proof.   □

10.11 THEOREM. *Let U be a connected open subset in $\mathbb{C}$, F a strictly convex complex Banach space, and $f \in H(U;F)$. Then f satisfies the strong maximum modulus theorem.*

Proof. From Theorem 10.2, we know that $\|f(z)\|$ is constant on U if $\|f(z)\|$ attains its maximum at a point in U. Thus we can assume that $\|f(z)\| = 1$ for all x in U. It follows from the preceding lemma that $\overline{co}(f(U))$ is contained in the unit sphere of F. Since each point of the unit sphere of F is a real extreme point of the closed unit ball, any convex set which lies on the unit sphere must consist of one point (why?). Thus $\overline{co}(f(U))$ is a one-point set, which shows that f is constant.   □

10.12 *Necessary Condition for the Strong Maximum Modulus Theorem.*
We shall show that if every holomorphic map f: U → F satisfies the
strong maximum modulus theorem, then the unit sphere of F consists
only of complex extreme points.

THEOREM. *Let F be a complex Banach space. If the unit sphere of F
contains a point which is not a complex extreme point of the closed
unit ball, then there exists a nonconstant holomorphic map f of the
open unit disc of* $\mathbb{C}$ *into F satisfying* $\|f(z)\| = 1$ *for all* $|z| < 1$.

Proof. Let $x \in F$ be a point that is not a complex extreme point of
the closed unit ball of F. Choose $y \neq 0$ such that $\|x + zy\| \leq 1$ for
all $|z| \leq 1$. Then $\|x + zy\| = 1$ for all $|z| \leq 1$. This can be shown
by the same argument given in the proof of Theorem 10.8. Now it is
obvious to define

$$f(z) = x + zy, \quad |z| < 1$$

Then $\|f(z)\| = 1$, which completes the proof.   □

10.13 To show the converse of the preceding theorem, we need the
following lemma due to Harris (1969) who simplified the original
proof of Thorp and Whitley (1967).

LEMMA. *Any holomorphic function* f: D → $\overline{D}$, *where* D = $\{z \in \mathbb{C}: |z| < 1\}$,
satisfies

$$|f(0)| + \frac{1 - |z|}{2|z|} \, |f(z) - f(0)| \leq 1$$

for all z $\in$ D\{0\}.
We first recall Schwarz's lemma in one variable before proving the
above lemma.

SCHWARZ'S LEMMA. *If* f: D → $\overline{D}$ *is holomorphic and* f(0) = 0, *then*
$|f(z)| \leq |z|$ *for all* z $\in$ D. *If* $|f(z_0)| = |z_0|$ *for some* $z_0 \in$ D, *then*
$f(z) = e^{ir}z$ *on D for some* r $\in \mathbb{R}$.

Proof. If $|f(z)| = 1$ for some z in D, then f is a constant map by the classical maximum modulus theorem and the lemma holds trivially. Hence we may assume that f: D → D. Then for any $\alpha \in D$, the Möbius transformation

$$\phi(z) = \frac{z-\alpha}{1-\bar{\alpha}z}$$

maps D onto D and α to 0. Hence the map

$$g(z) = \frac{f(z) - f(0)}{1 - \overline{f(0)}f(z)}$$

is a holomorphic mapping of D into D. Furthermore, $g(0) = 0$. By Schwarz's lemma, $|g(z)| \le |z|$ for all z in D; hence

$$|f(z) - f(0)| \le |z||1-\overline{f(0)}f(z)|$$

for all $z \in D$. By the triangle inequality,

$$|1-\overline{f(0)}f(z)| = |1-f(0)\overline{f(0)} + \overline{f(0)}[f(0) - f(z)]|$$

$$\le 1 - |f(0)|^2 + |f(0)||f(0) - f(z)|$$

so

$$|f(z) - f(0)| \le |z|(1 - |f(0)|^2) + |z||f(0) - f(z)||f(0)|$$

Hence

$$(1 - |z||f(0)|)|f(z) - f(0)| \le |z|(1 - |f(0)|^2)$$

Since $|f(0)| < 1$, we have $1 - |z| \le 1 - |z||f(0)|$, and $1 - |f(0)|^2 \le 2(1 - |f(0)|)$. Hence

$$(1 - |z|)|f(z) - f(0)| \le 2|z|(1 - |f(0)|)$$

which proves the lemma.   □

10.14 THEOREM. *Let f be a holomorphic mapping of the open unit*
disc in $\mathbb{C}$ *into a complex Banach space F and suppose that* $\|f(z)\| = 1$
for $|z| < 1$. *If f(p) is a complex extreme point of the closed unit*
ball of F for some p, $|p| < 1$, *then f is constant.*

Proof (Harris, 1969). We may assume without loss of generality that
f(0) is a complex extreme point of the unit ball of F. This follows
readily from a change of variables and the identity theorem (see
Ex. 9C).

   Let $\lambda \in \mathbb{C}$ be such that $|\lambda| \leq \dfrac{1-|z|}{2|z|}$. For any $A \in F^*$, $\|A\| \leq 1$,
since $|Af(z)| \leq 1$ for all $|z| < 1$, we have from the preceding lemma
that

$$\left| A\{f(0) + \lambda[f(z) - f(0)]\}\right| \leq 1$$

It follows from 2.12 that

$$\|f(0) + \lambda[f(z) - f(0)]\| \leq 1$$

for all z. Since f(0) is a complex extreme point, we must have
f(z) = f(0) for all z, $|z| < 1$. This completes the proof.   □

10.15 *The Strong Maximum Modulus Theorem*. Combining Theorem 10.12
and 10.14, we have the following statement.

THEOREM (Thorp and Whitley, 1967). *Let U be a connected open set*
in $\mathbb{C}$ *and F a complex Banach space. Then every map f in H(U;F) sat-*
isfies the strong maximum modulus theorem if and only if every point
in the unit sphere of F is a complex extreme point of the closed
unit ball.

Proof. Suppose that every point in the sphere of the unit ball of
F is a complex extreme point and $f \in H(U;F)$ such that $\|f(z)\| = a$,
a constant, for all $z \in U$. We may assume that $a \neq 0$. Then f(z)/A

has norm 1 on U. By the preceding theorem, f(z)/a is constant around some neighborhood in U. Then by the identity theorem, f is constant on U. ◻

10.15 REMARK. In light of the Thorp-Whitley theorem J. Globevnik (1975) introduced the following concepts. A complex normed space is *strictly c-convex* if its unit sphere consists of complex extreme points of its closed unit ball; a complex normed space is said to be *uniformly c-convex* if for every $\varepsilon > 0$ there exists $\delta > 0$ such that $x, y \in E$, $\|x + zy\| \leq 1$ ($|z| < 1$) and $\|y\| > \varepsilon$ implies $\|x\| < 1-\delta$. It is clear that uniform c-convexity implies strict c-convexity. Following Thorp and Whitley (1967), Globevnik showed that $L^1(I)$ is uniformly c-convex. It should be noted that recently Davis, Garling and Tomczak-Jaegermann (1984) studied the complex convexity of quasi-normed spaces. Globevnik's work was motivated by the strong maximal modulus theorem; Davis et al. was interested in finding complex versions of the modulus of convexity. ◻

EXERCISES

10A *Schwarz's Lemma.* Let f be a holomorphic map of the open unit disc D of $\mathbb{C}$ to a complex Banach space F. Suppose that $\|f(z)\| \leq M$ on D and f(0) = 0. Then $\|f(z)\| \leq M|z|$ on D.

10B *Extreme Points.* Let E be the Banach space of all bounded complex-valued mappings on a set X with the sup-norm. Then a function $f \in E$ with $\|f\| = 1$ is a real extreme point of the unit ball of E if and only if it is a complex extreme point.

10C *Hadamard's Three Circles Theorem.* Let f(z) be holomorphic for $|z| < R$ and let

$$M(r) = \max\{\|f(z)\| : |z| = r\}$$

for $0 \leq r < R$.

(1) M(r) is a nondecreasing function of r.

(2) If $M(r_1) = M(r_2)$ for $0 < r_1 < r_2$, then M(r) is a constant function.

(3)   If $f(z)$ is not identically zero for $|z| < R$, then $\log M(r)$
is a convex function of $\log r$ when $0 < r < R$; i.e., if
$r_1 < r_2 < r_3$, then

$$(\log r_3 - \log r_1)\log M(r_2) \leq (\log r_3 - \log r_2)\log M(r_1)$$

$$+ (\log r_2 - \log r_1)\log M(r_3)$$

10D   *Maximum Modulus Theorem for a Strip.*   Let $S = \{x + iy: x_0 < x < x_1, y \in \mathbb{R}\}$. If $f: U \to F$ is a bounded holomorphic map, where $S \subset U$, and $\|f(x_0 + iy)\| \leq M$, $\|f(x_1 + iy)\| \leq M$, then $\|f(z)\| \leq M$ for $x \in S$.

10E.   The closed unit ball of the Banach space c_o has no complex extreme points.   More generally, if X is a locally compact, non-compact Hausdorff space, then the Banach space $C_o(X)$ of all complex-valued functions vanishing at infinity on X with the sup norm has no complex extreme point.

11

Power Series

11.1 In this chapter the definition and basic properties of a power
series will be given. In particular, we will discuss various concepts
of the radius of convergence of a power series, namely, the radius
of uniform convergence and the radius of absolute convergence. Then
we will show that a function which is represented by a power series
is a C^{∞} function within the radius of uniform convergence. Normed
spaces appearing in this chapter are vector spaces over the field $\mathbb{K}$.

RADIUS OF UNIFORM CONVERGENCE

11.2 *Power Series.* Recall that a power series in classical analysis
is a formal series of polynomials. Maintaining a parallel with
classical analysis, we have the following definition of a power
series.

A *power series* from E to F about $\xi \in E$ is a formal series of
the form

$$\sum_{m=0}^{\infty} A_m (x - \xi)^m$$

where $A_m \in L_s(^m E;F)$, $m = 0,1,\dots$ If we prefer to use polynomials, a
power series from E to F about ξ is a formal series of the form

$$\sum_{m=0}^{\infty} P_m (x - \xi)$$

where $P_m \in P(^mE;F)$, $m = 0,1,\ldots$ Both A_m and P_m will be called the *mth coefficients of the power series.*

EXAMPLES. (1) An obvious example of a power series is a polynomial

$$P_0 + P_1 + \ldots + P_m$$

where $P_k \in P(^kE;F)$, $k = 0,1,\ldots,m$.

(2) For $f \in L(E;E)$, we introduced the exponential function

$$\exp f = \sum_{m=0}^{\infty} \frac{f^m}{m!}$$

in Exercise 2E. This is a power series from $L(E;E)$ to itself about the origin.

(3) If $E = \ell^2$ and $F = \mathbb{K}$,

$$\sum_{m=1}^{\infty} (x_m)^m \qquad x = (x_1, x_2, \ldots) \in \ell^2$$

is a power series around 0 since $P_m(x) = (x_m)^m$ is an m-homogeneous polynomial from E to F.

11.3 *Radius of Uniform Convergence.* We now explore the convergence properties of power series. To motivate the reader we first review the properties of the radius of convergence of a power series from $\mathbb{K}$ to $\mathbb{K}$.

THEOREM. *For a power series* $\sum_{m=0}^{\infty} a_m(z-a)^m$ *from* $\mathbb{K}$ *to* $\mathbb{K}$, *define the number* ρ, $0 \leq \rho \leq +\infty$, *by*

$$\frac{1}{\rho} = \lim \sup |a_m|^{1/m}$$

Then:

(a) *if* $|z-a| < \rho$, *the series converges absolutely;*

(b) if $|z-a| > \rho$, *the series diverges;*

(c) if $0 < r < \rho$, *the series converges uniformly on* $\{z: |z-a|$ $\leq r\}$. *Moreover, the number* ρ *is the only number satisfying properties (a) and (b), and* ρ *is called the radius of convergence of the power series.*

In other words, if ρ is the radius of convergence, then

(1) ρ is the largest number $(0 \leq \rho \leq + \infty)$ such that the series converges absolutely on $\{z: |z-a| < \rho\}$;

(2) ρ is the largest number $(0 \leq \rho \leq + \infty)$ such that if $0 < r < \rho$, the series converges uniformly on $\{z: |z-a| \leq r\}$.

Strictly speaking, the radius of convergence should be called the *radius of absolute convergence* and the *radius of uniform convergence* corresponding to the properties (1) and (2), respectively, stated above. Then the radius of absolute convergence is equal to the radius of uniform convergence. We will see later that this is not the case for power series defined on an infinite dimensional space (see 11.9).

We first introduce the radius of uniform convergence for a power series defined on a normed space.

For a given power series $\sum_{m=0}^{\infty} P_m(x-\xi)$ from E to F about $\xi \in E$, a number ρ, $0 \leq \rho \leq + \infty$, is said to be the *radius of uniform convergence* of the series if ρ is the supremum of all r, $0 \leq r \leq + \infty$, such that the power series converges uniformly on $\overline{B}_r(\xi)$. Since r can be 0, the radius of uniform convergence always exists.

The ball $B_\rho(\xi)$, where ρ is the radius of uniform convergence, is called the *open ball of uniform convergence*. When $\rho = + \infty$, we say that the power series is *entire*; that is, the power series converges uniformly on every bounded subset of E. For example, the exponential function exp f, f $\in$ L(E;E), is entire if E is a Banach space.

11.4 *Normal Convergence.* It will be useful to introduce the following concept of convergence. Let X be a set and F a Banach space. Consider the Banach space B(X;F) of bounded functions of X into F with the sup norm $\|f\| = \sup\{\|f(x)\|: x \in X\}$.

A series $\sum_{m=0}^{\infty} f_m$ of functions in $B(X;F)$ is said to be *normally convergent* if $\sum_{m=0}^{\infty} \|f_m\| < +\infty$. This implies that, for each $x \in X$, the series $\sum_{m=0}^{\infty} \|f_m(x)\|$ is convergent, and so the series $\sum_{m=0}^{\infty} f_m(x)$ is absolutely convergent. Since F is complete, $\sum_{m=0}^{\infty} f_m(x)$ converges (see 1.18). If $f(x)$ is the sum of this series, i.e.,

$$f(x) = \sum f_m(x) \quad x \in X$$

then

$$\|f\| \le \sum_{m=0}^{\infty} \|f_m\|$$

$$\lim_{p} \left\| f - \sum_{m=0}^{p} f_m \right\| = 0$$

Therefore, $\sum_{m=0}^{\infty} f_m$ converges to f uniformly on X. This shows that *a normally convergent series is absolutely and uniformly convergent.* This result is known as the *Weierstrass M-test.*

Recall that the limit of a uniformly convergent sequence of continuous functions on a topological space is always continuous, and hence the sum of a normally convergent series of continuous functions is also continuous.

We have the following diagram showing implications of different notions of convergence.

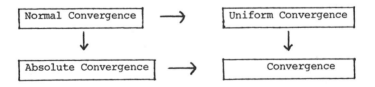

In general, uniform convergence does not imply normal convergence (see Exercise 11C).

11.5 *Cauchy-Hadamard Formula.* The radius of uniform convergence can be computed by the Cauchy-Hadamard formula if the range space F is complete. We first recall that if (a_n) is a sequence of real numbers, then the limit superior of (a_n) is defined by

$$\lim \sup_n a_n = \lim_n (\sup\{a_n, a_{n+1}, \dots\})$$

THEOREM. *Let F be a Banach space. Then the radius of uniform convergence ρ of a power series*

$$\sum_{m=0}^{\infty} P_m(x - \xi)$$

is given by

$$\frac{1}{\rho} = \lim \sup \|P_m\|^{1/m}$$

In this formula, we adopt the usual convention that $\rho = 0 \Leftrightarrow 1/\rho = +\infty$, $\rho = +\infty \Leftrightarrow 1/\rho = 0$.

Proof. Let $L = \lim \sup \|P_m\|^{1/m}$. We consider the following three cases.

(1) Suppose that $L = +\infty$. We claim that $\rho = 0$. In fact, let $A > 0$ be a given real number. Since $L > A$, there exists an infinite subset N_A of the natural numbers such that

$$\|P_m\|^{1/m} > A$$

for $m \in N_A$. Now let $r = 1/A$. We want to show that the series does not converge uniformly on $\bar{B}_r(\xi)$. This will show that $\rho = 0$ since $A > 0$ is fixed but arbitrary.

Using properties of the norm $\|P_m\|$, it is easy to find $y_m \in E$, $\|y_m\| = 1$, such that

$$\|P_m(y_m)\| > A^m$$

for each $m \in N_A$. Let $x_m = \xi + ry_m$. It follows that

$$\|P_m(x_m-\xi)\| = r^m\|P_m(y_m)\| > r^m\left(\frac{1}{r}\right)^m = 1$$

for $m \in N_A$. This proves that the power series does not converge uniformly on $\overline{B}_r(\xi)$.

(2) Suppose that $0 < L < + \infty$. Let $0 < r < \rho$ and

$$a_m = \|P_m\|r^m$$

for $m = 0,1,\ldots$ Then

$$\lim \sup_m (a_m)^{1/m} = (\lim \sup \|P_m\|^{1/m})r = Lr$$

By the root test on series of nonnegative numbers, $\Sigma\, a_m$ converges if $Lr < 1$. Therefore $\Sigma_{m=0}^{\infty} \|P_m\|r^m < + \infty$ for $r < 1/L$. This shows that $\Sigma_{m=0}^{\infty} P_m(x - \xi)$ converges normally and hence uniformly on the ball $\overline{B}_r(\xi)$ for each $r < 1/L$. Thus $1/L \leq \rho$.

On the other hand, if $r > 1/L$, then $L > 1/r$. Then for some infinite set F_r of natural numbers, we have

$$\|P_m\| > \frac{1}{r^m}$$

whenever $m \in F_r$. As in case (1), we can find $y_m \in E$, $\|y_m\| = 1$ for each $m \in F_r$, such that

$$\|P_m(y_m)\| > \frac{1}{r^m}$$

Then for $x_m = \xi + ry_m$, we have

$$\left\| P_m(x_m - \xi) \right\| = r^m \left\| P_m(y_m) \right\| > r^m \left(\frac{1}{r}\right)^m = 1$$

which shows that $\sum_{m=0}^{\infty} P_m(x-\xi)$ does not converge uniformly on $\overline{B}_r(\xi)$. This proves that $\rho \leq 1/L$; hence $\rho = 1/L$.

(3) For $L = 0$, let $r > 0$ and $\varepsilon = 1/2r$. Then there exists $N > 0$ such that

$$\left\| P_m \right\| \leq \varepsilon^m$$

for $m \geq N$. If $x \in \overline{B}_r(\xi)$, we have

$$\left\| P_m(x-\xi) \right\| \leq \left\| P_m \right\| \cdot \left\| x-\xi \right\|^m \leq (\varepsilon r)^m = \frac{1}{2^m}$$

for $m \geq N$. This shows that the power series converges uniformly on $\overline{B}_r(\xi)$. Since $r > 0$ is arbitrary, we conclude that $\rho = +\infty$. □

COROLLARY 1. *If ρ is the radius of uniform convergence for* $\sum_{m=0}^{\infty} P_m(x-\xi)$ *and* $0 < r < \rho$, *then*

$$\sum_{m=0}^{\infty} \left\| P_m \right\| r^m < +\infty$$

Proof. As in case (2) of the preceding proof, apply the root test with $a_m = \left\| P_m \right\| r^m$ to show the result. □

COROLLARY 2. *If F is a Banach space, the following statements are equivalent:*

(a) $\sum_{m=0}^{\infty} P_m(x-\xi)$ *converges uniformly on some* $\overline{B}_r(\xi)$, $r > 0$;

(b) $\lim \sup \left\| P_m \right\|^{1/m} < +\infty$;

(c) *the sequence* $(\left\| P_m \right\|^{1/m})$ *is bounded;*

(d) *there exist $C > 0$ and $c > 0$ such that*

$$\|P_m\| \le Cc^m$$

for all m = 0,1,2,...

In (d), to avoid $\|P_0\| \le 1$, we use the constant C.

Proof. The implications (a) $\Rightarrow$ (b) $\Rightarrow$ (c) $\Rightarrow$ (d) are clear. For (d) $\Rightarrow$ (a), let r > 0 be such that rc < 1. Then

$$\Sigma \ \|P_m(x-\xi)\| \le \Sigma \ C(rc)^m < + \infty$$

for $x \in \overline{B}_r(\xi)$, which shows that the series converges normally on $\overline{B}_r(\xi)$. Hence the radius of convergence is strictly positive. □

11.6 In the definition of the radius of uniform convergence and the Cauchy-Hadamard formula, we used the power series $\Sigma \ P_m(x-\xi)$ with polynomial coefficients. If

$$\sum_{m=0}^{\infty} A_m(x-\xi)^m$$

is a given power series with $A_m \in L_s(^mE;F)$, we also define the radius of uniform convergence as in 11.3. Let $\overline{\rho}$ be the radius of uniform convergence for this series. Then we can prove that

$$\frac{1}{\rho} = \lim \ \sup \ \|A_m\|^{1/m}$$

if F is complete.

We have the following relationship between ρ and $\overline{\rho}$.

THEOREM. *Let* F *be a Banach space and let*

$$\sum_{m=0}^{\infty} P_m(x-\xi) = \sum_{m=0}^{\infty} A_m(x-\xi)^m$$

where $P_m = \hat{A}_m$, $A_m \in L_s(^mE;F)$. *Then*

$$\frac{\rho}{e} \leq \bar{\rho} \leq \rho$$

Proof. Since

$$\|P_m\| \leq \|A_m\| \leq \frac{m^m}{m!} \|P_m\|$$

for all m (see 4.13), we have

$$\rho \geq \bar{\rho} \geq \rho \lim \frac{(m!)^{1/m}}{m}$$

By Stirling's formula

$$\lim \frac{m!}{m^m e^{-m} (2\pi m)^{1/2}} = 1$$

we have

$$\lim \frac{m}{(m!)^{1/m}} = e$$

and hence

$$\frac{\rho}{e} \leq \bar{\rho} \leq \rho$$

which completes the proof. □

The strict inequality $\bar{\rho} < \rho$ can be achieved as shown by the following example.

EXAMPLE. Let $E = \ell^1$ and $F = \mathbb{K}$. Then for each m = 0,1,2,..., there exists $A_m \in L_s(^mE;F)$ such that

$$\|A_m\| = 1 \qquad \text{and} \qquad \|A_m\| = \frac{m!}{m^m}$$

(see 4.14). Now by the Cauchy-Hadamard formula we obtain

$$\bar{\rho} = 1 \quad \rho = e$$

since

$$\lim \frac{m}{(m!)^{1/m}} = e$$

11.7 *Radius of Normal Convergence.* For a power series $\Sigma\, P_m(x-\xi)$
from E to F, where F is a Banach space, let ρ_n be the supremum of
all r, $0 \leq r < \infty$, such that the power series converges normally
on $\bar{B}_r(\xi)$. We call ρ_n the *radius of normal convergence* of the power
series. In the proof of the Cauchy-Hadamard formula for the radius
of uniform convergence we in fact proved the following theorem.

THEOREM. *If F is a Banach space, the radius of normal convergence
is equal to the radius of uniform convergence.*

RADIUS OF ABSOLUTE CONVERGENCE

11.8 *Radius of Absolute Convergence.* A number ρ_a, $0 \leq \rho_a \leq + \infty$,
is said to be the *radius of absolute convergence* for a power series

$$\sum_{m=0}^{\infty} P_m(x-\xi)$$

if ρ_a is the largest number such that the series converges absolutely
for $\|x-\xi\| < \rho_a$.

As remarked before in 11.3, the radius of absolute convergence
is equal to the radius of uniform convergence if E = **K** .

It is clear that $\rho \leq \rho_a$, but ρ can be strictly less than ρ_a for
some infinite dimensional spaces, as the following example shows.

Example. Let E = ℓ^2. For

$$P_m(x) = (x_m)^m$$

where $x = (x_1, x_2, \ldots) \in \ell^2$, we have

$$\|P_m\| = 1$$

for all m, and hence $\rho = 1$. On the other hand, since $x_m \to 0$ as $m \to \infty$, the series $\Sigma \; (x_m)^m$ converges absolutely for all $x \in \ell^2$. This shows that $\rho_a = +\infty$, and hence $\rho < \rho_a$.

11.9 *Formula for* ρ_a. Let F be a Banach space. For a power series

$$\sum_{m=0}^{\infty} P_m(x-\xi)$$

and $u \in E$, $\|u\| = 1$, we define $\rho(u)$ to satisfy

$$\frac{1}{\rho(u)} = \lim \; \sup \; \|P_m(u)\|^{1/m}$$

THEOREM. $\rho_a = \inf \{\rho(u): \|u\| = 1\}$.

Proof. Let $R = \inf \{\rho(u): \|u\| = 1\}$. Without loss of generality we may take $\xi = 0$. Suppose $R > 0$. Then if $\|x\| < R$, we can find $u \in E$, $\|u\| = 1$, and $\alpha \geq 0$ such that $x = \alpha u$. We now have

$$\lim \; \sup \; \|P_m(\alpha u)\|^{1/m} = \frac{\alpha}{\rho(u)} \leq \frac{\alpha}{R} < 1$$

It follows immediately by the root test that the series $\Sigma_{m=0}^{\infty} P_m(x)$ converges absolutely. Hence $R \leq \rho_a$.

On the other hand, if $\alpha > R$, we can find $\delta > 0$ such that $\alpha > R + \delta$. Since $R = \inf \{\rho(u): \|u\| = 1\}$, there exists $u \in E$, $\|u\| = 1$, such that

$$\alpha > R + \delta > \rho(u)$$

Then

$$\lim \sup \left\| P_m(\alpha u) \right\|^{1/m} = \frac{\alpha}{\rho(u)} > 1$$

and hence the series $\Sigma \left\| P_m(\alpha u) \right\|$ diverges by the root test. This shows that

$$\rho_a \leq R$$

which proves that $\rho_a = R$.   □

REMARK. If $E = \mathbb{K}$, the formula for ρ_a reduces to ρ.

11.10 *Uniqueness of Power Series.* If two power series $\Sigma P_m(x-\xi)$ and $\Sigma Q_m(x-\xi)$ converge absolutely and have the same sum on a neighborhood of ξ, we show that $P_m = Q_m$ for all $m = 0,1,2,\ldots$ This is an immediate consequence of the following theorem.

THEOREM. *If the power series* $\Sigma_{m=0}^{\infty} P_m(x-\xi)$ *converges absolutely and its sum is equal to zero on some neighborhood of* ξ, *then*

$$P_m = 0 \quad m = 0,1,2,\ldots$$

Proof. We first prove the theorem for $E = \mathbb{K}$. For simplicity assume $\xi = 0$, and let $\Sigma_{m=0}^{\infty} u_m x^m$ be a power series from $\mathbb{K}$ to F, where $u_m \in F$. Suppose that $\Sigma u_m x^m$ converges absolutely and its sum is equal to zero for $|x| \leq \delta$ for some $\delta > 0$. If we put $x = 0$, then $u_0 = 0$. Inductively, assume that $u_0 = u_1 = \cdots = u_{n-1} = 0$ for some $n > 1$. Since $\Sigma u_m \delta^m = 0$, $\left\| u_m \right\| \delta^m \to 0$ as $m \to \infty$; hence $\{ \left\| u_m \right\| \delta^m : m = 0,1,\ldots \}$ is bounded above, say, by C. Now for $x \neq 0$, $|x| \leq \delta$,

$$\left\| u_n \right\| \leq \sum_{m=n+1}^{\infty} \left\| u_m \right\| \cdot |x|^{m-n} \leq \frac{C|x|}{\delta^n (\delta - |x|)}$$

Thus we obtain $u_n = 0$ by letting $x \to 0$.

We are now ready to prove the theorem for the general case. Assume the power series converges absolutely to 0 for $\left\| x \right\| \leq r$, $r \neq 0$. Let $t \in E$ be fixed but arbitrary. Then for any $|\lambda| \leq r/\left\| t \right\|$, we have

$$\Sigma \ P_m(\lambda t) \ = \ \Sigma \ P_m(t)\lambda^m \ = \ 0$$

Hence from the case where $E = \mathbb{K}$ above, we have $P_m(t) = 0$ for all $m = 0,1,2,\ldots$ Since t is arbitrary, we can conclude that $P_m = 0$ for all $m = 0,1,2,\ldots$ □

DIFFERENTIABILITY OF POWER SERIES

11.11 THEOREM. *Let F be a Banach space and let* $\Sigma_{m=0}^{\infty} \ A_m(x-\xi)^m$ *be a power series whose radius of uniform convergence* ρ *is strictly positive, and let*

$$f(x) \ = \ \sum_{m=0}^{\infty} \ A_m(x - \xi)^m$$

be its sum on $B_\rho(\xi)$. *If* $\eta \in B_\rho(\xi)$, *then there exists a power series* $\Sigma_{k=0}^{\infty} \ B_k(x-\eta)^k$ *such that*

$$f(x) \ = \ \sum_{k=0}^{\infty} \ B_k(x - \eta)^k$$

for $\|x-\eta\| < \rho - \|\eta-\xi\|$, *and its radius of uniform convergence is* $\geq \rho - \|\eta-\xi\|$. *The coefficients* B_k *are given by*

$$B_k \ = \ \sum_{m=k}^{\infty} \ \binom{m}{k} \ A_m(\eta - \xi)^{m-k}$$

Proof. Let $0 < r < \rho$. Then $\Sigma_{m=0}^{\infty} \|A_m\| r^m < + \infty$ by Corollary 1 of 11.5, and $\Sigma_{m=0}^{\infty} A_m(x-\xi)^m$ converges uniformly on $\overline{B}_r(\xi)$. For $\eta \in B_r(\xi)$, apply the multinomial formula in 4.4 to $A_m[(\eta-\xi) + (x-\eta)]^m$ to get

$$A_m(x-\xi)^m \ = \ \sum_{k=0}^{\infty} \ \binom{m}{k} A_m(\eta-\xi)^{m-k}(x-\eta)^k$$

Then

$$\sum_{k=0}^{m} \binom{m}{k} \|A_m\| \cdot \|\eta - \xi\|^{m-k} \|x - \eta\|^k \leq \|A_m\| r^m$$

for $\|x - \eta\| < r - \|\eta - \xi\|$. Thus

$$\sum_{m=0}^{\infty} \|A_m (x - \xi)^m\| \leq \sum_{m=0}^{\infty} \sum_{k=0}^{\infty} \binom{m}{k} \|A_m\| \cdot \|\eta - \xi\|^{m-k} \|x - \eta\|^k$$

$$\leq \sum_{m=0}^{\infty} \|A_m\| r^m < +\infty$$

Since the iterated series in the preceding inequalities converges, we can interchange the order of summations to obtain

$$\sum_{k=0}^{\infty} \left(\sum_{m=k}^{\infty} \binom{m}{k} \|A_m\| \cdot \|\eta - \xi\|^{m-k} \right) \|x - \eta\|^k < +\infty$$

Therefore, if we set

$$B_k = \sum_{m=k}^{\infty} \binom{m}{k} A_m (\eta - \xi)^{m-k}$$

then it follows that

$$\|B_k\| < +\infty$$

Hence $B_k \in L_s(^k E; F)$. Thus we have

$$\sum_{m=0}^{\infty} A_m (x - \xi)^m = \sum_{k=0}^{\infty} B_k (x - \eta)^k$$

with the convergence in the last series being uniform for $\|x-\eta\| <$ $r - \|\eta-\xi\|$. This shows that the radius of uniform convergence for $\Sigma_{k=0}^{\infty} B_k (x-\eta)^k$ is at least $\rho - \|\eta-\xi\|$ since r was arbitrary. This completes the proof. $\square$

11.12 *Differentiability of Power Series.* Once we have the preceding theorem at our disposal we are ready to show that a function represented by a power series is a C^{∞} function within the radius of uniform convergence.

THEOREM. *Let F be a Banach space. If* $\Sigma_{m=0}^{\infty} A_m (x-\xi)^m$ *converges to* $f(x)$ *on* $B_\rho (\xi)$, *where ρ is the radius of uniform convergence for the series, then f is a C^{∞} function on $B_\rho (\xi)$ and*

$$A_m = \frac{1}{m!} d^m f(\xi)$$

for m = 0,1,2,...

Proof. Let $0 < r < \rho$. Then $\Sigma_{m=0}^{\infty} \|A_m\| r^m < + \infty$ by Corollary 1 of 11.5. Let

$$d = \sup\{\|A_m\| r^m: \ m = 0,1,2,... \ \}$$

Since $f(x) = \Sigma_{m=0}^{\infty} A_m (x-\xi)^m$, we have $f(\xi) = A_0$ and

$$f(x) - f(\xi) - A_1 (x-\xi) = \sum_{m=2}^{\infty} A(x-\xi)^m$$

Hence

$$\left\| f(x) - f(\xi) - A_1 (x-\xi) \right\| \leq \sum_{m=2}^{\infty} \frac{d}{r^m} \|x-\xi\|^m = \frac{d\|x-\xi\|^2}{r(r - \|x-\xi\|)}$$

for $x \in B_r(\xi)$. This proves that f is differentiable at ξ and $Df(\xi) = A_1$. Repeating this argument for $\eta \in B_\rho (\xi)$ and

$$f(x) = \sum_{k=0}^{\infty} B_k (x-\eta)^k$$

for $\|x-\eta\| < \rho - \|\eta-\xi\|$ (see 11.11), we conclude that f is differentiable at η, and

$$Df(\eta) = B_1 = \sum_{m=1}^{\infty} m A_m (\eta-\xi)^{m-1}$$

This proves that f is differentiable on $B_\rho(\xi)$ and

$$Df(x) = \sum_{m=1}^{\infty} m A_m (x-\xi)^{m-1} = \sum_{m=1}^{\infty} DA_m (x-\xi)^m$$

for $x \in B_\rho(\xi)$.

Since $\lim \sup \|m A_m\|^{1/m} = \lim \sup \|A_m\|^{1/m}$, the radius of uniform convergence of the power series $\sum_{m=0}^{\infty} m A_m (x-\xi)^{m-1}$ is equal to ρ. By repeating this argument, we can conclude that Df is differentiable on $B_\rho(\xi)$ and

$$d^2 f(x) = \sum_{m=2}^{\infty} m(m-1) A_m (x-\xi)^{m-2} = \sum_{m=2}^{\infty} d^2 A_m (x-\xi)$$

where the radius of uniform convergence is equal to ρ. (See 7.6 for the definition of $d^m f$.) Notice that

$$A_2 = \frac{1}{2!} d^2 f(\xi)$$

Inductively, we conclude that f is indefinitely differentiable on $B_\rho(\xi)$ and

$$A_m = \frac{1}{m!} d^m f(\xi)$$

for m = 0,1,2,...   □

COROLLARY 1. *Let f be as in the Theorem. Then*

$$d^k f(x) = \sum_{m=k}^{\infty} d^k A_m (x-\xi)^m = \sum_{m=0}^{\infty} d^k A_{m+k} (x-\xi)^{m+k}$$

$$\hat{d}^k f(x) = \sum_{m=k}^{\infty} \hat{d}^k P_m (x-\xi) = \sum_{m=0}^{\infty} \hat{d}^k P_{m+k} (x-\xi)$$

for $x \in B_\rho(\xi)$ *where* $P_m = \hat{A}_m$.

COROLLARY 2. *Let f be as in the theorem with* $P_m = \hat{A}_m$, *we have*

$$\frac{1}{k!} \hat{d}^k \left(\frac{1}{m!} \hat{d}^m f \right) (\xi) = \frac{1}{m!} \hat{d}^m \left(\frac{1}{(k+m)!} \hat{d}^{k+m} f(\xi) \right)$$

Proof. In the notation of the theorem with $P_m = \hat{A}_m$, we have

$$\frac{1}{k!} \hat{d}^k (\hat{d}^m f) (\xi) = \hat{d}^m P_{k+m} = \hat{d}^m \left(\frac{1}{(k+m)!} \hat{d}^{k+m} f(\xi) \right) \qquad \square$$

COROLLARY 3. *Let f be as in the theorem and* $\xi \in U$. *Then for any* k, m *we have*

$$\tau_{k, \hat{d}^m f, \xi} = \hat{d}^m [\tau_{k+m, f, \xi}]$$

EXERCISES

11A *Radius of Uniform Convergence.* Let F be a Banach space and ρ the radius of uniform convergence for a power series $\sum_{m=0}^{\infty} P_m (x-\xi)$.
Then (1) $\rho = \sup \{r: r > 0, \sum_{m=0}^{\infty} \|P_m\| r^m < + \infty\}$
     (2) $\rho = \lim \|P_m\| / \|P_{m+1}\|$

if the limit exists.

11B *Differentiability of Power Series.* Let ρ be the radius of uniform convergence for $\sum_{m=0}^{\infty} A_m (x-\xi)^m$ and let $f(x)$ be its sum for $x \in B_\rho(\xi)$. Then for any $r \in E$,

$$\frac{\partial f}{\partial r}(x) = \sum_{m=0}^{\infty} \frac{\partial A_m}{\partial r} (x-\xi)^m$$

for $x \in B_\rho(\xi)$ and the convergence is uniform on $\overline{B}_s(\xi)$ for $0 < s < \rho$.

11C *Normal versus Uniform Convergence.* Let $g_n : \mathbb{R}^+ \to \mathbb{R}$ be given by $g_n(x) = 1/n$ on $[n, n+1)$ and 0 elsewhere. Then $\sum_{n=1}^{\infty} g_n$ converges uniformly on $\mathbb{R}^+$ but it is not normally convergent.

11D *Radius of Absolute Convergence.* If $\sum P_m(x)$ diverges at some u, $\|u\| > \rho_a$, does $\sum P_m(x)$ diverge for any x with $\|x\| = \|u\|$?

12

Analytic Mappings

12.1   In this chapter we present the theory of analytic functions based on the Weierstrass approach, and study the principle of analytic continuation.   As we remarked in 9.9, the Weierstrass viewpoint of analytic functions is more general than the Cauchy viewpoint and unifies both real and complex cases in its development. We shall devote a later chapter to their equivalence in the complex case.

Throughout this chapter E and F will denote normed spaces over the field $\mathbb{K}$ and U will denote a non-empty open subset of E.   We will assume that F is always complete.

ANALYTIC MAPPINGS

12.2   *Taylor Series*.   We say that a function f: U → F has a *Taylor series expansion* at a point $\xi \in U$ if there exists a power series

$$\sum_{m=0}^{\infty} P_m(x - \xi) = \sum_{m=0}^{\infty} A_m(x - \xi)^m$$

from E to F about ξ such that the power series converges to f uniformly on the ball $B_r(\xi) \subseteq U$ for some r > 0.   The series, if it exists, is unique by 11.10.

If f has a Taylor series expansion at ξ, then the function is infinitely differentiable in a neighborhood of ξ as shown in 11.12; moreover, we have the following relations:

$$P_m = \frac{1}{m!} \hat{d}^m f(\xi)$$

$$A_m = \frac{1}{m!} d^m f(\xi)$$

for $m = 0,1,2,\ldots$ Thus if f has a Taylor series expansion at ξ,

$$f(x) = \sum_{m=0}^{\infty} \frac{1}{m!} \hat{d}^m f(\xi)(x - \xi) = \sum_{m=0}^{\infty} \frac{1}{m!} d^m f(\xi)(x - \xi)^m$$

in some neighborhood of ξ in U. Since the open ball of uniform convergence of the Taylor series at ξ and the domain of the function f could be different, we usually write

$$f(x) \cong \sum_{m=0}^{\infty} \frac{1}{m!} \hat{d}^m f(\xi)(x - \xi) \cong \sum_{m=0}^{\infty} \frac{1}{m!} d^m f(\xi)(x - \xi)^m$$

to indicate formally the relationship between f and its Taylor series expansion at ξ.

12.3 *Analytic and Holomorphic Mappings.* A function f: U → F is said to be *analytic* on U if f has a Taylor series expansion at each point ξ of U. If $\mathbb{K} = \mathbb{C}$, an analytic mapping will be specially called *holomorphic*.

It is obvious that any linear combination of analytic mappings is analytic, and hence we have a vector space of all analytic mappings from U to F. This space will be denoted by A(U;F). In the case that $\mathbb{K} = \mathbb{C}$, we write H(U;F) instead of A(U;F). For simplicity, we denote A(U) = A(U;$\mathbb{K}$) and H(U) = H(U;$\mathbb{C}$).

Since A(U;F) is a vector subspace of C^{∞}(U;F), the space of infinitely differentiable mappings from U to F, we have for $f \in A(U;F)$ the following differential mappings:

$$d^m f: x \in U \to d^m f(x) \in L_s(^m E;F)$$

$$\hat{d}^m f\colon x \in U \to \hat{d}^m f(x) \in P(^m E;F)$$

We also have from 11.11 the following differential operators:

$$d^m\colon f \in A(U;F) \to d^m f \in A(U;L_s(^m E;F))$$

$$\hat{d}^m\colon f \in A(U;F) \to \hat{d}^m f \in A(U;P(^m E;F))$$

for all m = 0,1,2,...

12.4 *Local Property*. It is clear that being analytic is a local property in the following sense: If f $\in$ A(U;F) and V is a non-empty open subset of U, then f$|_V \in$ A(V;F). Conversely, if U = $\cup_{\lambda \in I} V_\lambda$, where V_λ is an open non-empty subset of U, and f: U $\to$ F is such that for each $\lambda \in$ I, f$|_{V_\lambda} \in$ A(V$_\lambda$;F), then f $\in$ A(U;F).

12.5 THEOREM. *Let* $\Sigma_{m=0}^\infty P_m(x - \xi)$ *be a power series whose radius of uniform convergence* ρ *is positive, and let* f(x) *be its sum for* $\|x\| < \rho$. *Then* f *is analytic on* B$_\rho(\xi)$.

Proof. This is a restatement of 11.11.   □

12.6 *Infinitely Differentiable Functions*. Despite its generality, if $\mathbb{K} = \mathbb{R}$, A(U;F) is only a proper subset of C$^\infty$(U;F), the vector space of all infinitely differentiable mappings from U to F. The relation between these two classes is shown by the following theorem.

THEOREM (Pringsheim). *A function* f: U $\to$ F *has a Taylor series expansion at* $\xi \in$ U *if and only if there exist* C $\geq$ 0, c $\geq$ 0 *and* r > 0 *such that*

(a) f *is infinitely differentiable on* B$_r(\xi)$

(b) $\|d^m f(x)\| \leq Cc^m m!$

for x $\in$ B$_r(\xi)$ *and* m = 0,1,2,...

Proof. The necessary condition follows from 11.12 and Corollary 2 of 11.5.

Assume that f satisfies conditions (a) and (b). By Taylor's formula in 8.9, we have

$$\left\| f(x) - T_{m,f,\xi}(x) \right\| \leq C(rc)^{m+1}$$

for $x \in B_r(\xi)$. Since we may easily assume that $rc < 1$, $T_{m,f,\xi}(x) \rightarrow$ f(x) as $m \rightarrow \infty$ uniformly on $B_r(\xi)$. Hence f has a Taylor series expansion at ξ.   □

PRINCIPLE OF ANALYTIC CONTINUATION

12.7 *Analytic Continuation*. The problem of analytic continuation is the following: given an analytic map f in a connected open set U and a connected open set V containing U, find an analytic map g on V such that $g\big|_U = f$. The following theorems guarantee that such a mapping g is unique if it exists. This is the principle of analytic continuation.

12.8 THEOREM (Strong Form). *Let U be a connected open subset of E, f $\in$ A(U;F) and $\xi \in$ U. The following are equivalent:*

(a) $d^m f(\xi) = 0$ *for all* $m = 0,1,2,\ldots$;

(b) f *is identically zero in a neighborhood of* ξ;

(c) f *is identically zero in U.*

Proof. It is clear that (c) implies (a). It remains to show that (a) $\Rightarrow$ (b) $\Rightarrow$ (c).

(a) $\Rightarrow$ (b): Suppose (a) is satisfied. Since f $\in$ A(U;F), the Taylor series at ξ converges to f uniformly on some neighborhood of ξ. Thus f is zero identically in this neighborhood.

(b) $\Rightarrow$ (c): Consider the set W = {x $\in$ U: f(x) = 0}. Then by the assumption (b), W is non-empty and open since the Taylor series of f at each x $\in$ W represents f in a neighborhood of x. Furthermore, W is closed since f is continuous. Thus W = U since U is connected and W is a closed and open subset of U. This shows that f is identically zero on U.   □

12.9 THEOREM (Weak Form). *Let U be a connected open set and*
f,g ∈ A(U). *If f and g are identical on a non-empty open subset*
of U, then they are identical on U.

Proof. Let V be a non-empty subset of U and assume that f and g
are identical on V. It follows then from Theorem 12.8 that f = g
on U.   □

COROLLARY. *Let U and V be two connected open subsets of E, and*
f ∈ A(U;F). *Then there is at most one g* ∈ A(U;F) *satisfying f = g*
on U ∩ V.

12.10 REMARK. If $f,g \in A(U)$, where U is a connected open set in
$\mathbb{K}$, and if f = g on an infinite set A with a cluster point, where
$A \subset U$, then f = g on U. The analogous statement for dim E ≥ 2
does not hold. For example, let $E = \mathbb{K}^2$, $f(x_1,x_2) = x_1$ and $g(x_1,x_2)$
= 0. Then $f,g \in A(E)$ and f = g on $\{(x_1,x_2): x_1 = 0\}$, but $f \neq g$. In
general, if M is a vector subspace of E which is not dense in E,
then we can find a bounded linear functional f satisfying $\|f\| = 1$
and f = 0 on M.

12.11 *Analytic Continuations to the Complexification.* If E and F
are real normed spaces and f: U → F is analytic, we will show that
there is at most one holomorphic mapping g: U_c → F_c such that g = f
on U, where E_c is the complexification of E and U_c is an open set in
E_c containing U. The reader should recognize one very interesting
consequence of this idea for one complex variable: Any trigonometric
identity valid on $\mathbb{R}$ for the trigonometric functions on a real variable
is valid throughout $\mathbb{C}$.

THEOREM. *Let E and F be complex normed spaces and let S be a real*
vector subspace of E such that the complex subspace $S_c = S + iS$
generated by S is dense in E. If f,g ∈ H(U;F), *where U is a connected*
open set in E, and f = g on some non-empty open set V in U ∩ S,
then f = g on U.

We need the following lemma.

LEMMA. *Let S be as in the theorem and let* $P \in P(^mE:F)$ *be such that P vanishes on S. Then P = 0 identically.*

Proof. Let $A \in L(^mE;F)$ be such that $\hat{A} = P$; i.e., $P(x) = Ax^m$. By the multinomial formula in 4.4, we have

$$P(x + iy) = A(x + iy)^m = \sum_{k=0}^{m} i^k \binom{m}{k} x^{m-k} y^k$$

for $x,y \in E$. Since P = 0 on S, it follows from the polarization formula in 4.6 that

$$A(t_1, \ldots, t_m) = 0$$

for $t_1, \ldots, t_m \in S$. Hence $P(x + iy) = 0$ for $x,y \in S$. This shows that P = 0 on the dense set $S + iS$, and hence P = 0 on E since P is continuous.   □

Proof of Theorem. It is sufficient to show the result for g = 0 on U. Let $\xi \in V$ and consider the Taylor series $\sum P_m(x - \xi)$ of f at ξ. Since f = g on V and g = 0 on V, $P_m = 0$ on V for all m = 0,1,2,... It follows from the lemma that $P_m = 0$ on S, and hence $P_m = 0$ on E for all m = 0,1,... This proves that f is identically zero on U by Theorem 12.8.   □

12.12 *Integral Domain* A(U). Exercise 11E (multiplication of power series) in fact states that if $f,g \in A(U)$ then the multiplication $fg \in A(U)$. Thus A(U) is a ring.

THEOREM. A(U) *is an integral domain if and only if U is connected.*

Proof. The necessary condition is clear. Suppose that U is connected and $f,g \in A(U)$ are such that fg = 0. If $f \neq 0$, then there exists a non-empty open subset V of U such that $f \neq 0$ on V. Thus g must be zero on V. Since g is analytic, g must be zero identically on U by 12.9.   □

EXERCISES

12A *Open domain.* Let F be a Banach space and $f: U \to F$. Then the set of points in U at which f has a Taylor series expansion is open.

12B *The algebra* A(U). The vector space A(U) is an algebra with respect to the Cauchy product of series.

12C *Mappings into product spaces.* Let $f: U \to F$, where F is the Cartesian product of normed spaces $F_1, \ldots, F_m$. Let $f_k: U \to F_k$ be the coordinate map of f for $k = 1, \ldots, m$. Then f is holomorphic on U if and only if each f_k is holomorphic for $k = 1, \ldots, m$.

12D *Composite Mappings.* Let E, F and G be normed spaces, U a non-empty open subset of E, $f \in A(U;F)$ and $g \in L(F;G)$. Then $g \circ f \in A(U;G)$ and $d^m(g \circ f)(\xi) = g \circ (d^m f(\xi))$ for all $\xi \in U$ and $m = 0, 1, \ldots$

13

Holomorphic Mappings

13.1 From this chapter on we shall be concerned with holomorphic
maps defined on complex normed spaces E and taking values in a complex Banach space F. Theory of holomorphy is distinguished from
that of (real) analytic functions. One of the most important dis-
tinctions is that holomorphic mappings have the Cauchy integral
representation. Here it will be shown that a substantial portion
in the elements of complex function theory can be extended to
normed spaces in light of the Cauchy integral formula. We also
study here Cartan's uniqueness theorem, some aspects of homogeneous
domains, and fixed points of holomorphic mappings.

THE CAUCHY INTEGRAL FORMULA

13.2 For x,y in E, $x \neq y$, consider the 1-dimensional complex affine
subspace $L(x,y)$ passing through these points. If U is an open subset of E, then the intersection $L(x,y) \cap U$ is homeomorphic to the
open subset $U(x,y)$ defined by

$$U(x,y) = \{z \in \mathbb{C} : (1-z)x + zy \in U\}$$

Notice that $0 \in U(x,y)$ if $x \in U$; $1 \in U(x,y)$ if $y \in U$. If $\xi \in U$,
then it follows that there exist x in U and $r > 1$ such that

$$D_r = \{z \in \mathbb{C} : \ z \ \leq r\} \subset U(\xi,x)$$

since U is open.

13.3 THEOREM (Cauchy Integral Formula). *Let* $f \in H(U;F)$ *and* $\xi \in U$. *If for* $x \in U$ *there exists* $r > 1$ *such that* $D_r \subseteq U(\xi,x)$, *then*

$$f(x) = \frac{1}{2\pi i} \int_{|z|=r} \frac{f[(1-z)\xi + zx] \, dz}{z-1}$$

Proof. Let $g: U(\xi,x) \to F$ be defined by

$$g(z) = f[(1-z)\xi + zx]$$

Then g is holomorphic on the open subset $U(\xi,x)$ of complex numbers. Since D_r is a subset of $U(\xi,x)$, it follows from the Cauchy integral formula 9.12 that

$$g(1) = \frac{1}{2\pi i} \int_{|z|=r} \frac{g(z) \, dz}{z-1}$$

which gives the integral formula after replacing g with f. □

REMARK. The Cauchy integral formula is also valid for any normed space F although we assumed here that F is complete for simplicity. In fact, if F is not complete, we consider the completion $F^{\wedge}$ of F and treat f: $U \to F$ as a function of U into $F^{\wedge}$. Then the integral formula holds as in the Theorem.

13.4 *Cauchy Integral Formulas for Differentials.* For $\xi \in U$ and $x \in E$, we can find a $r > 0$ such that for $z \in \mathbb{C}$, $|z| \leq r$, we have $\xi + zx \in U$ because U is an open set.

THEOREM. *Let* $f \in H(U;F)$ *and* $\xi \in U$. *For* $x \in E$ *let* $r > 0$ *be such that* $\xi + zx \in U$ *for every* $z \in \mathbb{C}$, $|z| \leq r$. *Then*

$$\hat{d}^m f(\xi)(x) = \frac{m!}{2\pi i} \int_{|z|=r} \frac{f(\xi + zx) \, dz}{z^{m+1}}$$

for m = 0,1,... .

Proof. The identity holds trivially when x = 0. Otherwise, let

$$V = \{z \in \mathbb{C}: \xi + zx \in U, \ z \neq 0\}$$

Then V is an open subset of complex numbers which contains all of
z satisfying $0 < |z| < r$. Now define $g: V \to F$ by

$$g(z) = \frac{f(\xi + zx)}{z^{m+1}}$$

Since the origin is not in V, g is holomorphic on V. From the
Theorem 9.12(b), more precisely, from Exercise 9E we obtain

$$\int_{|z|=r} g(z) \ dz = \int_{|z|=s} g(z) \ dz$$

for any s in $(0,r)$. Therefore,

$$\int_{|z|=r} \frac{f(\xi + zx) \ dz}{z^{m+1}} = \int_{|z|=s} \frac{f(\xi + zx) \ dz}{z^{m+1}} \tag{$*$}$$

We now consider the Taylor series expansion of f at ξ:

$$f(t) = \sum_{k=0}^{\infty} P_k(t - \xi)$$

Choose $r > 0$ small enough such that this power series converges to
f uniformly on the closed ball around ξ with radius $r\|x\|$. Then we
get from $(*)$ that

$$\int_{|z|=r} f(\xi + zx) z^{-(m+1)} dz = 2\pi i P_m(x)$$

by integrating term by term the series

$$\sum_{k=0}^{\infty} P_k(zx) z^{-(m+1)}$$

with respect to z and by recognizing the integral of z^{-n} around $|z| = r$ is equal to 0 for $n > 1$. □

As a simple corollary of the theorem we obtain the Cauchy integral formula in a form which is slightly different from the formula in 13.3.

COROLLARY. *Let* $f \in H(U;F)$, $\xi \in U$, $x \in E$ *and* $r > 0$ *be such that* $\xi + zx \in U$ *for* $z \in \mathbb{C}$, $|z| \le r$. *Then*

$$f(\xi) = \frac{1}{2\pi i} \int_{|z| = r} \frac{f(\xi + zx) \, dz}{z}$$

13.5 The Cauchy integral formula for differentials

$$d^m f: x \in U \to d^m f(x) \in L_s(^m E;F)$$

can be obtained similarly. But its formulation is complicated and not as elegant as the formula 13.4.

THEOREM. *Let* $f \in H(U;F)$, $\xi \in U$, $m \in \mathbb{N}$, $k \in \mathbb{N}$, $1 \le k \le m$, $(x_1,\ldots,x_k)$ $\in E^k$ *and* $(r_1,\ldots,r_k) \in \mathbb{R}^k$ *be such that* $\xi + (z_1 x_1 + \ldots + z_k x_k) \in U$ *for every* $z_j \in \mathbb{C}$ *with* $|z_j| \le r_j$, $j = 1,\ldots,k$. *Then for any* $(n_1,\ldots,n_k)$ $\in \mathbb{N}^k$ *with* $m = n_1 + \ldots + n_k$, *we have*

$$\frac{1}{n_1! \ldots n_k!} d^m f(\xi) (x_1)^{n_1} \ldots (x_k)^{n_k}$$

$$= \frac{1}{(2\pi i)^k} \int_{\substack{|z_j|=r_j \\ 1 \le j \le k}} \frac{f[\xi + (z_1 x_1 + \ldots + z_k x_k)] dz_1 \ldots dz_k}{z_1^{n_1+1} \ldots z_k^{n_k+1}}$$

13.6 As a consequence of the Cauchy integral formulas we obtain
the following fundamental inequalities.

CAUCHY INEQUALITIES. Let $f \in H(U;F)$, $r > 0$ *and* $B_r(\xi) \subseteq U$. *Then*

(a) $\left\| \frac{1}{m!} \hat{d}^m f(\xi) \right\| \leq \frac{1}{r^m} \sup\{\|f(x)\| : \|x - \xi\| = r\}$

(b) $\left\| \frac{1}{m!} d^m f(\xi) \right\| \leq \frac{m^m}{m! \, r^m} \sup\{\|f(x)\| : \|x - \xi\| = r\}$

Proof. (a) In the integral formula 13.4, assume that $\|x\| = 1$.
Then it follows from the Cauchy integral formula 13.4 that

$$\left\| \frac{1}{m!} \hat{d}^m f(\xi) \right\| = \left\| \frac{1}{2\pi i} \int_{|z|=r} \frac{f(\xi + zx) \, dz}{z^{m+1}} \right\|$$

$$\leq \frac{1}{r^m} \sup \{\|f(t)\| : \|t - \xi\| = r\}$$

(b) Recall that if $A \in L_s(^m E ; F)$, then

$$\|\hat{A}\| \leq \|A\| \leq \frac{m^m}{m!} \|\hat{A}\|$$

(see 4.13). Hence from the Cauchy inequalities, we obtain the desired
result.   □

13.7 The following estimate for a Taylor polynomial (cf. 8.6) of
a holomorphic mapping is very useful in the study of the convergence
of a Taylor series.

THEOREM. *Let* $f \in H(U;F)$. *Given* $\xi \in U$, *let* $x \in E$ *and* $r > 1$ *be such
that* $D_r \subset U(\xi,x)$. *Then for any* $m \in \mathbb{N}$ *we have*

$$f(x) - T_{m,f,\xi}(x) = \frac{1}{2\pi i} \int_{|z|=r} \frac{f[(1-z)\xi + zx] \, dz}{z^{m+1}(z-1)}$$

$$\left\| f(x) - T_{m,f,\xi}(x) \right\| \leq \frac{1}{r^m(r-1)} \sup \left\{ \left\| f[(1-z)\xi + zx] \right\| : |z| = r \right\}$$

Proof. For z in $\mathbb{C} \backslash \{0,1\}$, we have

$$\frac{1}{z-1} = \sum_{k=0}^{m} \frac{1}{z^{k+1}} + \frac{1}{z^{m+1}(z-1)} \tag{*}$$

Multiplying $f[(1-z)\xi + zx]/2\pi i$ to both sides of (*), integrating the result term by term over $|z| - r$, and applying the Cauchy integral formulas 13.3 and 13.4 to this, we obtain the first relation.

The inequality in the theorem follows easily from the first relation by noticing inf $\{ |z-1| : |z| = r \} = r-1$. □

COROLLARY. *Let* $f \in H(U;F)$, $m \in \mathbb{N}$, $\xi \in U$ *and* $r > 0$ *be such that* $\overline{B}_r(\xi) \subset U$ *and* $x \in \overline{B}_r(\xi)$. *Then*

$$\left\| f(x) - T_{m,f,\xi}(x) \right\| \leq \frac{\|x-\xi\|^{m+1}}{r^m(r-\|x-\xi\|)} \sup \left\{ \|f(t)\| : \|t - \xi\| = r \right\}$$

Proof. If $x = \xi$, then the inequality holds trivially. Assume that $x \neq \xi$ and let $s = r\|x - \xi\|^{-1}$. Then $s > 1$ and $(1-z)\xi + zx \in U$ for all $z \in \mathbb{C}$ with $|z| \leq s$. Let $t = (1-z)\xi + zx$, where $|z| = s$. Then

$$\left\| t - \xi \right\| = \left\| z(x - \xi) \right\| = |z| \cdot \|x - \xi\| = s\|x - \xi\| = r$$

whence we have

$$\sup \left\{ \left\| f[(1-z)\xi + zx] \right\| : |z| = s \right\} \leq \sup \left\{ \|f(t)\| : \|t - \xi\| = r \right\}$$

In the inequality (2) of the preceding theorem, first replace r with s, and replace s with $r\|x - \xi\|^{-1}$. Finally we obtain the required inequality from (*). □

13.8 *Liouville's Theorem.* We now study a Liouville-type theorem
characterizing continuous polynomials.

THEOREM. *Let* $f \in H(E;F)$. *Then* f *is a continuous polynomial of
degree* $\leq m$ *if and only if there is* $M > 0$ *such that*

$$\|f(x)\| \leq M\|x\|^m$$

for all x, *with sufficiently large* $\|x\|$. *In particular, if* f *is
bounded, then* f *is a constant mapping.*

Proof. As in the classical case, this theorem follows from the
Cauchy inequalities and Theorem 13.7.   □

13.9 *Maximum Modulus Theorem.* In Chapter 10, we studied the maximum
modulus theorem for vector-valued functions on a domain in the complex plane and presented the strong maximum modulus theorem of Thorp
and Whitley (1967). We now generalize the weak maximum modulus
theorem 10.2 for functions defined on Banach spaces.

THEOREM. *Let* U *be a connected open subset of* E *and* $f \in H(U;F)$. *If*
$\|f(x)\|$ *has a maximum at a point in* U, *then* $\|f(x)\|$ *is a constant
function on* U.

Proof. Suppose that there exists a point ξ in U such that

$$\|f(\xi)\| = \sup\ \{\|f(x)\| : x \in U\}$$

Since U is open, we can find r such that $B_r(\xi) \subset U$. For x in $B_r(\xi)$,
$x \neq \xi$, we have the holomorphic mapping

$$g(z) = f[(1-z)\xi + zx]$$

on an open set V which contains the unit disk of $\mathbb{C}$. Then g takes
its maximum modulus at 0; thus $\|g\|$ must be constant on V. In par-
ticular, $\|g(0)\| = \|g(1)\|$; and hence $\|f(x)\| = \|f(\xi)\|$ holds for all x

in $B_r(\xi)$. We now consider the following set to complete the proof:

$$A = \{x \in U: \|f(x)\| = \|f(\xi)\|\}$$

Then A is closed in U. If $a \in A$, then in the above argument replace ξ with a, we show that a is an interior point of A. This proves that A is both closed and open in the connected set U; thus $A = U$.□

COROLLARY. *Let f satisfy the hypotheses of the theorem. If every point in the unit sphere of F is a complex extreme point of the closed unit ball, then f is a constant function on U.*

Proof. This is a consequence of Thorp-Whitley's theorem 10.15. In fact, in the proof above, we have $g(0) = g(1)$; consequently, we have that f is constant on the open ball $B_r(\xi)$. Therefore, f must be a constant function by the principle of analytic continuation 12.9.

□

CARTAN'S UNIQUENESS THEOREM

13.10. In this section we prove a sequence of results mainly due to Cartan originally proved for finite dimension. These include a characterization of those biholomorphic maps of a circular domain which fix the origin. There follows a proof of Poincaré's theorem that the ball and the polydisc are not biholomorphic. We first recall Schwarz's lemma from classical analysis.

Schwarz's Lemma. Let f be holomorphic on the open unit disk D of $\mathbb{C}$. If $f(0) = 0$ and $|f(z)| < 1$ for $|z| < 1$, then

(a) $|f(z)| \leq |z|$ *for* $|z| < 1$

(b) *if $|f(p)| = |p|$ for some point p in D, then for some t in $\mathbb{R}$, $f(z) = e^{it}z$*

In this section we give several generalizations of the lemma.

13.11 *Cartan's Uniqueness Theorem.* H. Cartan (1931) showed the following two theorems for bounded domains in $\mathbb{C}^2$. His proof holds for a bounded domain in a complex Banach space without substantial change (cf. Bochner and Martin (1948), p. 13).

THEOREM. *Let U be a bounded connected open set in E and let f: U → U be holomorphic. If f has a fixed point p in U (i.e., f(p) = p) and df(p) is the identity map on E, then f is the identity map on U.*

Proof. We may assume that U is a neighborhood of 0 and 0 is a fixed point for f. In the Taylor series expansion $f(x) = \Sigma\, P_n(x)$, $P_n \in P(^nE;F)$, the coefficient $P_0 = 0$ because $f(0) = 0$ and $P_1 = df(0)$. Hence

$$f(x) = x + P_k(x) + P_{k+1}(x) + \cdots$$

for some $k \geq 2$. Let $f^n = f \circ f^{n-1}$. By Chain Rule 5.9, we can easily show that the Taylor expansion of f^n around 0 is

$$f^n(x) = x + nP_k(x) + \cdots$$

Now, by the Cauchy inequalities, we have

$$\|nP_k\| \leq r^{-k}\, \sup\, \{\|f^n(x)\| : x \in U\}$$

where $r = d(0, E\backslash U)$. Since U is bounded,

$$\|P_k\| \leq M/(r^k n)$$

for some $M > 0$. Let $n \to \infty$ and conclude that $P_k = 0$. Repeating the same argument we obtain $P_n = 0$ for all $n \geq 2$, which proves that f is the identity map on U.   □

13.12 For the next theorem of Cartan, we define that an open set U of E is *circular* if, for any $x \in U$, $e^{ir}x \in U$ for all real numbers r.

A function f: U → U is said to be *biholomorphic* if both f and f^{-1} are holomorphic on U.

THEOREM. *Let U be a circular connected open set containing 0 of E and let f: U → U be biholomorphic and f(0) = 0. Then f is the restriction to U of a linear isometry of E onto itself.*

Proof. Since f(0) = 0, the Taylor expansion of f around 0 is of the form:

$$f(x) = P_1(x) + P_2(x) + \ldots$$

with $P_n \in P(^nE;F)$ and $P_1 = df(0)$. Since f is biholomorphic $df^{-1}(0)$ exists and $df^{-1}(0) = (P_1)^{-1}$. Thus P_1 is an isomorphism. For each real number r, let $E_r: U \to U$ be defined by $E_r(x) = e^{ir}x$. Then E_r is biholomorphic on U. Consider

$$g = f^{-1} \circ E_{-r} \circ f \circ E_r$$

Then g is biholomorphic on U and g(0) = 0 and dg(0) is the identity map of E. Now it follows from Theorem 13.11, g is the restriction to U of the identity mapping; i.e.,

$$E_r \circ f = f \circ E_r$$

or

$$e^{ir}[P_1(x) + P_2(x) + \ldots] = e^{ir}P_1(x) + e^{2ir}P_2(x) + \ldots$$

for any real number r. From the uniqueness theorem of power series (11.10), we infer that $P_n = 0$ for $n \geq 2$. It follows that f is the restriction to U of the isomorphism P_1. Since any linear isomorphism mapping the unit ball onto itself must be an isometry, f is an isometry. This completes the proof. □

13.13 L.A. Harris (1969) generalized Cartan's version of the Schwarz lemma for normed spaces.

THEOREM. *Let* U *and* V *be open unit balls of* E *and* F *respectively.* *Let* f: U$\to$V$^-$ *be holomorhic and suppose that* df(0) *is an isometry of* E *onto* F. *Then* f(x) = df(0)(x) *for all* x $\in$ U; *in particular,* f *is an isometry of* U *into* V.

Proof. Since df(0) is an isomorphism of E onto F, f cannot be a constant map. Hence f maps U into V by the maximum modulus theorem 13.9. Let g - df(0)$^{-1}$ o f. Then g maps U into U and dg(0) is the identity map on E. To apply the Cartan uniqueness theorem 13.11, we need to show that g(0) = 0.

Suppose otherwise, let g(0) = a. Then by the Hahn-Banach theorem, there exists u $\in$ E* such that $\|u\|$ = 1 and u(a) = $\|a\|$. Now consider the following map:

   h(z) = u(g(za/$\|a\|$))

Then h: D $\to$ D$^-$ where D is the open unit disk of the complex plane. Since h cannot be a constant holomorphic function, h maps D into D by the maximum modulus theorem. Then

$$\left|h^{'}(0)\right| + \left|h(0)\right|^2 \le 1$$

by a classical lemma which is stated below. But

   h$^{'}$(0) = u(a/$\|a\|$) = 1

Hence h(0) = 0, contrary to the fact that h(0) = $\|a\|$. Therefore g(0) = 0.

By the Cartan uniqueness theorem, g is the restriction to U of the identity map. Hence f = df(0) on U, which completes the proof. $\square$

Lemma. *Let* D *be the open unit disk of the complex plane. For any holomorphic mapping f: D → D, we have*

$$\left| f'(0) \right| + \left| f(0) \right|^2 \leq 1$$

Proof. The Moebius transformation $z \rightarrow (z-a)/(1-\bar{a}z)$ maps D onto D and a to 0. Hence the map

$$q(z) = \frac{f(z) - f(0)}{1 - \overline{f(0)}f(z)}$$

is a holomorphic mapping of D into D. Furthermore, $q(0) = 0$. By 13.10, $\left| q(z) \right| \leq \left| z \right|$ for all z in D. Hence $\left| q'(0) \right| \leq 1$. But $q'(0) = f'(0)/(1 - \left| f(0) \right|^2)$, which implies that

$$\left| f'(0) \right| + \left| f(0) \right|^2 \leq 1 \qquad\qquad \square$$

REMARK. L. A. Harris' generalization of Schwarz's lemma has appli-cations to Banach algebra theory, in particular, this yields a generalized Banach-Stone theorem for J*-algebra and a new proof of the Russo-Dye theorem (cf. Harris (1969, 1972)). We can also give a shorter proof for Theorem 13.12 using 13.13.

13.14 *Homogeneous Domain.* Connected open sets U and V are said to be *holomorphically equivalent* if there exists a biholomorphic map-ping of U onto V. A connected open set U is said to be a *homogeneous domain* if for each pair of points x,y $\in$ U there exists a biholo-morphic mapping f: U → V with f(x) = y.

H. Poincaré (1907) showed that in $\mathbb{C}^2$, the unit balls corresponding to the maximum norm and the natural norm are not holomorphically equivalent. Therefore, the Riemann mapping theorem does not hold for higher dimensional spaces. [The Riemann mapping theorem states that any two proper simply connected open sets in $\mathbb{C}$ are holomor-phically equivalent.] This opened the study of holomorphic equiv-alency; in particular, classification of homogeneous domains.

We record the following two theorems for infinite dimension
which were first proved by Kaup and Upmeier (1976) and simplified
by Harris (1979).

13.15 THEOREM. *Let* U *and* V *be open unit balls on* E *and* F *respectively. If* U *is a homogeneous domain, then* U *and* V *are holomorphically equivalent if and only if* E *and* F *are isometrically isomorphic normed spaces.*

Proof. Suppose U and V are holomorphically equivalent. Then there
is a biholomorphic mapping f: U $\to$ V. Since U is homogeneous, there
is a biholomorphic mapping g: U $\to$ U such that g(O) - f^{-1}(O). Hence
h = f o g maps U onto V biholomorphically and h(O) = O. Hence by
Theorem 13.12, h extends to a linear isometry of E onto F. The converse is obvious.   □

13.16 THEOREM. *If the unit ball* U *of* E *is homogeneous, then* U *is
holomorphically equivalent to the product of the open unit balls of
normed spaces* F *and* G *if and only if* E *is isometrically isomorphic
to the space* F × G *with the norm* $\| (x,y) \|$ = max $\{ \|x\|, \|y\| \}$.

Proof. Note that the unit ball of F × G is the product of the
respective unit balls. Therefore, this theorem is an obvious
corollary of the preceding one.   □

FIXED POINTS OF HOLOMORPHIC MAPPINGS

13.17. S. Kakutani (1943) has shown that the Brouwer fixed point
theorem fails to hold for infinite dimensional spaces. He exhibited
a homeomorphism of the closed ball of the real Hilbert space ℓ^2 onto
itself with no fixed point. His example can be easily modified to
be a diffeomorphism. This might indicate that holomorphy is strong
enough to insure fixed points. However, the following example shows
the contrary in general.

EXAMPLE. Let E = c_o and define A: E $\to$ E to be the linear shift

$$A(x_1, x_2, \ldots) = (0, x_1, x_2, \ldots)$$

Then A is holomorphic on E, and for any a in E the translate of A by a (that is, A + a) is also holomorphic. In particular, the mapping

$$f(x_1, x_2, \ldots) = (1/2, x_1, x_2, \ldots)$$

is holomorphic on E. It is easy to see that f has no fixed point since a fixed point of f must be $(1/2, 1/2, \ldots)$, which is not in E. But notice that f maps the closed unit ball into itself.

13.18 THEOREM (Earle-Hamilton, 1970). *If U is a bounded open subset of a Banach space E and if* f: U → U *is a holomorphic mapping which maps U strictly inside U (i.e.,* f(U) *lies at a positive distance from* E \ U*), then f has a unique fixed point in U.*

For finite dimensional spaces, the theorem is in Herve (1963), p. 84 and Reiffen (1965), p. 322. For infinite dimension, Earle and Hamilton (1970) proved the theorem using a CRF-pseudometric (where CRF stands for Caratheodory-Reiffen-Finsler) satisfying the Schwarz-Pick lemma that holomorphic mappings do not increase distances. The following proof is adapted from Harris (1979). We will use implicitly the concept of CRF-pseudometrics.

Proof. We may assume that U is connected without losing generality. Let α: U × E → $\mathbb{R}$ be defined by

$$\alpha(x, y) = \sup \{\|Dg(x)y\| : g \in H(U; U)\}$$

where H(U;U) denotes the space of all holomorphic mappings of U into itself. Then α is locally bounded on U × E by the Cauchy inequalities, and it is lower semicontinuous on U × E since it is a supremum of continuous functions there. It is immediate that

$$\alpha(x, ky) = |k|\alpha(x, y)$$

for any real number k.

Let γ be a curve in U with continuous derivative. Call such a function *admissible*. Then $\alpha(\gamma(t),\gamma'(t))$ is a bounded measurable function on $[0,1]$. Hence we may define

$$L(\gamma) = \int_0^1 \alpha(\gamma(t),\gamma'(t))\,dt$$

Given $x,y \in U$, we define

$$\rho(x,y) = \inf \{L(\gamma): \gamma \text{ is admissible and } \gamma(0) = x,\ \gamma(1) = y\}$$

Then ρ is a pseudometric on U. In fact, it is a metric on U since

$$\|x - y\| \leq \rho(x,y) \tag{1}$$

for all x and y in U. To show this relation, first notice that

$$\alpha(x,y) \geq \|y\|$$

since the identity mapping is holomorphic. Hence for any admissible curve γ connecting x and y, we have

$$L(\gamma) = \int_0^1 \alpha(\gamma(t),\gamma'(t))\,dt \geq \int_0^1 \|\gamma'(t)\|\,dt \geq \left\|\int_0^1 \gamma'(t)\,dt\right\|$$

$$= \|\gamma(1) - \gamma(0)\| = \|y - x\|$$

Hence the relation (1) is verified. (The metric ρ is a CRF-metric.)

If g and h are in $H(U;U)$, so is $g \circ h$. It follows from the Chain Rule that for any $u \in U$ and $v \in E$

$$\|Dg[\,h(u)\,]Dh(u)v\| = \|D(g \circ h)(u)v\| \leq \alpha(u,v)$$

Fixing h and taking supremum for all g in $H(U;U)$, we get

$$\alpha(h(x), DH(x)y) \le \alpha(x,y) \tag{2}$$

for all x in U and y in E.

We now assume that f maps U strictly inside U. We claim that there exists a constant k in (0,1) such that

$$\alpha(f(x), Df(x)y) \le k\alpha(x,y) \tag{3}$$

and

$$\rho(f(x), f(y)) \le k\rho(x,y) \tag{4}$$

for all x,y in U.

In fact, since f(U) lies strictly inside U, there exists r > 0 such that the distance between f(U) and E\U is greater than r. Let s = r/d, where d denotes the diameter of the bounded set U. Then for x in U and y in E, the mapping

$$g(u) = f(u) + s[f(u) - f(x)]$$

is holomorphic from U into itself. It then follows from (2) that

$$\alpha(g(x), Dg(x)y) \le \alpha(x,y)$$

Hence

$$\alpha(f(x), (1+s)Df(x)y) \le \alpha(x,y)$$

This proves that the relation (3) holds with k = 1/(1+s).

To show the relation (4), first notice that f ∘ γ is an admissible curve connecting f(x) and f(y) whenever γ is an admissible curve connecting x and y. Thus

$$\rho(f(x),f(y)) \leq L(f \circ \gamma) = \int_0^1 \alpha(f[\gamma(t)],Df[\gamma(t)]\gamma´(t))dt$$

$$\leq k\int_0^1 \alpha(\gamma(t),\gamma´(t))dt \leq kL(\gamma)$$

Hence the relation (4) holds.

The inequality (4) shows that f is a contraction for the metric ρ. Let a be an arbitrary point in U and set $a_n = f^n(a)$ for all natural numbers n. The standard proof of the Banach fixed point theorem for a contraction in a complete metric space shows that (a_n) is a ρ-Cauchy sequence. Hence it is a Cauchy sequence for the norm because of the relation (1). Let b be the limit of (a_n). Then b is a point in U since f maps U strictly inside U. Then f(b) = b by the continuity of f. The contraction inequality (4) shows that f cannot have two distinct fixed points in U. This completes the proof.   □

13.19 We state without proof the following remarkable theorem.

THEOREM (Hayden-Suffridge, 1976). *Let B be the closed unit ball of a separable and reflexive Banach space and let f: B → B be continuous on B and holomorphic on the interior of B. Then for almost all (a.e.) t in* [0,2π] *the mapping* $e^{it}f$ *has a fixed point in B.*

EXERCISES

13A. Let r > 0 be such that the closed ball around a with radius r is contained in U. Let C be the smallest number satisfying the following inequality for all f $\in$ H(U;F)

$$\left\|\hat{d}^m f(a)\right\| \leq m!C \sup\{\left\|f(x)\right\| : \left\|x - a\right\| = r\}$$

Then $C = r^{-m}$. (This shows that r^{-m} is the best universal coefficient for the Cauchy inequalities.)

13B *Liouville's Theorem.* (1) Let $f \in H(U;F)$. Then f is the restriction to U of a continuous polynomial of degree $\leq$ m if and only if there is a point a $\in$ U such that $\hat{d}^n f(a) = 0$ for all $n > m$.

(2) A continuous polynomial P is an m-homogeneous polynomial if and only if $\hat{d}^k P(0) \neq 0$ for only $k = m$.

13C *Hadamard's Three Circles Theorem.* If U contains the annulus $\{x \in E: r_1 \leq \|x\| \leq r_2\}$, then for any $f \in H(U;F)$, the function $r \to \log M(r)$ is a convex function on $[r_1, r_2]$; i.e.,

$$\log M(r) \leq t \log M(r) + (1 - t) \log M(r_2)$$

for all $0 \leq t \leq 1$. In particular,

$$M(r)^{\log (r_2/r_1)} \leq M(r_1)^{\log (r_2/r_3)} M(r_2)^{\log (r/r_1)}$$

for $r_1 \leq r \leq r_2$.

13D *Schwarz's Lemma.* Let U and V be the open unit balls of E and F respectively. If $f: U \to V$ is holomorphic and $f(0) = 0$, then

(a) $\|f(x)\| \leq \|x\|$ for $x \in U$;

(b) if, for some point $p \in U$, the equality $\|f(p)\| = \|p\|$, then $\|f(zp)\| = \|zp\|$ for all $|z| < \|p\|^{-1}$

Furthermore, if the set $\{e^{it} f(p)/\|f(p)\|: t \in \mathbb{R}\}$ contains a complex extreme point of V, then $f(zp) = zf(p)$ for all $|z| < \|p\|^{-1}$.

13E. Let U be a bounded circular open connected set containing 0. Suppose that for each a $\in$ U there exists a biholomorphic mapping $L_a: U \to U$ with $L_a(0) = a$. Then for each biholomorphic mapping $f: U \to U$ there is an invertible continuous linear map $M: E \to E$ such that $M(U) = U$ and $f = L_{f(0)} \circ M$.

13F. Theorem 13.13 no longer holds when the condition that $df(0)$ is onto is omitted, even when $E = F$. (Hint: Let E be the space of all bounded sequences of complex numbers with the sup norm, and take $f(x) = ([x_1]^2, x_1, x_2, \ldots).$)

13G. Let U be a convex open set in E and let $\alpha(x,y) = \|y\|$ for x in U and y in E. Define L and ρ as in the proof of Theorem 13.18. Then ρ is the metric associated with the norm.

13H. Theorem 13.18 holds for any function $f \in H(U;U)$ such that f^n maps U strictly inside U.

13I. Let U be the open unit ball of a Banach space E and let $f \in H(U;U)$ be such that f maps U strictly inside U. For each $z \in \mathbb{C}$, $|z| < 1$, let x(z) be the unique fixed point of zf. Then the map $z \to x(z)$ is holomorphic.

14

Gâteaux Holomorphy

14.1 The theory of holomorphic mappings between infinite dimensional spaces that we developed in the previous chapters was based on the Weierstrass method of power series. It was shown that in one complex variable theory the Weierstrass concept of holomorphy is equivalent to differentiability. In this chapter we will show that this is also the case for infinite dimensional spaces. Although it is possible to present a direct proof of this fact, we introduce the concept of Gâteaux holomorphy (G-holomorphy) whose definition is directly based on one complex variable. The main theme of this chapter is to study various conditions making G-holomorphic mappings holomorphic. In particular, we study the Graves-Taylor-Hille-Zorn theorem which states that a locally bounded G-holomorphic mapping is holomorphic; then we prove Zorn's theorem stating that the set of continuities of a G-holomorphic mapping is closed and open in its domain. We also prove Hartogs' theorem concerning separately holomorphic mappings.

14.2 We start with the following theorem.

THEOREM. *Let* E, F *and* G *be complex Banach spaces, and* U *be a non-empty open subset of* E. *Then for any* f ∈ H(U;F) *and* u ∈ L(F;G), *we have* u ∘ f ∈ H(U;G). *Furthermore,*

$$d^m(u \circ f)(a) = u \circ (d^m f(a))$$

for m ∈ ℕ *and* ξ ∈ U.

Proof. For a ∈ U there exists r > 0 such that the Taylor series

$$f(x) = \sum_{m=0}^{\infty} \frac{1}{m!} d^m f(a)(x-a)^m$$

converges uniformly on $\overline{B}_r(a) \subset U$. Since u: F → G is continuous uniformly on E, we have

$$u[f(x)] = \sum_{m=0}^{\infty} \frac{1}{m!} u[d^m f(a)](x-a)^m$$

and the convergence is uniform on $\overline{B}_r(a)$. It remains to show that
$u[d^m f(a)]$ is a continuous symmetric m-linear mapping of E^m into G.
But this is an easy exercise (cf. Ex. 4B). Hence u o f ∈ H(U;G).

The differential identity in the theorem is a consequence of the
chain rule and the identity theorem of power series.   □

HOLOMORPHIC MAPPINGS ON $\mathbb{C}^n$

14.3 *Hartogs' Theorem.* Historically, holomorphic functions on an
open set in $\mathbb{C}^n$ were defined to be functions that are holomorphic in
each variable separately and bounded on compact sets, then the Cauchy
integral formula for polydisks implies that they are in fact C^∞-
functions. It was F. Hartogs (1906) who finally proved that the
hypothesis of boundedness on compact sets is superfluous. We state
Hartogs' theorem here without giving a proof (cf. Hormander, (1966),
p. 28).

THEOREM (Hartogs). *Let* U ⊂ $\mathbb{C}^n$ *be an open set and* f: U → $\mathbb{C}$. *If for
each fixed set of* $z_1, \ldots, z_{j+1}, \ldots, z_n$, *the function*

$$z \to f(z_1, \ldots, z_{j-1}, z, z_{j+1}, \ldots, z_n)$$

is holomorphic for each j, *then* f *is continuous on* U. *In particular,*
f *is holomorphic on* U.

14.4 We now consider mappings of $U \subset \mathbb{C}^n$ into a Banach space F, and generalize the Dunford theorem 9.12.

THEOREM. *A function* f: U → F *is holomorphic if and only if it is weakly holomorphic; i.e., for every* u ∈ F*, u o f *is holomorphic.*

Proof. By 14.2, every holomorphic map is weakly holomorphic. To show the reverse implication for a ∈ U let r > 0 be such that $\overline{B}_r(a)$ ⊆ U. If $x \in \mathbb{C}^n$ with $\|x-a\| \le r$, then for all u ∈ F*, the mapping

$$z \to u \circ f(a + zx)$$

is holomorphic on an open set in $\mathbb{C}$ containing the unit disk. Hence $z \to f(a + zx)$ is a holomorphic mapping of z by theorems 9.12 and 9.17. Thus there exists a sequence $(P_m(x))$ in F such that

$$f(a+zx) = \sum_{m=0}^{\infty} P_m(x) \frac{z^m}{m!}$$

for $|z| \le 1$. Here we note that

$$P_m(x) = \frac{d^m}{dz^m} f(a+zx) \Big|_{z=0}$$

Hence we have, in particular,

$$f(a+x) = \sum_{m=0}^{\infty} \frac{1}{m!} P_m(x)$$

It remains to show that P_m are homogeneous polynomials. Since u o f is holomorphic

$$\frac{d^m}{dz^m}[u \circ f(a+zx)]\Big|_{z=0} = u \circ \frac{d^m}{dz^m} f(a+zx) \Big|_{z=0}$$

are homogeneous polynomials of degree m (from $\mathbb{C}^n$ to $\mathbb{C}$). Thus, for

all u ∈ F*, u ∘ P_m(x) is a homogeneous polynomial of $\mathbb{C}^n$ into $\mathbb{C}$. It
follows from Exercise 4C that P_m(x) is an m-homogeneous polynomial. □

14.5 HARTOGS' THEOREM. *Let* dim E < ∞, *and* U *an open subset of* E.
If a mapping f: U → F *is holomorphic in each variable separately,*
then f *is holomorphic on* U.

Proof. Since f is holomorphic in each variable separately, for each
u ∈ F*, u ∘ f is holomorphic in each variable separately. Hence
the classical Hartogs theorem insures that u ∘ f is holomorphic.
Hence f is holomorphic since f is weakly holomorphic.   □

GRAVES-TAYLOR-HILLE-ZORN'S THEOREM

14.6. Let U now be an open set in a Banach space E. Motivated by
the Hartogs theorem, we say that a mapping f: U → F is *Gâteaux*
holomorphic or *G-holomorphic* on U if for every a ∈ U and x ∈ E, the
mapping

   z → f(a+zx)

is holomorphic on U(a,x) = {z ∈ $\mathbb{C}$: a+zx ∈ U} in the classical one
variable sense. This definition is due to R. Gâteaux (1919). Our
definition of Gâteaux holomorphy is equivalent to the following:

THEOREM. *A mapping* f: U → F *is said to be G-holomorphic on* U *if for*
every a ∈ U, x ∈ E *and* u ∈ F*, *the mapping*

   z → u[f(a+zx)]

is holomorphic on U(a,x).

   This is a consequence of Theorem 9.12 claiming that f: U(a,x) → F
is holomorphic if and only if f is weakly holomorphic.

   It is clear that every holomorphic map is G-holomorphic. However,
the converse is not necessarily true.

EXAMPLE. If dim E = ∞, then there is a linear functional u which is not continuous. But u is a G-holomorphic mapping of E into $\mathbb{C}$.

This follows from the fact that every linear functional on a finite dimensional space is continuous.

14.7 *Local Boundedness.* The classical Hartogs theorem shows that a separately holomorphic function is continuous, from which we can prove that it is holomorphic by the Cauchy integral formula. We have shown in the preceding example that a G-holomorphic mapping may not be continuous. It is natural to ask when a G-holomorphic mapping is continuous. This leads us to consider the concept of local boundedness.

A function f from a topological space into a normed space is said to be *locally bounded* if f is bounded on a neighborhood of each point in the domain.

THEOREM. *If X is a metric space and F is a normed space, then a function* f: X → F *is locally bounded if and only if f is bounded on each compact subset of X.*

Proof. It is clear that a locally bounded function is bounded on each compact subset. Conversely, suppose that f: X → M is not locally bounded at a point x in X. Then we can find a sequence (x_m) in X converging to x such that $f(x_m)$ is unbounded. But the sequence and its limit forms a compact subset of X on which f is not bounded. This is absurd. Hence a function bounded on compact subsets is locally bounded.   □

14.8 *Graves-Taylor-Hille-Zorn's Theorem.* In passing from the notion of G-holomorphy to holomorphy, L. Graves (1935) and A. E. Taylor (1937) proved independently that the added assumption of continuity was sufficient. Recognizing the original definition of holomorphic mapping on an open set in $\mathbb{C}^n$ [see (14.3)], and the equivalency of local boundedness and boundedness on compact subsets (14.7), E. Hille (1949) has shown that the continuity in A. E. Taylor (1957) can be replaced by local boundedness. This result was further generalized by M. A. Zorn (1945a) replacing local boundedness with continuity in the sense of Baire (see M. A. Zorn (1945a, 1945b)).

14.9   THEOREM (Graves-Taylor-Hille-Zorn). *Let U be an open subset of a Banach space E, and let* f: U → F. *The following are equivalent.*

   (a)   f *is holomorphic on* U.

   (b)   f *is G-holomorphic and continuous on* U.

   (c)   f *is G-holomorphic and locally bounded on* U.

   (d)   f *is G-holomorphic and Baire continuous on* U *(i.e.,* f *is continuous off a set of the first category).*

Proof. We prove here only the implication (c) ⇒ (a) and leave the remaining properties to the reader to verify (cf. Zorn (1945a) or Hille and Phillips (1957), pp. 778-779). Our proof follows essentially M. A. Zorn (1945a). Let f be a locally bounded G-holomorphic mapping of U into F. Then for any a ∈ U and x ∈ E, the mapping $z \to f(a+zx)$ is holomorphic on $U(a,x) = \{z \in \mathbb{C}: a + zx \in U\}$; and hence it is continuous on $U(a,x)$. We claim that there exists a power series from E to F such that

$$f(x) = \sum_{m=0}^{\infty} P_m(x-a)$$

converges uniformly on an open ball around a, where P_m are continuous m-homogeneous polynomials. We divide the proof into five steps.

   (1)   Since f is locally bounded, we can find r > 0 and M > 0 such that $\overline{B}_r(a) \subseteq U$ and

$$\|f(x)\| \leq M$$

for all x in $\overline{B}_r(a)$. We shall fix this r and M for the entire proof. If $\|x\| = 1$, then it is easy to see that the set $U(a,x)$ contains the closed complex disk $|z| \leq r$.

   Define $P_m: E \to F$ to be $P_m(0) = 0$ and

$$P_m(x) = \frac{c^m}{2\pi i} \int_{|z|=r} \frac{f(a+zy)}{z^{m+1}} dz \qquad (*)$$

where $x \in E\backslash\{0\}$, $y = x/\|x\|$ and $c = \|x\|$.

The mapping P_m is well-defined since $f(a+zy)$ is continuous for $|z| \leq r$. Also notice that the integral is taking the value over any circle $|z| = s$, $0 < s < r$ since $z \to f(a+zy)$ is holomorphic on $U(a,x)$. From the definition of P_m, it follows immediately that P_m is m-homogeneous; i.e., $P_m(tx) = t^m P_m(x)$ for any $t \in \mathbb{C}$.

(2) P_m is uniformly bounded on the unit ball. In fact, taking the norm on the both sides of (*) we obtain

$$\|P_m(x)\| \leq M/r^m$$

for $\|x\| \leq 1$.

(3) We now claim that P_m is an m-homogeneous polynomial. First we show that P_m is an m-homogeneous polynomial if $E = \mathbb{C}^n$. By the Hartogs theorem (14.5), in this case, f is holomorphic on U and we have the uniform convergent Taylor series expansion of f around a:

$$f(x) = \sum_{m=0}^{\infty} \frac{1}{m!} \hat{d}^m f(a)(x-a)$$

Let $s > 0$ be such that the series converges uniformly on the closed ball $\overline{B}_s(a)$. From the Cauchy integral formula we have

$$\frac{1}{m!} \hat{d}^m f(a)(x) = \frac{1}{2\pi i} \int_{|z|=s} \frac{f(a+zx)}{z^{m+1}} dz$$

We may assume $s < r$ and substitute $y = x/\|x\|$ for x and $c = \|x\|$ to conclude that

$$P_m(x) = \frac{1}{m!} \hat{d}^m f(a)(x)$$

Hence P_m is an m-homogeneous polynomial from $\mathbb{C}^n$ into F.

(4) Now we consider the general case where dim E = ∞. From the preceding case (3), we know that P_m is a mapping of E into F such that for any finite dimensional subspace S of E, the restriction to S of P_m is an m-homogeneous polynomial on S. To show that P_m is an m-homogeneous polynomial from E to F, we must show that there is a symmetric m-linear map A_m such that $\hat{A}_m = P_m$. Let A_m be a mapping of E^m into F such that for any finite dimensional subspace S of E, A_m is the symmetric m-linear map on S^m. We use induction to show A_m is m-linear. Notice the following fact:

> If A: E → F is a mapping such that A is linear on every two
> dimensional subspace of E, then A is linear.

Therefore we can easily prove by induction that $A_m : E^m \to F$ is m-linear. This and (2) together show that P_m is an m-homogeneous continuous polynomial of E into F.

(5) Finally we prove that the power series $\Sigma\ P_m(x - a)$ converges uniformly to f(x) around a. Let s > 1 and t > 0 be such that st < r. Then t < r. For $x \in \bar{B}_t(a)$ and for all $z \in \mathbb{C}$, $|z| \leq s$, we have

$$(1-z)a + zx = a + z(x - a) \in \bar{B}_r(a)$$

Mimicking the proof of Theorem 13.7, we obtain

$$f(x) - \sum_{k=0}^{m} P_k(x-a) = \frac{1}{2\pi i} \int_{|z|=s} \frac{f[(1-z)a+zx]}{z^{m+1}(z-1)}\, dz$$

and

$$\left\| f(x) - \sum_{k=0}^{m} P_k(x-a) \right\| \leq \frac{M}{s^m(s-1)}$$

for all $x \in \bar{B}_t(a)$. Since $s > 1$, $\Sigma P_m(x-a)$ converges to f uniformly on $\bar{B}_t(a)$. This shows that f is holomorphic at a; hence f is holomorphic on U since a is arbitrary. This completes the proof. □

14.10 *Balanced Set.* In the preceding proof, steps (3) and (4) actually show that if $f: U \to F$ is G-holomorphic, f can be represented by a power series $\Sigma P_m(x - a)$ on a neighborhood of each point a in U. The homogeneous polynomials in this representation need not be continuous.

It is natural to ask the size of the domain of convergence of the power series. To answer this we introduce the following concept.

A subset A of a vector space over $\mathbb{K}$ is said to be *balanced* (or *equilibrated*) if $\lambda A \subseteq A$ for every $\lambda \in \mathbb{K}$ such that $|\lambda| \leq 1$. It is clear that a nonempty balanced set contains the origin.

The union of an arbitrary family of balanced sets is balanced. Therefore, given any subset A of E, there exists a largest balanced set B contained in A. The set B is called the *balanced core* of A. It is nonempty if and only if A contains the origin.

Also notice that a point x in A is in the balanced core of A if and only if the balanced set $C(x) = \{zx: |z| \leq 1\}$ is contained in A.

In a normed space, the balanced core of a neighborhood of the origin is always open (why?).

If V is a balanced set and $\xi \in E$, then the set $\xi + V$ is called a ξ-*balanced set*. Hence, if U is a nonempty open set and $\xi \in U$, then U contains the largest ξ-balanced set (or the ξ-balanced core) which is open.

14.11 THEOREM. *A function* $f: U \to F$ *is G-holomorphic if and only if for any* $\xi \in U$ *there is a unique sequence* (P_m), *where* $P_m \in P_a(^mE;F)$ *(not necessarily continuous) such that*

$$f(x) = \sum_{m=o}^{\infty} P_m(x-\xi)$$

for all x in the largest ξ-balanced set in U.

Proof. As we remarked before in 14.10, for each $\xi \in U$ there corresponds a unique power series $\Sigma \, P_m(x-\xi)$ which represents $f(x)$. Since
f is G-holomorphic on U, the restriction of f to the affine line
passing ξ and x is holomorphic in the sense of one complex variable.
Hence the power series restricted to the same affine line converges
on the largest one-dimensional disk centered at ξ lying in U. This
shows that the domain of convergence of the power series is the
largest ξ-balanced set in U.

The converse is clear since

$$f[(1-z)\xi+zx] = \Sigma \, P_m(x-\xi) z^m$$

for all $|z| \le 1$ and x in the largest ξ-balanced set in U.   □

14.12 *Cauchy and Weierstrass Viewpoints.* As we remarked in 9.9,
the theory of holomorphic function based on complex differentiability
is referred to as the *Cauchy viewpoint*, and the theory based on
power series is called the *Weierstrass viewpoint*. These viewpoints
are equivalent in the one or several complex variable theory. We
have adopted the Weierstrass viewpoint since this approach is generally more suitable for both real and complex cases. The theorems
by Graves, Taylor, Hille, and Zorn (14.9) show that both viewpoints
are indeed identical for complex Banach spaces. We summarized this
fact in the following theorem.

14.13 THEOREM. *The following are equivalent for a function* f: U → F.

   (a) f *is Fréchet differentiable on* U.

   (b) f *is holomorphic on* U.

COROLLARY. *A composite mapping of two holomorphic mappings is
holomorphic.*

Proof. It is immediate from the Chain Rule.   □

14.14 *WEAK HOLOMORPHY.* A mapping f: U → F is said to be *weakly
holomorphic* if for every u $\in$ F*, u ∘ f: U → C is holomorphic.

THEOREM. *A function f: U → F is holomorphic if and only if f is weakly holomorphic.*

Proof. Since every holomorphic mapping is weakly holomorphic, it remains to show that a weakly holomorphic mapping is holomorphic. If f: U → F is weakly holomorphic, then by Theorem 14.4 or Theorem 9.12, f is G-holomorphic. We claim that f is locally bounded. Suppose the contrary. Then there exists a $\in$ U and a sequence (x_n) in U such that $x_n \to a$ and $\|f(x_n)\| > n$ for all $n \in \mathbb{N}$. But this is impossible because $\{u[f(x_n)]: n = 0,1,2,\ldots\}$ is bounded for all $u \in F^*$; hence $\{f(x_n): n = 0,1,2,\ldots\}$ is bounded (cf. Theorem 2.18). □

VITALI'S CONVERGENCE THEOREM

14.15. The property of being holomorphic is not necessarily preserved under pointwise convergence. In 6.10 we have studied that differentiability is preserved under pointwise convergence for differentiable maps with some mild restrictions on differentials. However, Vitali's convergence theorem for holomorphic mappings requires no restrictions on differentials. We first recall Vitali's convergence theorem for one complex variable from Titchmarsh (1939, p. 168).

THEOREM. *Let (f_n) be a sequence of holomorphic functions in a connected open set D in $\mathbb{C}$ satisfying*

(a) (f_n) is uniformly bounded on D;

(b) $(f_n(x))$ converges to a function f(x) at each point x in a subset of D having a cluster point in D.

Then f is holomorphic on D; the convergence is uniform on each compact subset of D.

This theorem is also true for functions taking values in a Banach space since holomorphy and weak holomorphy are equivalent by Dunford's theorem 9.12. Assuming this result we are ready to state and prove an infinite dimensional version of Vitali's convergence theorem. However, we have to modify here the assumptions and conclusions of the one dimensional case.

14.16 THEOREM. *Let* U *be a connected open subset of a normed space*
E, *and let* (f_n) *be a sequence of holomorphic mappings of* U *into* F
satisfying

(a) (f_n) *is locally uniformly bounded on* U;

(b) ($f_n(x)$) *converges at each point* x *in a nonempty open subset*
of U.

Then (f_n) *converges to a holomorphic mapping on* U.

Proof. We first show that (f_n) converges to a G-holomorphic mapping
f on U. To show this we first need the following topological lemma
which is also quite useful later; its proof is left to the reader.

LEMMA. *Let* U *be a connected subset of a normed space and* W *a non-*
empty subset of U. *If we have* a + V ⊂ W *for any* a *in* W *and for any*
balanced open set V *with* a + V ⊂ U, *then* W = U.

We now continue our proof that (f_n) converges to a G-holomorphic
mapping on U. Let W be the set of all points a in U such that the
sequence (f_n) converges on a neighborhood of a. Then W is a nonempty
subset of a connected set U. Now let's apply the lemma. For a point
a in W, consider a balanced open set V satisfying

(1) V(a) = a + V ⊂ U;

(2) The sequence (f_n) converges on V(a).

Since V(a) is an open a-balanced subset, for each x in V(a) there
exists r = r(x) > 1 such that the sequence ($f_n([1-z]a+zx)$) converges
to a holomorphic mapping, say, f([1-z]a+zx) for $|z|$ < r by Vitali's
convergence theorem 14.15. In particular, taking z = 1, we conclude
that ($f_n(x)$) converges to f(x) for all x in V(a). Hence, V(a) is a
subset of W. It follows from the lemma above that W = U; hence f is
G-holomorphic on U.

Finally we use the Graves-Taylor-Hille-Zorn theorem to prove that
f is holomorphic on U. It suffices to show that f is locally bounded
on U. Since (f_n) is locally uniformly bounded on U, for each point
a in U there exist M and a neighborhood V of a such that $\|f(x)\| \leq M$
for all x in V, which certainly shows that f is locally bounded.
This completes the proof.   □

14.17 In the preceding theorem we did not require that E be a
Banach space. If E is a Banach space, using the Baire category
theorem we obtain the following variation of Vitali's convergence
theorem.

THEOREM. *Let U be an open subset of a Banach space E. If (f_n) is*
a sequence of holomorphic mappings of U into F such that $(f_n(x))$
converges to f(x) at each point of U, then f is holomorphic on an
open dense subset of U.

We first prove the following lemma.

LEMMA. *Let X be a Baire space. If (f_n) is a sequence of continuous*
mappings of X into a normed space F which converges to a function
at each point in X, then the set of local boundedness for f is open
and dense in X.

Proof. Let V be a non-empty open subset of X. For each natural
number k, the set

$$W(k,n) = \{a \in V: \|f_n(a)\| \leq k\}$$

is closed in V; hence the following set

$$W_k = \cap\{W(k,n): n \in \mathbb{N}\}$$

is also closed in V. It is clear that

$$V = \cup \{W_k: k \in \mathbb{N}\}$$

Since V is a Baire space, some W_k must have non-empty interior. Let
$W = \text{int } W_k$. Then (f_n) is uniformly bounded on W.   □

Proof of Theorem. Since U is a Baire space, by the lemma above the
set of local boundedness for f is open and dense in U. Therefore,
f is holomorphic on an open dense subset of U by Vitali's convergence
theorem 14.16.   □

14.18 THEOREM. *Let* U *be an open subset of a Banach space* E *and let*
f_n: U → F *be holomorphic such that* $\Sigma f_n(x)$ *converges to a function*
f(x) *at each point of* U. *Then the sequence* (f_n) *is locally uniformly*
bounded on an open dense subset of U; *in particular,* f *is holomorphic*
on this dense set.

Proof. We need the following theorem whose proof is identical,
mutatis mutandis, to the proof of Banach-Steinhaus Theorem 2.17.

OSGOOD'S THEOREM. *Let* X *be a Baire space and* F *a normed space. If*
a sequence (f_n) *of continuous functions of* X *into* F *is uniformly*
bounded at each point of X, *then* (f_n) *is locally uniformly bounded on*
an open dense subset of X.

For each x in U, since $\Sigma f_n(x)$ converges, we have $\|f_n(x)\| \to 0$ as
n → ∞. Hence there exists M_x such that

$$\|f_n(x)\| \leq M_x$$

for all n. Now it follows from Osgood's Theorem that (f_n) is locally
uniformly bounded on an open dense subset. Vitali's convergence
theorem now takes care of the rest.   □

14.19 *Banach-Steinhaus Theorem.* We have the following theorem as a
corollary of Osgood's Theorem which was given already in 4.17.

THEOREM. *Let* E *be a Banach space. If* (P_k) *is a sequence in* $P(^mE;F)$
which converges to a function P *at each point of* E, *then* $P \in P(^mE;F)$.

ZORN'S THEOREM

14.20. In his paper (1945a), M. Zorn proved that if U is an open and
connected subset of a Banach space E, the continuity or the local
boundedness at each point of U in the Graves-Taylor-Hille-Zorn
Theorem can be replaced with continuity or local boundedness at a
single point. In this striking theorem, we must assume that E is a

Banach space; otherwise, the result is not true in general as shown in Example 14.25.

To prove this, we need the following sequence of theorems and lemmas all of which are interesting in their own right.

14.21 LEMMA. *Let* P: E → F *be a G-holomorphic mapping satisfying the following conditions:*

(1) *For* $|z| = 1$, $\|P(zx)\| = \|P(x)\|$;

(2) *There exists a point a in E, M > 0, and a balanced open set V such that* $\|P(x)\| \leq M$ *for all x in a + V.*

Then

$$\|P(x)\| \leq M$$

for all x in S·a + V where S = $\{z \in \mathbb{C}: |z| = 1\}$.

Proof. For x in V, the function g(z) = P(za+x) is holomorphic on $\mathbb{C}$. By the maximum modulus theorem, the restriction to S of g takes its maximum modulus at some point w, $|w| = 1$. Then for any z in S, we have

$$\|P(za+x)\| \leq \|P(wa+x)\| = \|P(a+w^{-1}x)\| \leq \sup\ \{\|P(a+v)\|: v \in V\} \leq M$$

which completes the proof. □

14.22 THEOREM. *Let U be an open balanced subset of a normed space E and let* $\Sigma\ P_m$ *be a power series converging to a function f at each point of U. If the sequence* (P_m) *is uniformly bounded on U, then for any r in (0,1) f is bounded on rU = $\{rx: x \in U\}$. Conversely, if f is bounded on U, the sequence* (P_m) *is uniformly bounded on U.*

Proof. Let $\|P_m(x)\| \leq M$ for all x in U and m. For r in (0,1) if x is in rU, then x' = x/r is in U; and hence

$$\|P_m(x')\| = r^{-m}\|P_m(x)\| \leq M$$

From this we obtain

$$\|P_m(x)\| \leq r^m M$$

for all x in rU.   Hence

$$\|f(x)\| \leq \Sigma \|P_m(x)\| \leq M/(1-r)$$

for all x in rU.   Since f is G-holomorphic on U and f is uniformly
bounded on rU, we conclude that f is holomorphic on rU.

Conversely, if $\|f(x)\| \leq M$ on U, then f is holomorphic on U.   Hence
for any x in U we have

$$P_m(x) = \frac{1}{2\pi i} \int_{|z|=1} \frac{f(zx)}{z^{m+1}} \, dz$$

Therefore

$$\|P_m(x)\| \leq M \qquad\qquad\qquad \square$$

COROLLARY. *Let E be a Banach space and let* $f(x) = \Sigma\, P_m(x)$ *be defined
on an open balanced subset* U *where* $P_m \in P(^m E; F)$. *Then f is holomorphic on some open balanced subset of* U.

Proof. By Theorem 14.18, there are a point a in U and an open
balanced subset V such that $a + V \subset U$ and (P_m) is uniformly bounded
on a + V. Then by Lemma 14.21, (P_m) is uniformly bounded on V.
Hence f is holomorphic on rV for any r in (0,1). $\square$

14.23 THEOREM. *Let U be an open balanced subset of a Banach space
E on which a power series* $\Sigma\, P_m$ *of continuous homogeneous polynomials
converges to a function f. Then f is holomorphic on* U.

Proof. The mapping f is clearly G-holomorphic on U. Furthermore,
it is locally bounded on an open dense subset of U. Let a be an

arbitrary point in U. Then the function can be represented by a power series

$$f(a+x) = \sum_{m=0}^{\infty} Q_m(x) \tag{1}$$

which converges for all x in some balanced open set.

We now show that Q_m must be continuous. First find A_m in $L(^mE;F)$ such that $\hat{A}_m = P_m$. Then by the hypothesis we have

$$f(a+x) = \sum_{m=0}^{\infty} A_m(a+x)^m \tag{2}$$

Since the power series expansion must be unique, comparing (1) and (2) we have for each k

$$Q_k(x) = \sum_{m=0}^{\infty} \binom{m}{k} A_m a^{m-k} x^k$$

by the binomial formula 4.4. This identity holds for all x in E. Since

$$A_m a^{m-k} : E^k \to F$$

is a continuous k-linear mapping for each m, we conclude that Q_k is a k-homogeneous polynomial by the Banach-Steinhaus Theorem 14.19 for homogeneous polynomials. Now it follows from Corollary 14.22 that f(a+x) is holomorphic on a neighborhood of O. □

14.24 Zorn's Theorem. *Let U be a connected open subset of a Banach space and* f: U → F *a G-holomorphic mapping. Then the following are equivalent:*

   (a) *f is holomorphic.*

   (b) *f is continuous at one point in U.*

   (c) *f is locally bounded at one point in U.*

Proof. The forward implications are clear. It remains to show that
(c) implies (a).

Let W be the set of local boundedness of f in U. Then W is non-
empty. For a in W, choose the largest open balanced set V satisfying
a + V ⊂ U. Since f is G-holomorphic, the Taylor expansion

$$f(a+x) = \sum_{m=0}^{\infty} P_m(x)$$

converges for each x in V. Since a is in W, $\sum P_m(x)$ is holomorphic
on a neighborhood of 0 by the Graves-Taylor-Hille-Zorn theorem.
Hence each P_m is continuous. It follows immediately from Theorem
14.23 that $\sum P_m$ is holomorphic on V. Thus a + V ⊂ W. Applying
Lemma 14.16, we can conclude that W = U. Therefore f must be holomorphic on U by the Graves-Taylor-Hille-Zorn theorem.   □

14.25. In Zorn's Theorem we assumed that E is a Banach space. We
now show that this assumption is essential for the theorem.

EXAMPLE (Noverraz, 1973). Let c_o be the vector space of all complex
sequences converging to 0. Then c_o is a vector subspace of ℓ^{∞} which
is given in 1.6. The space c_o is a closed subspace of ℓ^{∞}; hence c_o
is a Banach space. Let E be the set of all complex sequences having
only a finite number of non-zero terms. Then E is a dense subspace
of c_o. Therefore E is not a Banach space.

For each natural number m, define $P_m: E \to \mathbb{C}$ to be the polynomial

$$P_m(x) = [x_1]^{m-1} x_m$$

where $x = (x_1, x_2, \ldots)$. Then P_m is an m-homogeneous polynomial with
$\|P_m\| = 1$. We now introduce a function f: E → ℂ by setting

$$f(x) = \sum_{m=0}^{\infty} P_m(x)$$

Then f is well-defined since for each x in E the series $\Sigma\, P_m(x)$ contains only a finite number of non-zero terms. Hence f is G-holomorphic on E.

We want to show that f is locally bounded in a certain open set, say U, but f cannot be locally bounded on the outside of U. In fact, denote

$$U = \{x \in E: \left|x_1\right| < 1\}$$

and let a be a point in U. Find r in $(0,1)$ such that $B_{2r}(a) \subset U$. We now claim that f is uniformly bounded on $B_r(a)$.

Let $M = \|a\|+r$ and $L = \left|a_1\right|+r$. Then $L < 1$. For any x in the ball $B_r(a)$, we have

$$\left|f(x)\right| \leq \left|x_1\right|+\left|x_1 x_2\right|+\left|[x_1]^2 x_3\right|+\ldots+\left|[x_1]^{n-1} x_n\right|+\ldots$$

$$\leq M(1+L+L^2+\ldots\,) = M/(1-L)$$

Thus f is locally bounded on U, and hence f is holomorphic on U. On the other hand, take $x = (1,0,\ldots\,)$ and

$$h_n = (1/n^{\frac12},0,\ldots,0,1/n^{\frac12},0,\ldots)$$

where non-zero terms appear at the first and $(n+1)$th places. Then $h_n \to 0 = (0,0,\ldots\,)$. But

$$f(x+h_n) = (1+1/n^{\frac12}) + (1/n^{\frac12})(1+1/n^{\frac12})^n$$

tends to ∞ while $f(x) = 1$. Therefore, there is no neighborhood of x on which f can be bounded. Thus f is not holomorphic on the entire space. □

HARTOGS' THEOREM

14.26. We need the following lemma in the proof of the Hartogs theorem for infinite dimensional spaces.

LEMMA. *Let X be a Baire space, Y a metric space, and F a normed*
space. If a function f: X × Y → F is continuous with respect to
each variable separately, then f is bounded on a non-empty open sub-
set of X × Y.

Proof. For any $m \in \mathbb{N}$ and $y \in Y$, let $A_y(m) = \{x \in X: \|f(x,y)\| \le m\}$.
For $b \in Y$ and $n \in \mathbb{N}$, set $B_n = \{y \in Y: d(b,y) < 1/n\}$. Then the
following set is closed:

$$A_{m,n} = \cap\{A_y(m) : y \in B_n\}$$

We claim that X is the union of $A_{m,n}$ for all m and n in $\mathbb{N}$. In
fact, for each x, the function $y \to f(x,y)$ is continuous at b, and
hence there exist m and n such that $\|f(x,y)\| \le m$ for $y \in B_n$. This
shows that $x \in A_{m,n}$.

Since X cannot be a countable union of nowhere dense sets, there
exists a non-empty open set U of X which is contained in $A_{p,q}$ for
some p and q. Then f is bounded on the open set $U \times B_q$, which
completes the proof.   □

Notice that the preceding argument also gives another proof of
Theorem 3.5.

14.27 HARTOGS' THEOREM. *Let E, F and G be Banach spaces and let U*
be an open connected subset of E × F. If f: U → G is separately
holomorphic on U, then f is holomorphic.

Proof. Let $(a,b) \in U$ and $(x,y) \in E \times F$. Then the mapping

$$(z,w) \to f(a+zx,b+wy)$$

is separately holomorphic on a subset V of $\mathbb{C}^2$. Hence by Hartogs'
Theorem 14.5, the mapping is holomorphic on V. In particular, the
mapping

$$z \to f(a+zx,b+zy)$$

is holomorphic on a suitable subset of $\mathbb{C}$. Hence f is G-holomorphic.

To complete the proof, we need to show that f is locally bounded at some point since U is connected. But this is a consequence of Lemma 14.26.   □

EXERCISES

14A *Finitely Open Sets.* A subset U of a normed space E is said to be *finitely open* if

$$U(a;x_1,x_2,\ldots,x_m) = \{(z_1,\ldots,z_m) \in \mathbb{C}^m : a+z_1x_1+\ldots+z_mx_m \in U\}$$

is open in $\mathbb{C}^m$ for any a in U and $x_1,\ldots,x_m$ in E.

(a)   Every open subset of E is finitely open.

(b)   Some finitely open sets are not open.

(c)   Every G-holomorphic mapping on a finitely open set is continuous if we endow E with the topology consisting of all finitely open subsets (we call this the finite topology on E).

(d)   The scalar multiplication $(\lambda,x) \in \mathbb{C} \times E \to \lambda x \in E$ is continuous for the finite topology on E.

(e)   The addition $(x,y) \in E \times E \to x + y \in E$ is continuous for the finite topology when one of the terms is restricted to vary in a finite dimension subspace of E.

(f)   The finite topology on a vector space E is a vector space topology (i.e., both scalar multiplication and addition are continuous) if and only if the dimension of E is at most countably infinite (see Kakutani and Klee (1963)).

14B *Baire Continuity.* A function f on a metric space X is said to be *Baire continuous* if removal of a set P of the first category yields a continuous function on $X \setminus P$.

(a)   Let S be an open sphere of X and F a Banach space. If f: S → F is Baire continuous, then the extended function f': X → F defined by f' = f on S and f' = 0 off S is Baire continuous.

(b)   Let E and F be Banach spaces. If f: E → F is linear and
      Baire continuous, then f is continous on E.

(c)   Let U be an open subset of a Banach space. Then f: U → F
      is holomorphic if and only if f is G-holomorphic and
      Baire continuous.

14C   *Weak Holomorphy.*

(a)   The mapping g: U → F* is holomorphic if and only if the
      function x ∈ U → f(x)(y) ∈ ℂ is holomorphic for every y
      in F.

(b)   Let F and G be Banach spaces. The mapping f: U → L(F;G) is
      holomorphic if and only if the function x ∈ U → φ[f(x)(y)]
      is holomorphic for every y ∈ F and φ ∈ G*.

14D   *Algebraic Weak Holomorphy.*

(a)   Let f be a mapping from U to the algebraic dual $F' = L_a(F;ℂ)$
      such that, for every y ∈ F, the function f_y: x ∈ U → f(x)(y)
      ∈ ℂ is holomorphic. Let U be connected and assume that
      there is some a ∈ U such that, for every x in E and m, the
      linear functional

      $$y ∈ F → \hat{d}^m f_y(a)(x) ∈ ℂ$$

      is continuous. Then f(U) ⊂ F* and f: U → F* is holomorphic.

(b)   Let f: U → F' = $L_a(F;ℂ)$ be such that, for every y ∈ F, the
      function x ∈ U → f(x)(y) ∈ ℂ is holomorphic. Let U be
      connected and assume that f(V) ⊂ F*. The f(U) ⊂ F* and
      f: U → F* is holomorphic.

14E   *Absolute Convergent Power Series.* Let f(x) = Σ P_m(x) be a power
series on a Banach space with the radius of absolute convergence
$r_a > 0$. Then f is holomorphic for $\|x\| < r_a$ if and only if f is continuous at one point.

14F   *Taylor Series for Differentials.*

(a)   If f ∈ H(U;F) and

      $$f(x) = \sum_{k=o}^{∞} P_k(x-a)$$

is the Taylor series of f at a in U, then the Taylor series of $d^m f$ and $\hat{d}^m f$ at a are

$$d^m f(x) \approx \sum_{k=o}^{\infty} d^m P_{m+k}(x-a)$$

$$\hat{d}^m f(x) \approx \sum_{k=o}^{\infty} \hat{d}^m P_{m+k}(x-a)$$

(b)   If $f \in H(U;F)$ and $a \in U$, then for $k,m = 0,1,\ldots,$

$$\frac{1}{k!} \hat{d}^k \left[\frac{1}{m!} \hat{d}^m f \right](a) = \frac{1}{m!} \hat{d}^m \left(\frac{1}{(k+m)!} \hat{d}^{k+m} f(a) \right)$$

(c)   If $f \in H(U;F)$ and $a \in U$ then for $k,m = 0,1,\ldots,$ we have

$$\hat{d}^m (T_{k+m,f,a}) = T_{k,\hat{d}^m f,a}$$

15

Radius of Boundedness

15.1. It is a great mistake to think of the subject of holomorphic mappings of infinite dimensional spaces as being a mere generalization of one or several complex variables. Deep new phenomena and profound and as yet unsolved problems present themselves naturally in the study of infinite dimensional holomorphy. The concept of radius of boundedness which we study in this chapter has no meaningful counterpart in finite dimensional holomorphy. We first study the convergence of Taylor series before introducing the concept of radius of boundedness. We also present a brief introduction to subharmonic and plurisubharmonic functions and their relationship with radius of boundedness.

CONVERGENCE OF TAYLOR SERIES

15.2 As before we assume again that E is a normed space, U a non-empty open subset of E, and F a Banach space.

Let $f \in H(U;F)$. It was shown in 14.11 that for any point ξ in U the Taylor series of f at ξ converges to f pointwisely on the largest ξ-balanced subset of U. If E is one dimensional, then the Taylor series converges to f uniformly on any closed disk around ξ which is contained in U (cf. 9.17). For the infinite dimensional case, we have the following theorem.

15.3 THEOREM. *If U is a ξ-balanced open set and $f \in H(U;F)$, then the Taylor series of f at ξ converges to f uniformly on some neighborhood of every compact subset of U.*

We first prove the following lemma.

LEMMA. *Let U be a ξ-balanced open set and f ∈ H(U;F). If K is a compact subset of U, then there correspond an open set V with K ⊂ V ⊂ U and a number r > 1 such that f is bounded on the open set*

$$V_\xi = \{(1-z)\xi + zx : x \in V, \ |z| \le r\} \subset U$$

Proof. Consider a mapping A of $\mathbb{C} \times E$ into E defined by

$$A(z,x) = (1-z)\xi + zx$$

Then A is continuous; and hence $A(D_1 \times K)$ is compact in U where D_r denotes the closed disk centered at the zero with radius r. Since f is holomorphic on U, f must be bounded on $A(D_1 \times K)$. Since f is also locally bounded and $D_1 \times K$ is compact, we can find a number r > 1 and an open subset V with $K \subset V \subset U$ such that f is bounded on $A(D_r \times V) \subset U$. But $A(D_r \times V) = V_\xi$, which completes the proof. □

Proof of Theorem. For each compact K, let V and V_ξ be the open sets in the lemma above. It then follows from Theorem 13.7 that

$$\left\| f(x) - \sum_{m=0}^{k} \frac{1}{m!} \hat{d}^m f(\xi)(x-\xi) \right\| \le \frac{M}{r^k(r-1)}$$

for all x in V, where $M = \sup \{\|f(x)\| : x \in V_\xi\}$. Since the right side of the above estimate converges to 0 as $k \to \infty$, the Taylor series converges to f uniformly on V, which completes the proof. □

15.4. In the preceding theorem we have not used the fact that f is G-holomorphic. As a consequence, we can obtain Theorem 14.11 for f ∈ H(U;F) without using the fact that f is G-holomorphic.

THEOREM. *Let f ∈ H(U;F) and ξ ∈ U. The nth Taylor series of f at ξ converges to f on the largest ξ-balanced set U_ξ in U, and the convergence is uniform on every compact subset of U_ξ.*

RADIUS OF BOUNDEDNESS

15.5 A phenomenon in infinite dimensional holomorphy which has no
counterpart in several complex variables is that an entire function
may be unbounded on a bounded set as shown by an example below. But
in a finite dimensional space, every closed bounded set is compact;
and hence every holomorphic mapping is bounded on bounded sets.

EXAMPLE. We now show an example showing that an entire function may
not be bounded on a bounded set.

Let E be the Banach space c_o of all sequences $x = (x_1, x_2, \ldots)$ of
complex numbers converging to 0 with the norm $\|x\| = \sup |x_m|$. We
define the m-homogeneous polynomial P_m of E into $\mathbb{C}$ by

$$P_m(x) = x_1 \ldots x_m$$

For convenience we assign $P_o(x) = 1$. Then the power series $\Sigma \, P_m(x)$
converges absolutely at any point x in E. Since $\|P_m\| = 1$ for all m,
by Theorem 14.23 we conclude that the power series $f(x) = \Sigma \, P_m(x)$ is
holomorphic on E.

For any r in $(0,1)$ and $\|x\| \le r$ we have

$$\|f(x)\| \le \Sigma \, \|P_m\| r^m = 1/(1-r)$$

Thus f is bounded on the closed ball, $\|x\| \le r$. However,

$$\sup \, \{\|f(x)\| : \|x\| = 1\} = \infty$$

since

$$f((1,\ldots,1,0,0,\ldots)) = m+1$$

where 1 appears in the first m terms. This shows that f is unbounded
on every ball centered at O with radius $r > 1$.   □

15.6 *Radius of Boundedness*. The above example led L. Nachbin (1969)
to consider the concept of radius of boundedness.

For f ∈ H(U;F) and ξ ∈ U, the *radius of boundedness* of f at ξ is defined to be the supremum of all r > 0 such that $B_r(\xi) \subset U$ and f is bounded on $\overline{B}_s(\xi)$ for all s in (0,r). We denote the radius of boundedness of f at ξ by $r_f(\xi)$.

15.7. In Example 15.5, we have $r_f(0) = 1$, and radius of uniform convergence at 0 is also 1. We will see that this is not a coincidence.

For f ∈ H(U;F) and ξ ∈ U, let $R_f(\xi)$ denote the radius of uniform convergence of f at ξ and $d_U(\xi) = \inf\{\|\xi-x\|: x \in E\backslash U\}$ the distance from ξ to the boundary of U.

THEOREM (Nachbin, 1969). *For f ∈ H(U;F) and ξ ∈ U, we have*

$$r_f(\xi) = \inf\{d_U(\xi), R_f(\xi)\}$$

Proof. It is clear that $r_f(\xi) \leq d_U(\xi)$. Let $f(x) = \Sigma\, P_m(x-\xi)$ be the Taylor series of f at ξ.

We first show that $r_f(\xi) \leq R_f(\xi)$; hence

$$r_f(\xi) \leq \inf\{d_U(\xi), R_f(\xi)\}$$

Let s in $(0, r_f(\xi))$ and set

$$M_s = \sup\{\|f(x)\|: \|x-\xi\| \leq s\}$$

Then M_s is finite, and it follows from the Cauchy inequalities that

$$\|P_m\| \leq M_s/s^m$$

for all m. Hence

$$\limsup \|P_m\|^{1/m} \leq 1/s$$

which shows that $s \leq R_f(\xi)$ by the Cauchy-Hadamard formula 11.5. Let $s \to r_f(\xi)$, and we get $r_f(\xi) \leq R_f(\xi)$.

To show that $r_f(\xi) \geq \inf \{d_U(\xi), R_f(\xi)\}$, consider a real number r satisfying $0 < r < \inf \{d_U(\xi), R_f(\xi)\}$. The Taylor series of f at ξ then converges uniformly on $\overline{B}_r(\xi)$. We claim that $r \leq r_f(\xi)$; i.e., f is bounded on $\overline{B}_r(\xi)$. In fact, from Corollary 1 in 11.5 we obtain

$$\|f(x)\| \leq \Sigma \|P_m\| r^m < \infty$$

for all x in $\overline{B}_r(\xi)$. Hence $r < r_f(\xi)$, which shows that

$$r_f(\xi) > \inf \{d_U(\xi), R_f(\xi)\} \qquad \qquad \square$$

REMARK. In a finite dimensional space, we have

$$r_f(\xi) = d_U(\xi) \leq R_f(\xi)$$

but in infinite dimensional holomorphy as we have seen in Example 15.5, it may occur that

$$r_f(\xi) = R_f(\xi) < d_U(\xi)$$

For this reason the radius of boundedness was never considered in finite dimensions.

HOLOMORPHIC MAPPINGS OF UNBOUNDED TYPE

15.8. The following theorem is used frequently in constructing an entire function.

THEOREM. *Let E be a normed space and* (u_m) *a sequence in E*. Then*

$$f(x) = \sum_{m=0}^{\infty} [u_m(x)]^m$$

is an entire function on E if and only if $u_m(x) \to 0$ *for every x in E; i.e.,* (u_m) *converges weakly to 0.*

Proof. For each x in E, since the function $z \rightarrow f(zx)$ is an entire function on $\mathbb{C}$ and

$$f(zx) = \sum_{m=0}^{\infty} [u_m(x)]^m z^m$$

we have from the Cauchy-Hadamard formula $\lim \sup |u_m(x)| = 0$. Since

$$0 \leq \lim \inf |u_m(x)| \leq \lim \sup |u_m(x)| = 0$$

we obtain $\lim |u_m(x)| = 0$

Conversely, if $u_m(x) \rightarrow 0$ for each x in E, then $|u_m(x)| \rightarrow 0$. Thus

$$\sum_{m=0}^{\infty} |u_m(x)|^m < \infty$$

for each x in E. This shows that f is a limit of an absolutely convergent series of homogeneous polynomials; hence f is G-holomorphic on E. Since $[u_m]^m \in P(^mE;F)$ for each m, we conclude that f is holomorphic on E by Theorem 14.23. □

15.9 *Unbounded Type.* An entire function on E is said to be a holomorphic mapping of *unbounded type* if it is unbounded on a bounded neighborhood of each point of E.

Of course, if E is finite dimensional, there is no entire function of unbounded type. It turns out that in the case of finite dimension there is always a holomorphic mapping of unbounded type (see 15.11). We first present a simple example of such a mapping.

EXAMPLE. Let $E = \ell^P (p > 1)$. For each m, let $u_m(x) = x_m$ where $x = (x_1, x_2, \ldots)$ in ℓ^P. Then $u_m \in E^*$ and $u_m(x) \rightarrow 0$ as $m \rightarrow \infty$. Hence the mapping

$$f(x) = \sum_{m=0}^{\infty} [x_m]^m$$

is an entire function. Since $\|u_m\| = 1$, we have $r_f(0) = R_f(0) = 1$.
Therefore, for any ξ in E, $r_f(\xi) \leq \|\xi\| + r_f(0) < \infty$.   □

15.10. We now show that for every infinite dimensional Banach space
E, there is an entire function of unbounded type. In proving this
theorem, it is sufficient to know that for any infinite dimensional
Banach space E there is a sequence (u_n) in E* such that $\|u_n\| = 1$
and $u_n(x) \to 0$ for each x in E (see Dineen (1972)). Josefson (1975) and
Nissenzweig (1975) independently affirmed that such a sequence indeed
exists in the dual E* of every infinite dimensional Banach space E.
We state this deep theorem of functional analysis here without
giving a proof (for a new proof of the theorem, see Hagler and
Johnson (1977)). This is a good example demonstrating a close
interaction between functional analysis and infinite dimensional
holomorphy.

THEOREM (Josefson-Nissenzweig, 1975). *If E is an infinite dimensional*
Banach space, then there exists a sequence (u_m) *in E* such that*
$\|u_m\| = 1$ *for all m and* $u_m(x) \to 0$ *for each x on E; i.e., the weak*
sequential convergence in the dual of a Banach space does not imply
the norm convergence.

15.11 THEOREM (Josefson, 1975). *If E is an infinite dimensional*
Banach space, then there exists an entire function f: E → F *with*
$r_f(\xi) < \infty$ *for every point* ξ *in E.*

Proof. It is sufficient to consider F = $\mathbb{C}$. Let (u_m) be as in the
Josefson-Nissenzweig theorem and consider

$$f(x) = \sum_{m=0}^{\infty} [u_m(x)]^m$$

Then f is an entire function of unbounded type on E with $r_f(0) = 1$. □

THEOREM (Aron, 1974). *Let B be the unit ball of a Banach space E.*
Then there exists an entire function on E such that for any r > 0
there corresponds x_r *in B with*

$$\sup \ \{\|f(x)\| : \ \|x - x_r\| < r\} = \infty$$

In particular,

$$\inf \ \{r_f(x) : \ x \in B\} = 0$$

SUBHARMONIC FUNCTIONS

15.13 *Harmonic Functions*. If U is an open subset of $\mathbb{C}$, then a function f: $U \to \mathbb{R}$ is *harmonic* if f has a continuous second partial derivatives and

$$f_{xx} + f_{yy} = 0$$

where f_{xx} denotes the second partial derivative with respect to x. This equation is called *Laplace's equation*.

A continuous function f: $U \to \mathbb{R}$ has the *mean value property* if whenever $\overline{B}_r(a) \subset U$, we have

$$f(a) \ = \ \frac{1}{2\pi} \int_0^{2\pi} f(a + re^{it}) \, dt$$

It is a classical result that a harmonic function f: $U \to \mathbb{R}$ has the mean value property. Conversely, a continuous function defined on a connected open set and which has the mean value property is a harmonic function. It is also well-known that a harmonic function on a simply connected open set U is a real part of a holomorphic mapping on U.

15.14 *Subharmonic Functions*. The theory of subharmonic functions, introduced by F. Riesz in 1922, has become an important adjunct to the theory of complex variables. We present here some properties of subharmonic functions.

Keeping in mind that a function defined on a connected set is harmonic exactly when it has the mean value property, the choice of terminology in the following definition becomes appropriate.

Let U be an open subset of $\mathbb{C}$. A function f is called *subharmonic* on U if the following conditions are satisfied:

(1) $-\infty \leq f(x) < \infty$;

(2)   f is upper semicontinuous; i.e., $\{x \in U: f(x) < a\}$ is open for each real number a:

(3)   Whenever $\overline{B}_r \subset U$, we have

$$f(a) \leq \frac{1}{2\pi} \int_0^{2\pi} f(a+re^{it})\ dt$$

(4)   None of the integrals in (3) are $-\infty$.

Note that the integrals in (3) always exist and are not ∞, since (1) and (2) jointly imply that f is bounded above on every compact subset K of U. [In fact, if $K_n = \{x \in K: f(x) \geq n\}$, then K_n is empty for some n or $\cap \{K_n: n \in \mathbb{N}\}$ is non-empty, in which case $f(x) = \infty$ for some x in K.] The integral in (3) should be considered as the Lebesgue integral since f could be badly discontinuous.

15.15   We first mention some simple properties of subharmonic functions; their proofs are left to the reader as exercises.

THEOREM.   (1)   *If f and g are subharmonic functions, then af + g is also a subharmonic function for any real number a > 0.*

(2)   *If f is subharmonic, then* $f^*(z) = \limsup_{x \to z} f(x)$ *is also a subharmonic function.*

(3)   *If both f and -f are subharmonic, then there exists a holomorphic function g on U such that* Re $g = f$.

15.16   THEOREM.   *If f is subharmonic on U, and if u is a monotonically increasing convex function on $\mathbb{R}$, then u o f is subharmonic.*

Proof. Since u is increasing and continuous, u o f is upper semicontinuous.   If $\overline{B}_r(a) \subset U$, we have

$$u[f(a)] \leq u\left(\frac{1}{2\pi} \int_0^{2\pi} f(a+re^{it})\,dt\right) \leq \frac{1}{2\pi} \int_0^{2\pi} u[f(a+re^{it})]\,dt$$

The first of these inequalities holds since u is increasing and f is subharmonic; the second follows from the convexity of u. Thus u o f is subharmonic. □

COROLLARY. *If f is subharmonic on U then* e^f *is also subharmonic.*

15.17 THEOREM. *If f: U → ℂ is holomorphic and not identically 0, then the following functions are subharmonic:*

$$\log |f| \qquad \log^+|f| = \max \{0, \log |f|\} \qquad |f|^p \ (p > 0)$$

Proof. It is understood that $\log|f(z)| = -\infty$ if $f(z) = 0$. Then $\log |f|$ is clearly upper semicontinuous. If $\log |f(z)| = -\infty$, there is nothing to show. Otherwise, $\log |f(z)|$ is the real part of $\log f(z)$. Hence $\log |f(z)|$ is harmonic; thus $\log |f|$ is subharmonic.

The remaining assertions follow if we apply Theorem 15.16 to $\log |f|$ in place of f, with $u(x) = \max \{0, x\}$ and $u(x) = e^{px}$. □

15.18 THEOREM. *Let* $(f_w)_{w \in I}$ *be a collection of subharmonic functions on U and let*

$$f^*(z) = \lim_{x \to z} \sup \left(\sup_{w \in I} f_w(x)\right)$$

If f* is upper semicontinuous, then f is subharmonic.

Proof. By the monotone convergence theorem, $t \to f^*(z+re^{it})$ is integrable on $[0,2\pi]$ and the following inequalities hold:

$$f^*(z) = \lim_{x \to z} \sup \left(\sup_{w \in I} f_w(x)\right) \leq \frac{1}{2\pi} \lim_{x \to z} \sup \left(\sup_{w \in I} \int_0^{2\pi} f_w(x+re^{it})\,dt\right)$$

$$\leq \frac{1}{2\pi} \lim_{x \to z} \sup \int_0^{2\pi} \sup_{w \in I} f_w(x+re^{it})\,dt \leq \frac{1}{2\pi} \int_0^{2\pi} f^*(z+re^{it})\,dt \qquad \square$$

COROLLARY. If (f_n) is a monotone increasing sequence of subharmonic
functions and if $f(x) = \lim f_n(x)$, then f is subharmonic.

15.19 THEOREM (Maximum Principle). Let f be a subharmonic function
on a connected open subset U of $\mathbb{C}$. If f takes a maximum at a point
in U, then f is a constant function on U.

Proof. The proof is similar to the proof given in 13.9. Suppose
that f takes a maximum at $p \in U$, and denote

$$A = \{z \in U: f(z) = f(p)\}$$

Since U is connected, it suffices to show that A is a closed and
open subset of U; and hence A = U. Since f is upper semicontinuous,
A is obviously closed in U.

Let $a \in A$ and choose r such that $\overline{B}_r(a) \subseteq U$. Suppose that there
is a point b in $B_r(a)$ such that $f(b) \neq f(a)$; then $f(b) < f(a)$. By
upper semicontinuity, $f(z) < f(a)$ for all z in a neighborhood of b.
If $s = |b-a|$, then $b = a+se^{it}$ for some t in $[0,2\pi]$. For any $\varepsilon > 0$
we can find an interval $[u,v]$ in $[0,2\pi]$ on which $f(a+se^{it}) < f(a) - \varepsilon$
for t in $[u,v]$. Then

$$\frac{1}{2\pi}\int_0^{2\pi} f(a+se^{it})\,dt = \frac{1}{2\pi}\left(\int_0^u + \int_u^v + \int_v^{2\pi}\right) f(a+se^{it})\,dt$$

$$\leq \frac{1}{2\pi}f(a)u + \frac{1}{2\pi}[f(a)-\varepsilon](v-u) + \frac{1}{2\pi}f(a)(2\pi-v) = f(a)-\varepsilon\frac{v-u}{2\pi} < f(a)$$

which shows a contradiction, and hence $B_r(a) \subseteq A$.   □

15.20 THEOREM. Let f be a subharmonic function in the open set U
and assume that f is not $-\infty$ identically in any component of U.
Then f is integrable on any compact subset of U.

Proof. Suppose the contrary. Then there exists a point a in U such
that f is not integrable on any closed ball containing a in its
interior. Let $s > 0$ be such that $\overline{B}_{2s}(a) \subseteq U$ and let $D = \overline{B}_s(a)$. Then

$$\iint_D f(z)\,dxdy = -\infty$$

Since f is subharmonic in U, we have

$$\frac{1}{2}\,f(a)s^2 = \int_0^S f(a)\,rdr \le \frac{1}{2\pi}\int_0^S rdr \int_0^{2\pi} f(a+re^{it})\,dt$$

$$= \frac{1}{2\pi}\iint_D f(z)\,dxdy = -\infty$$

Hence $f(a) = -\infty$. If $b \in D$, then $a \in B_s(b)$. This shows that

$$\iint_{B_s(b)} f(z)\,dxdy = -\infty$$

and hence $f(b) = -\infty$. Therefore f takes $-\infty$ on a component of U, which is a contradiction. This completes the proof. □

The following two theorems are quite important but their proofs are rather involved and we refer the reader to Hormander (1966), pp. 21-22 and Rado (1937), p. 15, respectively.

15.21 A $\mathbb{C}^2$-*function f is subharmonic on* U *if and only if*

$$f_{xx} + f_{yy} \ge 0$$

15.22 THEOREM (Rado, 1937). *If f is a function on* U, *then the following are equivalent:*

(1) The mapping $z \to \log f(z)$ *is subharmonic;*

(2) The mapping $z \to |e^{az}|f(z)$ *is subharmonic for any* $a \in \mathbb{C}$.

15.23 We now return to our study of holomorphic mappings.

THEOREM. *If* U *is an open subset of* $\mathbb{C}$ *and* $f \in H(U;F)$ *then the mapping* $z \to \|f(z)\|$ *is subharmonic in* U.

Proof. For any a in U and r > 0 with $\overline{B}_r(a) \subseteq U$, we have

$$\|f(a)\| \le \frac{1}{2\pi}\int_0^{2\pi} \|f(a+re^{it})\|\,dt$$

Hence f is subharmonic.   □

From 15.19 and the preceding theorem, we obtain the weak maximum modulus theorem again.

WEAK MAMIMUM MODULUS THEOREM.   Let U be a connected open subset of $\mathbb{C}$ and $f \in H(U;F)$.   If $\|f(\xi)\| = \sup \{\|f(z)\| : z \in U\}$ for some $\xi \in U$, then the mapping $z \to \|f(z)\|$ is constant.

15.24   THEOREM.   If $f \in H(U;F)$, then the mapping $z \to \log\|f(z)\|$ is subharmonic in U.

Notice that this theorem is not a trivial consequence of Theorem 15.17 where $F = \mathbb{C}$.

Proof.   In this proof we use Theorem 15.22. Since f is holomorphic, for each $a \in \mathbb{C}$, the mapping $z \to e^{az}f(z)$ is holomorphic. By Theorem 15.23, the mapping $z \to |e^{az}|\|f(z)\|$ is subharmonic in U. Therefore, it follows from Theorem 15.22 that $z \to \log\|f(z)\|$ is a subharmonic function on U.   □

PLURISUBHARMONIC FUNCTIONS

15.25   *Plurisubharmonic Functions*.   Let U be an open subset of a normed space E. A function f is called *plurisubharmonic*, in short, *plush* if the following conditions are satisfied:

   (1) $-\infty \leq f(x) < \infty$;

   (2)   f is upper semicontinuous;

   (3)   For each a in U and x in E there corresponds a neighborhood V of 0 in $\mathbb{C}$ such that $z \to f(z+zx)$ is subharmonic in V. This means that the restriction of f to each complex line in U should be subharmonic.

EXAMPLE.   If U is an open subset of E and if $f \in H(U;F)$, then $\log \|f\|$ and $\|f\|^p$ (p > 0) are plush by Theorem 15.24 and Theorem 15.23, respectively.

Plush functions play important roles in the study of convexity of domains; in particular, they are closely related to the characterization of domains of holomorphy or the problem of analytic continuation of holomorphic mappings (see Chapter 20). At present, our aim is to show that $-\log r_f$ is a plush function.

15.26 *Properties of Plush Functions.* The following properties of plush functions are immediate consequences of the corresponding properties for subharmonic functions.

(1) If f and g are plush, then af + g is also plush by any a > 0.

(2) If f is plush, so is $f^*(z) = \lim\sup_{x \to z} f(x)$.

(3) If f is plush and u: $\mathbb{R} \to \mathbb{R}$ is a monotone increasing convex function, then u o f is also plush.

(4) If f_w is plush for each w $\in$ I and if

$$f^*(x) = \lim_{y \to x} \sup \left[\sup_{w \in I} f_w(y) \right]$$

is upper semicontinuous, then f* is plush.

(5) If f is plush on a connected open set U and f takes its maximum in U, then f is a constant function.

15.27 THEOREM. *If both f and -f are plush functions on U, then there exists a holomorphic mapping* g: U $\to$ $\mathbb{C}$ *such that* Re g = f.

Proof (Noverraz, 1970). We assume that U contains the origin O. For any complex line L passing through O, the restriction f_L of f to L $\cap$ U is harmonic since both f_L and $-f_L$ are subharmonic. Hence there corresponds a unique holomorphic mapping g_L on L such that Re g_L = f_L with Im $g_L(O)$ = 0. If L does not pass through the origin, we consider the finite dimensional subspace generated by L and the origin. On this subspace, f and -f are plush; and hence there exists a holomorphic mapping, denoted again by g_L, on this subspace with Re g_L = f_L and Im $g_L(O)$ = 0.

We must show that this gives a well-defined holomorphic function g: U $\to$ $\mathbb{C}$ such that its restriction to L produces g_L for each L. In fact, let L and L' be two arbitrary complex lines and g_L and $g_{L'}$ the functions defined in the preceding paragraph. Consider the finite dimensional subspace A generated by L, L' and the origin O and the unique holomorphic mapping g_1 on $A \cap U$ such that Re $g_1 = f|_{A \cap U}$ and Im $g_1(0) = 0$. Then $g_1|_L = g_L$ and $g_1|_{L'} = g_{L'}$. Hence the function g is a well-defined G-holomorphic mapping.

We now show that g is holomorphic. It is sufficient to prove that g is locally bounded; then g is holomorphic by the Graves-Taylor-Hille-Zorn Theorem. Since Re $g = f$ is continous, it is locally bounded. It follows that e^g is locally bounded G-holomorphic mapping since $|e^g| = e^{\text{Re } g}$. Therefore, e^g is holomorphic, and hence g is locally bounded. This completes the proof.   □

15.28 *Pseudoconvex Domain.* A connected open subset U of E is said to be *pseudoconvex* if the $-\log d_U$ is a plush function on U, where d_U is the boundary distance function defined by $d_U(x) = d(x, E \backslash U)$. This concept will be studied in Chap. 20.

PROPERTIES OF RADIUS OF BOUNDEDNESS

15.29 THEOREM. *Let U be a connected open subset of E. If $f \in H(U;F)$ is of unbounded type, i.e., $r_f(a) < \infty$ for all a in U, then for any a and b in U the following inequality holds:*

$$|r_f(a) - r_f(b)| \le \|a - b\|$$

Proof. Let a, b in U. Since the conclusion is vacuously true if $r_f(a) = r_f(b)$, we assume that $r_f(a) > r_f(b) > 0$. We further assume that $\|a-b\| < r_f(a)$, otherwise the result is trivial.

Let r be a number between $\|a-b\|$ and $r_f(a)$ and let $s = r - \|a-b\|$. Then f is bounded on $\overline{B}_s(b)$. Hence $r_f(b) > r - \|a-b\|$. Letting $r \to r_f(a)$, we obtain

$$r_f(a) > r_f(b) > r_f(a) - \|a - b\|$$

which completes the proof.   □

15.30 THEOREM (Lelong, 1968). *If $f \in H(U;F)$, then the mapping $x \to -\log R_f(x)$ is a plush function, where $R_f(a)$ denotes the radius* *of uniform convergence of the Taylor series of f at a.*

Proof. It is clear from 15.29 that the mapping $x \to \log R_f(x)$ is lower semicontinuous; hence $x \to -\log R_f(x)$ is upper semicontinuous. For any x in U, we have

$$\frac{1}{R_f(x)} = \lim \sup \left\| \frac{1}{m!} \hat{d}^m f(x) \right\|^{1/m}$$

by the Cauchy-Hadamard formula. Since the mapping $\hat{d}^m f: U \to L(^m E; F)$ is holomorphic, $x \to \log \left\| \hat{d}^m f(x) \right\|$ is plush on U. Hence

$$-\log R_f(x) = \lim \sup \frac{1}{m} \log \left\| \frac{1}{m!} \hat{d}^m f(x) \right\|$$

is a plush function. ☐

15.31 THEOREM. *If U is a pseudoconvex domain and* $f \in H(U;F)$, *then the mapping* $x \to -\log r_f(x)$ *is a plush function in U.*

Proof. Recall from 15.7 that r_f satisfies $r_f(a) = \inf \{d_U(a), R_f(a)\}$ From this, we have

$$-\log r_f(a) = \sup \{-\log d_U(a), -\log R_f(a)\}$$

Since U is pseudoconvex, $x \to -\log r_f(x)$ is a plush function. ☐

15.32 REMARK. In Theorem 15.29 we have shown that the function r_f is a Lipschitz function with Lipschitz constant less than or equal to 1. In fact, Kiselman (1977) proved that for a uniform convex Banach space the inequality in 15.29 is always strict.

15.33 *Prescribing the Radius of Boundedness.* Motivated by Theorem 15.29 and Theorem 15.31, Kiselman (1977), and later, Coeure (1977) studied the problem of prescribing the radius of boundedness and produced the following remarkable theorem.

THEOREM. *Let E be an infinite dimensional separable Banach space and let g be a function satisfying the following conditions:*

(1) $0 < g(x) \le \infty$;

(2) $|g(a) - g(b)| \le \|a - b\|$ *for all a and b in E;*

(3) $-\log g$ *is a plush function on E.*

then

(4) *For* $E = \ell^1$, *there exists* $f \in H(E)$ *with* $r_f = g$.

(5) *For E with a Schauder basis there exists* $f \in H(E)$ *such that*

$$g/3 \le r_f \le g$$

Part (4) is due to Kiselman (1977) and Part (5) is due to Coeure (1977). See Schottenloher (1977) for a more accessible proof of the theorem.

EXERCISES

15A. If $E = \ell^\infty$ and

$$f(x) = \sum_{n=1}^{\infty} [x_n]^n$$

then $f \in H(E)$; but the radius of uniform convergence is 1.

15B. Let $E = F = \ell^\infty$, and let $f: E \to F$ be defined by

$$f(x) = \sum_{n=1}^{\infty} (0, \ldots, 0, [x_n]^n, 0, \ldots)$$

where $[x_n]^n$ appears in the nth coordinate in the nth term of the series. Then the radius of uniform convergence is equal to 1, but f is not an entire function. The series converges at x if and only if $\lim \sup |x_n| < 1$.

15C. Let $E = c_o$ and let

$$f(x) = \sum_{n=2}^{\infty} [x_1 x_n]^n$$

Then the radius of uniform convergence of the series is given by

$$R_f(x) = 1/2[(|x_1|^2 + 4)^{1/2} + |x_1|]$$

(see Kiselman, 1977).

15D.   Let $E = \ell^1$.   If $f \in H(E)$ is defined by

$$f(x) = \sum_{n=1}^{\infty} \exp(-nx_1)[x_n]^n$$

we have

$$R_f(x) = \begin{cases} 1 + \operatorname{Re} x_1 & \text{if} \quad \operatorname{Re} x_1 \geq 0 \\ \exp(\operatorname{Re} x_1) & \text{if} \quad \operatorname{Re} x_1 \leq 0 \end{cases}$$

Hence $|R_f(x) - R_f(y)| = \|x-y\|$ when x and y are positive multiples of $e_1 = (1,0,0,\ldots)$ (see Kiselman, 1977).

15E.   If $f \in H(U;F)$ and v is a plush function on a neighborhood of $f(U)$, then $v \circ f$ is a plush function.

PART III

TOPOLOGIES ON SPACES OF HOLOMORPHIC MAPPINGS

16

The Compact-Open Topology on H(U;F)

16.1 We come now to one of the most important topics in infinite dimensional holomorphy. This is the investigation of various topologies on the vector space H(U;F) of all holomorphic mappings of U into F. Looking back on the theory of several complex variables, it is quite natural to consider the compact-open topology τ_o which is also called the topology of uniform convergence on compact sets.

The purpose of this chapter is to present motivation for considering some other locally convex topologies on H(U;F) which are better suited in proving facts about the space H(U;F) when U is an open set in an infinite dimensional normed space. However, we should note that the compact-open topology is adequate for the study of finite dimensional holomorphy.

We first briefly discuss the bare bones of locally convex spaces. We will frequently refer the reader to Dunford and Schwartz (1958), Horváth (1966), Robertson and Robertson (1964), and other sources in the subject for properties of locally convex spaces which might take too much space to develop in this book. Our standard reference for terminology for locally convex spaces will be Horváth (1966).

LOCALLY CONVEX SPACES

16.2 *Seminorms.* Recall the definition of a seminorm from (1.2). A real-valued function p on a vector space E is called a seminorm on E if

(1) $p(x + y) \leq p(x) + p(y)$ for all x, y in E;

(2) $p(\lambda x) = |\lambda| p(x)$ for all x in E and for all scalars λ.

For example, if f is a linear functional on E, then we can define a seminorm p_f on E by setting

$$p_f(x) = |f(x)|$$

for all x in E. Hence, on any vector space we can find a family of seminorms.

It follows from (2) that $p(0) = 0$. Conversely, if $p(x) = 0$ implies x = 0, then p is a norm. For any seminorm p on E, we have

$$|p(x) - p(y)| \leq p(x - y)$$

for all x and y in E.

16.3 *Topology Induced by Seminorms.* Let E be a vector space, and let Γ be a family of seminorms on E. The topology τ_Γ induced on E by the collection Γ is the smallest topology on E making each semi-norm in Γ continuous. It evidently is that topology on E for which the sets

$$V(p;\varepsilon) = \{x: p(x) < \varepsilon\}$$

for $p \in \Gamma$ and $\varepsilon > 0$, form a subbase for the neighborhood system of the origin. Hence, a neighborhood base of the origin consists of sets of the form

$$V(p_1,\ldots p_n;\varepsilon_1,\ldots \varepsilon_n) = \{x: p_i(x) < \varepsilon_i, \ 1 \leq i \leq n\}$$

for any finite subset $p_1,\ldots,p_n$ and positive numbers $\varepsilon_1,\ldots,\varepsilon_n$.

Since E is a vector space, it is sufficient to describe a topology on E by a neighborhood system of the origin since a neighborhood of an arbitrary point ξ is then of the form

$\xi + V$

where V is a neighborhood of the origin.

It is easy to show that an equivalent neighborhood base of the origin is of the form

$$V(p_1,\ldots,p_n;\epsilon) = V(p_1,\ldots,p_n;\epsilon,\ldots,\epsilon)$$

taking $\epsilon = \epsilon_i, i=1,\ldots,n$.

16.4 *Topological Vector Spaces*. A topological vector space E is a topological space such that the vector operations

$$(x,y) \in E \times E \to x + y \in E$$

$$(\lambda,x) \in \mathbb{K} \times E \to \lambda x \in E$$

are continuous where $E \times E$ and $\mathbb{K} \times E$ are endowed with their respective product space topologies.

It is clear that a normed space with its normed topology is a topological vector space. More generally, if Γ is a family of seminorms on a vector space E, then E becomes a topological vector space under the topology τ_Γ induced by Γ.

16.5 *Locally Convex Spaces*. We may ask whether the topology of each topological vector space is defined by a family of seminorms. The answer to this questions is in general negative (see Köther (1969), p. 156). Locally convex spaces, which we introduce, are a characterization of those topological vector spaces for which the answer is affirmative.

A topological vector space E with a topology τ is called a *locally convex space* if every τ-neighborhood of the origin 0 contains a convex τ-neighborhood of 0.

It is easy to see that if τ is defined by a family of seminorms, then E with τ is a locally convex space. Conversely, if τ is the topology of a locally convex space E, then there is at least one

nonempty family Γ of seminorms on E such that τ is defined by Γ; that
is $\tau = \tau_\Gamma$. In fact, if we take Γ to be the set of all continuous
seminorms on E, then $\tau = \tau_\Gamma$; this Γ is the largest family for which
$\tau = \tau_\Gamma$. We summarize this remark in the following theorem.

THEOREM. *A topological vector space is a locally convex space if
and only if the topology is defined by a family of seminorms.*

Since the topology of a locally convex space is generated by a
family of seminorms at sight, it is easy to work with locally convex
spaces. Most important topological vector spaces are locally convex.

16.6 *Weak and Weak* Topologies.* If E is a normed space and E* is
its dual space, then the family of seminorms of the form

$$p_f(x) = \left| f(x) \right| \qquad\qquad (f \in E^*, \; x \in E)$$

defines a locally convex topology on E. This topology is called the
weak topology on E, and denoted by $\sigma(E,E^*)$.

The weak* topology on E* is induced by the family of seminorms of
the form

$$p_x(f) = \left| f(x) \right| \qquad\qquad (x \in E, \; f \in E^*)$$

It is clear that the norm topology on E is finer (stronger) than
the weak topology and the norm topology on E* is stronger than the
weak* topology.

16.7. We use frequently the following form of the Hahn-Banach theorem
(see Robertson and Robertson (1964), p. 30 or Dunford and Schwartz
(1958), p. 418).

THEOREM (Hahn-Banach Separation Theorem). *Let E be a locally convex
space. If K is a convex balanced closed subset of E and a $\notin$ K, then
there exists a continuous linear functional f satisfying*

$$\sup \left\{ \left\| f(x) \right\| : \; x \in K \right\} < f(a)$$

METRIZABILITY OF H(U;F)

16.8 *Compact-Open Topology on* C(U;F). Let U be an open subset of a normed space, and let F be a Banach space. We assume throughout the book $F \neq \{0\}$. We shall write C(U;F) for the vector space of continuous functions of U into F. Then the vector space H(U;F) is a vector subspace of C(U;F). The *compact-open topology* on C(U;F) is the locally convex topology generated by the seminorms of the form

$$p_K(f) = \|f\|_K = \sup \{\|f(x)\| : x \in K\}$$

where K ranges over the compact subsets of U. We denote this topology by τ_o which is separated (or Hausdorff). The compact-open topology is also called the *topology of compact convergence*.

Naturally the compact-open topology restricted on H(U;F) will be called again the compact-open topology τ_o on H(U;F).

In the following theorem, we take $F = \mathbb{C}$ for simplicity.

16.9 THEOREM. *Let E be a Banach space and let U be an open subset of E. Then* (H(U),τ_o) *is metrizable if and only if E is finite dimensional.*

Proof. If E is finite dimensional, we claim (C(U),τ_o) is metrizable. Then H(U) will be metrizable as a subspace of a metrizable space. In fact, there is a sequence (K_n) of compact subsets of U with the following properties:
(1) $U = \bigcup K_n$; (2) $K_n \subset K_{n+1}$, $n \in \mathbb{N}$; (3) any compact subset of U is contained in some K_n. The compact-open topology is then generated by seminorms

$$p_n(f) = \sup \{|f(x)| : x \in K_n\}$$

For each $f \in C(U)$ set

$$d(f) = \sum_{n=1}^{\infty} 2^{-n} \inf \{1, p_n(f)\}$$

The $d(f - g)$ defines the distance between f and g; hence $C(U)$ is a metrizable space.

Conversely, assume that $H(U)$ is metrizable. It is a first countable space; hence we find a countable neighborhood basis (U_n) at the origin for the topology τ_0. For each n, we can choose a compact subset K_n and $r_n > 0$ such that

$$\{f \in H(U) : \|f\|_{K_n} < r_n\} \subset U_n \tag{*}$$

If we let L_n be the closed convex hull of K_n (i.e., the smallest closed convex set containing K_n), then L_n is compact by the Mazur theorem, which states that if K is a compact subset of a Banach space E then the closed convex hull of K is compact (see Dunford and Schwartz (1964), p. 416). Let D be the closed unit disk of the complex field and let $\phi: D \times L_n \to E$ be defined by $\phi(\lambda,x) = \lambda x$. Then ϕ is continuous and the image B_n of the compact set $D \times L_n$ under ϕ is compact, convex and balanced.

Suppose that E is not finite dimensional, then B_n must be nowhere dense by Riesz's theorem (1.15). We now claim that U is the union of all B_n's, which is an obvious contradiction since E is a Baire space, and U as an open subset of E cannot be of the first category.

Let $x \in U$ be arbitrary but fixed for the argument. Consider $V = \{f: |f(x)| < 1\}$. Then V is a neighborhood of the origin for τ_0. Hence $U_n \subset V$ for some n. Hence by (*) if $f \in H(U)$ satisfies $\|f\|_{K_n} < r_n$, we get $|f(x)| < 1$. This shows that

$$|f(x)| \leq (r_n)^{-1}\|f\|_{K_n} \tag{**}$$

for all f in $H(U)$. If we apply this to f^m, take the mth root, and let $m \to \infty$ to obtain

$$|f(x)| \leq \|f\|_{K_n}$$

for all f in $H(U)$.

Since we can consider E* as a vector subspace of H(U), for any A ∈ E*, we have

$$|A(x)| \leq \|A\|_{K_n} \leq \|A\|_{B_n} \qquad (***)$$

But by the Hahn-Banach separation theorem 16.7, we must have x in B_n since B_n is compact, convex and balanced. This proves that U is contained in the union of B_n's. This completes the proof. [The second half of the proof is due to Alexander (1968).]   □

16.10 The preceding theorem is a special case of the following theorem which shows that H(U;F) is never a metrizable space if E is infinite dimensional for any topology we study in this book.

THEOREM. *Let* τ *be a locally convex topology on* H(U;F) *which is finer than the topology of pointwise convergence (i.e., the topology generated by seminorms* $p_{\{x\}}(f) = |f(x)|$ *for all x in U). If* (H(U;F),τ) *is metrizable, then E is finite dimensional.*

Proof. If τ is a metrizable topology for H(U;F), we choose a countable neighborhood basis (U_n) for 0 in the τ-topology. For each x in U define

$$V_x = \{f \in H(U;F) : \|f(x)\| < 1\}$$

Then V_x is a neighborhood of 0; hence $U_m \subset V_x$ for some m. For each n if we set

$$K_n = \{x \in U : U_n \subset V_x\}$$

then K_n is closed in U; in fact, if (x_k) is a sequence in K_n with $x_k \to x$ in U, we claim that x must be in K_n. Notice that if $f \in U_n$, we have $\|f(x_k)\| \leq 1$ for all k; hence $\|f(x)\| \leq 1$. This shows that x is in K_n.

It is clear that U is the union of all K_n's. From the Baire category theorem, it follows that some K_p must contain a nonempty open set W. Thus, if $f \in U_p$, then $\|f\|_W \leq 1$. Since U_p is a neighborhood of the origin, it is absorbing; i.e., for any f in H(U;F) there is r > 0 such that f is in rU_p. Then $\|f\|_W \leq r$. Notice that this is an apparent contradiction to the results of Josefson and Aron (see 15.11 and 15.12) which implies that there exists f in H(E;F) with $\|f\|_W = \infty$. This completes the proof.   □

COMPLETENESS OF H(U;F)

16.11 *Complete Locally Convex Spaces.* As we have seen in the foregoing section, H(U;F) is not metrizable for an infinite dimensional Banach space E. Thus it is not sufficient to describe the completeness using Cauchy sequences; we need Cauchy nets or filters.

A net $(x_\alpha)_{\alpha \in A}$ in a topological vector space E is said to be a *Cauchy net* if for each neighborhood V of 0 there is α in A such that if $\beta, \gamma \geq \alpha$ in the ordering of A, then $x_\beta - x_\lambda \in V$. Every convergent net in E is a Cauchy net.

A topological vector space E is called *complete* if every Cauchy net in the space converges to a point of the space. Evidently each closed subspace of a complete space is complete. If E is a separated (i.e., Hausdorff) topological vector space and G is a complete subspace of E, then G is closed in E.

In order to show $(H(U;F), \tau_o)$ is complete for a Banach space F, we need the following theorem.

16.12 THEOREM. *Let E be a normed space and F a Banach space. Then the space* H(U;F) *is a closed subspace of* C(U;F) *for the compact-open topology.*

Proof. Let f be a cluster point of H(U;F) in C(U;F). We claim that f is holomorphic. By the Graves-Taylor-Hille-Zorn theorem 14.9 it suffices to show that f is G-holomorphic on U; i.e., for any a in U, x in E, and λ in F*, the mapping

$z \rightarrow \lambda[f(a + zx)]$

is holomorphic on the open set

$U(a,x) = \{z \in \mathbb{C}: a + zx \in U\}$

Since the mapping is already continuous, by the Morera theorem it is enough to show that for any simple rectifiable closed curve γ with its interior completely contained in $U(a,x)$, we have

$$\int_\gamma \lambda[f(a + zx)]dz = 0 \qquad\qquad (*)$$

Since γ is compact, the image K of γ by the continuous map $z \rightarrow a + zx$ must be compact in U. Hence for any $\varepsilon > 0$ we can find f_ε in H(U;F) satisfying

$$\|f - f_\varepsilon\|_K < \varepsilon$$

Then for each z in γ we obtain

$$\left|\lambda[f(a + zx) - f_\varepsilon(a + zx)]\right| \leq \|\lambda\|\|f - f_\varepsilon\|_K \leq \varepsilon\|\lambda\|$$

Applying the Cauchy integral theorem to the holomorphic mapping $\lambda \circ f_\varepsilon$, we get

$$\left|\int_\gamma \lambda[f(a + zx)]dz\right| \leq \left|\int_\gamma \lambda[f(a + zx) - f_\varepsilon(a + zx)]dz\right|$$

$$+ \left|\int_\gamma \lambda[f_\varepsilon(a + zx)]dz\right| \leq \int_\gamma \left|\lambda[f(a + zx) - f_\varepsilon(a + zx)]\right|\,|dz|$$

$$\leq \varepsilon\|\lambda\|\int_\gamma |dz|$$

Since $\varepsilon > 0$ is arbitrary, by letting $\varepsilon \to 0$ we obtain the relation (*), which completes the proof.   □

16.13 THEOREM. $(H(U;F), \tau_o)$ *is complete.*

Proof. By Theorem 16.12, $H(U;F)$ is closed in $C(U;F)$ for τ_o, so it suffices to show that $C(U;F)$ is complete for τ_o.

Let (f_α) be a Cauchy net in $(C(U;F), \tau_o)$. Then the net $(f_\alpha(x))$ is clearly a Cauchy net in F for each x in U; hence it converges to, say, $f(x)$. We now claim that $f_\alpha \to f$ in $(C(U;F), \tau_o)$. Let K be a compact subset of U and $\varepsilon > 0$. Then there is for some α such that whenever $\beta, \gamma \geq \alpha$, we have

$$\left\| f_\beta - f_\gamma \right\|_K < \varepsilon$$

Thus we obtain for $\beta \geq \alpha$,

$$\left\| f - f_\beta \right\|_K \leq \varepsilon \qquad\qquad\qquad (*)$$

Thus $f_\beta|_K \to f|_K$ uniformly on K, which shows that $f|_K$ is continuous on K. Since E is metrizable (hence E is a k-space), and f is continuous on every compact subset of U, we conclude that f is continuous on U. The relation (*) shows that $f_\alpha \to f$ in $(C(U;F), \tau_o)$, which completes the proof.   □

BOUNDED SETS IN H(U;F)

16.14 *Boundedness.* A set B in a topological vector space E is said to be *bounded* if for any neighborhood U of the origin there exists a positive number λ such that

$$B \subset \lambda U$$

Boundedness in the sense of the above definition may easily be seen to coincide in a normed space with boundedness in norm.

It is easy to see that if E is a locally convex space and Γ is a family of seminorms on E which generates the topology on E, then a set B is bounded in E if and only if each seminorm in Γ is bounded on B.

16.15 *Three Inequalities.* In the study of holomorphic mappings we often use the following three inequalities which is due to L. Nachbin (1968).

First Inequality. Given any $k \in \mathbb{N}$, $k \leq m$, $P \in P(^mE;F)$, we have

$$\left\| \frac{1}{k!} \hat{d}^k P(x) \right\| \leq 2^m \|P\| \|x\|^{m-k}$$

Proof. If $x = 0$, then the inequality holds trivially. For $x \neq 0$, set $r = \|x\|$. Since P is holomorphic, we can use the Cauchy inequalities to obtain

$$\left\| \frac{1}{k!} \hat{d}^k P(x) \right\| \leq \frac{1}{r^k} \sup \{ \|P(y)\| : \|y - x\| = r \}$$

Since

$$\|P(y)\| \leq \|P\| \|y\|^m \leq \|P\| (\|x\| + \|y - x\|)^m = 2^m \|P\| r^m$$

we conclude that

$$\left\| \frac{1}{k!} \hat{d}^k P(x) \right\| \leq 2^m \|P\| \|x\|^{m-k}$$

Second Inequality. If $B_r(a) \subset U$, $x \in B_r(a)$, and $f \in H(U;F)$, then

$$\left\| \frac{1}{k!} \hat{d}^k f(\bar{x}) \right\| \leq \sum_{m=k}^{\infty} 2^m r^{m-k} \left\| \frac{1}{m!} \hat{d}^m f(a) \right\|$$

Proof. Let $P_m = \frac{1}{m!} \hat{d}^m f(a)$. We have the following

$$\frac{1}{k!}\,\hat{d}^k f(x) = \sum_{m=k}^{\infty} \frac{1}{k!}\,\hat{d}^k P_m(x - a)$$

from Corollary 11.12.   Therefore

$$\left\|\frac{1}{k!}\,\hat{d}^k f(x)\right\| \le \sum_{m=k}^{\infty} 2^m r^{m-k}\left\|\frac{1}{m!}\,\hat{d}^m f(a)\right\|$$

by the first inequality.   □

Third Inequality.   Let $f \in H(U;F)$, $X \subset U$, and $B_r(X) \subset U$ with $r > 0$.
Then

$$\sum_{m=0}^{\infty} \varepsilon^m \sup_{x\in B_r(X)} \left\|\frac{1}{m!}\,\hat{d}^m f(x)\right\| \le \sum_{m=0}^{\infty} [2(r + \varepsilon)]^m \sup_{x\in X} \left\|\frac{1}{m!}\,\hat{d}^m f(x)\right\|$$

Proof.   Let $a \in X$.   For each $x \in B_r(a)$ and $\varepsilon > 0$

$$\sum_{m=0}^{\infty} \varepsilon^m \left\|\frac{1}{m!}\,\hat{d}^m f(x)\right\| \le \sum_{k=0}^{\infty} \varepsilon^k \left[\sum_{m=k}^{\infty} 2^m r^{m-k}\left\|\frac{1}{m!}\,\hat{d}^m f(a)\right\|\right]$$

$$= \sum_{m=0}^{\infty} \sum_{k=0}^{m} \varepsilon^k 2^m r^{m-k}\left\|\frac{1}{m!}\,\hat{d}^m f(a)\right\| \le \sum_{m=0}^{\infty} [2(r + \varepsilon)]^m\left\|\frac{1}{m!}\,\hat{d}^m f(a)\right\|$$

since $(r^m + \varepsilon r^{m-1} + \ldots + \varepsilon^m) \le (r + \varepsilon)^m$.   □

16.16   THEOREM.   *Let* X *be a subset of* $H(U;F)$. *Then the following
are equivalent.*

(1) X is bounded for τ_0;

(2)   If K is a compact subset of U, then

$$\sup \{\|f\|_K : f \in X\} < \infty$$

(3) For any compact subset K of U *there exists an open subset* V
of U *containing K such that*

$$\sup \ \{\|f\|_V : \ f \in X\} < \infty$$

(4) For any point ξ in U *there exist C ≥ 0 and c ≥ 0 such that*

$$\sup \ \{\|\frac{1}{m!} \ \hat{d}^m f(\xi)\| : \ f \in X\} \leq Cc^m$$

for every m ∈ ℕ;

(5) For every point ξ in U *there exist C ≥ 0 and c ≥ 0, and an
open neighborhood* V *of ξ in* U *such that*

$$\sup \ \{\|\frac{1}{m!} \ \hat{d}^m f\|_V : \ f \in X\} \leq Cc^m$$

for every m ∈ ℕ;

(6) For every compact subset K of U *there exist C ≥ 0 and c ≥ 0
such that*

$$\sup \ \{\|\frac{1}{m!} \ \hat{d}^m f\|_K : \ f \in X\} \leq Cc^m$$

for every m ∈ ℕ ;

(7) For every compact subset K of U *there exist C ≥ 0 and c ≥ 0,
and an open subset* V *of* U *containing K such that*

$$\sup \ \{\|\frac{1}{m!} \ \hat{d}^m f\|_V : \ f \in X\} \leq Cc^m$$

for every m ∈ ℕ.

Proof. The equivalency between (1) and (2) is clear.
(2) ⇒ (3). There is no loss of generality to assume that K is a

singleton $\{\xi\}$. We claim that there exists $r > 0$ such that $B_r(\xi) \subset U$ and X is uniformly bounded on $B_r(\xi)$.

Suppose the contrary. Then we can find a sequence (x_n) in U with $x_n \to \xi$ and another sequence (f_n) in X such that $\|f_m(x_m)\| > m$. Since the sequence (x_n) and its limit ξ form a compact set, say L, we conclude that X is not bounded on the compact set L, which contradicts the statement (2).

(3) $\Rightarrow$ (4). Let $\xi \in U$ and $r > 0$ be such that $B_r(\xi) \subset U$ and X is bounded on this ball. It follows from the Cauchy inequalities that

$$\left\| \frac{1}{m!} \hat{d}^m f(\xi) \right\| \leq \frac{1}{r^m} \sup \ \{\|f(x)\| : x \in B_r(\xi)\}$$

for every $m \in \mathbb{N}$ and $f \in X$. Set $c = r^{-1}$ and $C = \sup \ \{\|f(x)\| : x \in B_r(\xi)\}$ to obtain the statement (4).

(4) $\Rightarrow$ (5). Let $\xi \in U$. By (4), we can find $C \geq 0$ and $c \geq 0$ such that

$$\left\| \frac{1}{m!} \hat{d}^m f(\xi) \right\| \leq Cc^m$$

for every $m \in \mathbb{N}$ and $f \in X$. Choose $r > 0$ to satisfy $\overline{B}_{2r}(\xi) \subset U$ and $2cr < 1$. Let

$$f(x) = \sum_{k=0}^{\infty} P_k(x - \xi)$$

be the Taylor series of f at ξ. Using Corollary 11.12, we have

$$\frac{1}{m!} \hat{d}^m f(x) = \sum_{k=m}^{\infty} \frac{1}{m!} \hat{d}^m P_k(x - \xi)$$

where

$$P_k = \frac{1}{k!} \hat{d}^k f(\xi)$$

This series converges for x in $B_r(\xi)$ to the indicated sum in the sense of the Banach space $P(^mE;F)$. Since $P_k \in P(^kE;F)$, by the First Inequality of 16.15 we obtain

$$\left\|\frac{1}{m!}\hat{d}^mP_k(x - \xi)\right\| \leq 2^k\|P_k\|\|x-\xi\|^{k-m} \leq C(2cr)^kr^{-m}$$

Hence

$$\left\|\frac{1}{m!}\hat{d}^mf(x)\right\| \leq Cr^{-m}\frac{(2cr)^m}{1 - 2cr} = \left(\frac{C}{1 - 2cr}\right)(2c)^m$$

for all x in $B_r(\xi)$.

(5) $\Rightarrow$ (6). The statement (6) follows immediately from (5) since K can be covered by a finite number of open balls satisfying the inequality in (5).

(6) $\Rightarrow$ (7). For (7), let V be the union of open balls mentioned in the proof (5) $\Rightarrow$ (6). Then the results follows.

(7) $\Rightarrow$ (1). This is obviously true.   □

BARRELLEDNESS OF H(U;F)

16.17. One of the most important problems in the theory of holo-morphic mappings is the characterization of domains and envelopes of holomorphy (see Chap. 19). In the classical several complex variables, $(H(U),\tau_o)$ is a complete metrizable locally convex space; i.e., a Fréchet space. On Fréchet spaces the open mapping theorem and the uniform boundedness principle are available (see Horváth (1966)). In the case of infinite dimension, $(H(U;F),\tau_o)$ is not a Fréchet space. Since there is no Fréchet space structure in H(U;F) in the infinite dimensional case, classical arguments of several complex variables cannot be readily applicable in infinite dimensions. We study a property which generalizes the Fréchet space structure, the barrelledness of H(U;F).

16.18 *Barrelled Spaces*. Let E be a locally convex space. An absorbing, balanced, convex, and closed subset of E is said to be a *barrel* (*tonneau* in French)[Recall that a subset A is absorbing if for any $x \in E$ there is $r > 0$ such that $x \in rA$]. A locally convex space E is said to be *barrelled* (tonnelé in French) if every barrel in E is a neighborhood of 0. A locally convex space E which is a Baire space is barrelled; in particular, every Fréchet space [a complete metrizable locally convex space] is barrelled (see Horváth (1966), pp. 211-224, for other properties of a barrelled space). We will see that $(H(U;F), \tau_o)$ is not barrelled if E is infinite dimensional. We need the following lemma.

16.19 LEMMA. *If τ is the compact-open topology on $P(^mE;F)$ then the compact-open topology τ_o on* H(U;F) *induces on $P(^mE;F)$ the topology τ.*

Proof. Since H(U;F) and H(U − ξ;F) are topologically isomorphic, we may assume $0 \in U$. Since E has more compact subsets than U, it is clear that τ is finer than τ_o. Conversely, if K is a compact subset of E, then we can find $r > 0$ such that $L = rK \subset U$. Then for $P \in P(^mE;F)$ and $x \in K$, we have

$$\left\| P(x) \right\| = r^{-m} \left\| P(rx) \right\| \leq r^{-m} \left\| P \right\|_L$$

i.e., $\left\| P \right\|_K \leq r^{-m} \left\| P \right\|_L$; this implies that τ_o is finer than τ. Hence $\tau = \tau_o$.   □

16.20 THEOREM. *The vector space $P(^mE;F)$ is a closed complementary subspace of $(H(U;F), \tau_o)$ for each* m.

Proof. It is clear that we can consider $P(^mE;F)$ as a subspace of H(U;F) since the restriction of any continuous m-homogeneous polynomial to the open set U is holomorphic. We assume that $0 \in U$ as in the proof of (16.19).

    To show that $P(^mE;F)$ is complementary we must find a continuous linear map

T: H(U;F) → H(U;F)

such that T^2 = T (i.e., T is a projection) and T maps H(U;F) onto P(^mE;F). In fact, consider a mapping T defined by

$$T(f) = \frac{1}{m!} \hat{d}^m f(0)$$

Then T is clearly a linear map satisyfing T^2 = T since T(P) = P for any P ∈ P(^mE;F). Hence T is a projection onto the subspace P(^mE;F). We now prove that T is continuous. Let K be a compact subset of U. Then there is r > 0 such that

$$L = \{zx: z \in \mathbb{C}, |z| \leq r, x \in K\} \subset U$$

since 0 ∈ U. Since the map (z,x) ∈ $\mathbb{C}$ × K → zx ∈ E is continuous, L is compact. If f ∈ H(U;F), then for any x ∈ K,

$$\|T(f)(x)\| \leq r^{-m}\|f\|_L$$

that is, $\|T(f)\|_K \leq r^{-m}\|f\|_L$, which shows that T is continuous with respect to the compact-open topology.

It remains to show that P(^mE;F) is closed in H(U;F). Notice that

$$P(^mE;F) \to H(U;F) \overset{T}{\to} P(^mE;F) \tag{*}$$

Since the inclusion P(^mE;F) → H(U;F) is continuous, the composite mapping in (*) is continuous; furthermore, it is the identity mapping on P(^mE;F). Hence P(^mE;F) is closed. This completes the proof. □

16.21 THEOREM. $(H(U),\tau_o)$ *is barrelled if and only if* E *is finite dimensional.*

Proof. If $(H(U;F),\tau_o)$ is barrelled, then every closed complementary subspace is also barrelled (see Horváth (1966), pp. 214-215). (Note

that a closed subspace of a barrelled space may not be barrelled; but
the complementary closed subspaces are.) Since E* is a closed complementary subspace of H(U), E* must be barrelled with respect to the
compact open topology.

We claim in Lemma 16.22 that if (E^*, τ_o) is barrelled, then E is
finite dimensional. This will complete the proof of the theorem. □

This theorem motivates us to find a new topology on H(U) which is
barrelled for the infinite dimensional case as well as the finite
dimensional case.

16.22 LEMMA. *If* (E^*, τ_o) *is barrelled, then* E *is finite dimensional.*

Proof (Alexander, 1968). Let B and D be the closed unit balls of E
and E*, respectively. We claim that D is a barrel for τ_o; i.e., D
is absorbing, convex, balanced and closed for τ_o. The first three
properties are clear; it remains to show that D is closed for τ_o.
Since D is weak* closed and the weak* topology on E* is weaker
(coarser) than the compact-open topology on E*, D is closed for τ_o.
This shows that D is a barrel for the topology τ_o; hence D is a
neighborhood of 0 in E* for the topology τ_o. It follows that there
exist a compact set K in E and r > 0 such that

$$V = \{f \in E^*: |f|_K < r\} \subset D$$

Now take the polar operation ^{o} to obtain

$$B = D^o \subset V^o$$

where $X^o = \{x \in E; |f(x)| \le 1, f \in X\}$.

If $\xi \in V^o$, then we first show that

$$|f(\xi)| \le r^{-1}|f|_K$$

for all $f \in E^*$. In fact, let $g = rf/(|f|_K + s)$ where s > 0. Then
$|g|_K < r$; hence $g \in V$ and $|g(\xi)| \le 1$. This implies that

$\left| f(\xi) \right| \leq r^{-1}(\left| f \right|_K + s)$ for all $s > 0$.

If L is the closed convex balanced hull of K, then L is compact and

$$\left| f(r\xi) \right| \leq \left| f \right|_K \leq \left| f \right|_L$$

for all $f \in E^*$. Since L is closed, convex and balanced, it follows from the Hahn-Banach separation theorem that $r\xi \in L$; hence $\xi \in r^{-1}L$. This shows that

$$B \subset V^o \subset r^{-1}L$$

Since $r^{-1}L$ is compact, B is compact. By the Riesz theorem 1.15, E is finite dimensional. □

16.23 *Nachbin-Shirota Theorem.* In 1954, L. Nachbin and T. Shirota independently proved that if X is completely regular, then $(C(X),\tau_o)$ is barrelled if and only if for any closed noncompact subset K of X there is a function $f \in C(X)$ with $\left| f \right|_K = \infty$. Hence for any open subset U of a normed space E, $(C(U),\tau_o)$ is barrelled. We have thus produced a closed subspace of a barrelled space which is not barrelled. Our example H(U) is much more natural than those examples given in the literature (see Köther (1966), p. 434).

16.24 *Differential Operators* $\hat{d}^m$. We close this chapter with the following fact which shows the inadequacy of the compact-open topology in infinite dimensional cases.

THEOREM. *Let* $\hat{d}^m$: $H(U;F) \rightarrow H(U;P(^mE;F))$ *be the differential operator defined by* $\hat{d}^m(f) = \hat{d}^m f$. *If* $\hat{d}^m$ *is continuous for the compact-open topologies on both spaces for some* $m \geq 1$, *then E is finite dimensional.*

Proof. Without loss of generality, we may take $m = 1$ and $U = E$, $F = \mathbb{C}$. Then the composite map

$$f \in H(E) \rightarrow \hat{d}f \in H(E;E^*) \rightarrow \left\| \hat{d}f(0) \right\| \in \mathbb{R}^+$$

is a continuous seminorm on $H(E)$ for τ_o. Therefore, we can find a number $C > 0$ and a convex balanced compact subset K of E such that

$$\left\| \hat{d}f(0) \right\| \leq C \left\| f \right\|_K$$

for all f in $H(E)$. In particular, if $\phi \in E^*$, then

$$\left\| \phi \right\| \leq C \left\| \phi \right\|_K$$

As in (***) in the proof of (16.9), by the Hahn-Banach separation theorem, $B_1(0) \subseteq K$. This shows that K has an interior point; hence E must be finite dimensional by the Riesz theorem 1.15.   □

EXERCISES

16A. The compact-open topology on $P(^mE;F)$ is coarser (i.e., weaker) than the norm topology.

16B. The differential operator

$$\hat{d}: H(U) \rightarrow H(U;E^*)$$

defined by $\hat{d}(f) = \hat{d}(f)$ is not continuous for τ_o topologies if E is infinite dimensional.

16C *Topology of Pointwise Convergence.* Let τ_p be the topology of pointwise convergence on $C(X)$, the space of continuous function of a topological space into $\mathbb{C}$.

(1) If K is a compact topological space and if $C(K)$ has a count-able sequence (f_n) which separates points of K, then K is metrizable; hence separable.

(2) If S is a separable topological space, then every compact subset of $(C(S), \tau_p)$ is metrizable.

(3) If U is an open subset of a separable normed space E then every compact subset of $(H(U),\tau_p)$ is metrizable.

16D *Compact Holomorphic Maps* (Aron and Schottenloher, 1976). Let E and F be Banach spaces. A mapping $f: E \to F$ is called *compact* at a point ξ if there exists a neighborhood V_ξ of ξ such that $f(V_\xi)$ is relatively compact in F; i.e., the closure of $f(V_\xi)$ in F is compact. Then the following are equivalent for $f \in H(E;F)$:

(a) f is compact at each point of E;

(b) f is compact at some point of E;

(c) There exists a compact subset K of F such that f(E) is contained in the vector space spanned by K

(d) The transpose $f^*: (F^*,\tau_o) \to (H(E),\tau_o)$ defined by $f^*(u) = u \circ f$ is continuous.

(e) For each m, $\hat{d}^m f(0)$ is compact.

17

The Nachbin Topology on H(U;F)

17.1 As we have seen in the preceding chapter, if E is infinite
dimensional and U is an open subset of E, the compact-open topology
τ_o on H(U;F) does not fully extend those properties which are normally
valid when E is considered to be finite dimensional. Hence we are
interested in finding a natural topology on H(U;F) which generalizes
those essential topological properties of H(U;F) which were valid
for several complex variables with respect to τ_o. This natural
topology τ, if there is one, must satisfy at least the following
conditions:

   (1) $\tau \geq \tau_o$; $\tau = \tau_o$ if E is finite dimensional.

   (2) $\hat{d}^m$: H(U;F) $\to$ H(U;P(^{m}E;F)) is continuous for τ.

   (3) τ-bounded subsets of H(U;F) are τ_o-bounded sets.

   (4)   (H(U;F),τ) is a barrelled space.

The Nachbin topology τ_ω on H(U;F) which we study in this chapter
satisfies the first three conditions easily and the fourth condition
if E has an unconditional basis (see 18.16); in general, it is
unknown that τ_ω is a barrelled topology.

   In this chapter we also study various other topologies which lie
between topologies τ_o and τ_ω.

TOPOLOGY τ_∞

17.2 *Topology τ_m.* As a bridge from the topology τ_o to τ_ω, we first
introduce the topology τ_m which makes the differential operator

258

$\hat{d}^m$: $H(U;F) \to H(U;P(^mE;F))$ continuous for $m \in \mathbb{N}$. The locally convex topology τ_m on $H(U;F)$ is defined by seminorms of the form:

$$P_{K,n}(f) = \left\|\hat{d}^n f\right\|_K = \sup \left\{\left\|\hat{d}^n f(x)\right\| : x \in K\right\}$$

where K is a compact subset of U and $n \in \{0,1,\ldots,m\}$. It is clear that each τ_m is a Hausdorff (or separated) topology on $H(U;F)$.

The *topology* τ_∞ is the locally convex topology defined by all seminorms $P_{K,m}$ where K ranges over all compact subsets of U and $m \in \mathbb{N}$. It immediately follows that

$$\tau_0 \leq \tau_1 \leq \ldots \leq \tau_m \leq \ldots \leq \tau_\infty$$

We also have that the differential operator

$$\hat{d}^m: H(U;F) \to H(U;P(^mE;F))$$

is continuous for the topology τ_∞ on both spaces. This follows from the following inequality which can be obtained by combining Corollary 2, 11.12 and the first inequality of 16.15:

$$\left\|\frac{1}{k!}\hat{d}^k\left[\frac{1}{m!}\hat{d}^m f\right](x)\right\| = \left\|\frac{1}{m!}\hat{d}^m\left[\frac{1}{(k+m)!}\hat{d}^{k+m}f(x)\right]\right\| \leq 2^{k+m}\left\|\frac{1}{(k+m)!}\hat{d}^{k+m}f(x)\right\|$$

17.3 THEOREM. *If E is infinite dimensional, then*

$$\tau_0 < \tau_1 < \cdots < \tau_m < \tau_{m+1} < \cdots < \tau_\infty$$

Proof. We may assume that $0 \in U$. Let $m \in \mathbb{N}$ and

$$P_m(f) = \left\|\hat{d}^m f(0)\right\|$$

Then it is clearly τ_m-continuous on $H(U;F)$. We now show that P_m is

not τ_n-continuous if $n < m$. This fact was proved for $m = 1$ in
Theorem 16.24. The general case can be proved similarly.   □

17.4 THEOREM. $\tau_o = \tau_\infty$ *if and only if E is finite dimensional.*

Proof. If $\tau_o = \tau_\infty$, then $\tau_o = \tau_1$. Hence E is finite dimensional by
Theorem 16.24.

  Conversely, suppose that E is finite dimensional and assume that
$U = E$ for simplicity. Then we can find a sequence (K_n) of compact
sets in E with

$$K_1 \subset \mathrm{Int}K_2 \subset K_2 \subset \mathrm{Int}K_3 \subset \ldots$$

such that any compact subset K of E is contained in some K_n. It
follows from the Cauchy inequalities that

$$p_{K,m}(f) \leq \|\hat{d}^m f\|_{K_n} \leq c(m,n)\|f\|_{K_{n+1}}$$

for some $c(m,n) > 0$. Hence each continuous seminorm for τ_∞ is continuous for τ_o; this shows that $\tau_o = \tau_\infty$.   □

17.5 THEOREM. $(H(U;F),\tau_\infty)$ *is complete.*

Proof. We have shown that τ_o is complete in (16.13). Let $(f_\alpha)_{\alpha \in A}$
be a Cauchy net in $(H(U;F),\tau_\infty)$. Since (f_α) is then a Cauchy net
for τ_o, there is some $f \in H(U;F)$ such that $f_\alpha \to f$ for τ_o. It is
sufficient to show that for each compact set K in U and for each m

$$\|\hat{d}^m f - \hat{d}^m f_\alpha\|_K \to 0$$

Let $\xi \in K$. The net $(\hat{d}^m f_\alpha(\xi))_{\alpha \in A}$ is a Cauchy net in $P(^mE;F)$; hence
it converges to an element in $P^m(E;F)$, say P_m. On the other hand,
for each $x \in E$

$$\frac{1}{2\pi i} \int_{|z|=r} \frac{f(\xi+zx) - f_\alpha(\xi+zx)}{z^{m+1}} \, dz \;\to\; 0$$

if we choose $r > 0$ small enough. By the Cauchy integral formulas
for differentials (13.4),

$$\frac{1}{m!} \hat{d}^m f_\alpha(\xi)(x) \;\to\; \frac{1}{2\pi i} \int_{|z|=r} \frac{f(\xi+zx)}{z^{m+1}} \, dz$$

Hence $P_m = \hat{d}^m f(\xi)$. This completes the proof.   □

COROLLARY. $(H(U;F), \tau_m)$ *is complete for* $m \in \mathbb{N}$.

Proof. The proof of the corollary is a part of the preceding
proof.   □

THE NACHBIN TOPOLOGY τ_ω

17.6 *Compact-Ported Seminorms*. L. Nachbin (1967) noticed that the
compact-open topology on $H(U;F)$ is not the largest natural locally
convex topology (unlike what happens when E is finite dimensional
and thus locally compact); he introduced the topology τ_ω which is
the main topic of this chapter. The *Nachbin topology* τ_ω on $H(U;F)$
is defined by all seminorms ported by compact subsets of U as defined
below. His definition was motivated by certain properties of analytic functionals of several complex variables studied by the late
A. Martineau (1963, 1966).

A seminorm p on $H(U;F)$ is said to be *ported by a compact subset*
K of U if, given any open subset V of U containing K, there exists
a real number $c(V) > 0$ such that

$$p(f) \leq c(V) \|f\|_V$$

holds for every $f \in H(U;F)$.

The *Nachbin topology* τ_ω on H(U;F) defined by all compact-ported seminorms on H(U;F) is clearly finer than τ_∞ by the Cauchy inequalities, i.e.,

$$\tau_o \leq \tau_m \leq \tau_\infty \leq \tau_\omega$$

REMARK. It is to be noted that f is not necessarily bounded on any V; however, once $f \in$ H(U;F) and the compact subset K of U are given, we can always find an open subset V of U containing K on which f is bounded since every holomorphic mapping is locally bounded (see 14.7). In other words, the above estimate imposes a restriction on p for every f, although its right-hand side may occasionally be infinite.

We can see clearly that the Nachbin topology is really the topology of *"local uniform convergence,"* in the sense that if (f_n) converges f in τ_ω then (f_n) converges locally uniformly.

17.7 *Ported vs. Supported.* The term "compact-ported" should not be confused with "compact-supported". In fact, if p is a compact-ported seminorm on H(U;F), it is not always true that among the compact subsets of U porting p there exists a smallest one. As a matter of fact, if p is ported by K_1 and K_2, it does not follow necessarily that p is ported by their intersection $K_1 \cap K_2$. For example, let p be a seminorm on H($\mathbb{C}$) defined by

$$p(f) = |f'(0)|$$

Then p is ported by any circle $|z|=r$. This is the reason why we use the term "ported" rather than "supported" which is reserved for the case in which there is a smallest support.

17.8 The following characterization of compact-ported seminorms is very useful and we shall use this frequently in the future.

THEOREM. *For a seminorm* p *on* H(U;F), *the following are equivalent:*

(a) p *is ported by a compact subset K of U;*

(b) *Corresponding to every* $\varepsilon > 0$ *there is* $c(\varepsilon) > 0$ *such that,*
for every $f \in H(U;F)$

$$p(f) \leq c(\varepsilon) \sum_{m=0}^{\infty} \varepsilon^m \sup_{x \in K} \left\| \frac{1}{m!} \hat{d}^m f(x) \right\|$$

(c) *Corresponding to every* $\varepsilon > 0$ *and every open subset V of U*
containing K there is $c(\varepsilon,K) > 0$ *such that, for every* $f \in H(U;F)$

$$p(f) \leq c(\varepsilon,K) \sum_{m=0}^{\infty} \varepsilon^m \sup_{x \in V} \left\| \frac{1}{m!} \hat{d}^m f(x) \right\|$$

Proof. We will establish the theorem in the following order: (b)
$\Rightarrow$ (a) $\Rightarrow$ (c) $\Rightarrow$ (b).

(b) $\Rightarrow$ (a). Let V be an open subset of U containing K and let $r > 0$
be such that $\overline{B}_r(\xi) \subset V$ for every $\xi \in K$. It then follows from the
Cauchy inequalities that

$$\sup_{x \in K} \left\| \frac{1}{m!} \hat{d}^m f(x) \right\| \leq r^{-m} \|f\|_V \tag{*}$$

Choose $\varepsilon > 0$ to satisfy $\varepsilon < r$, and let $c(V) = \dfrac{rc(\varepsilon)}{r-\varepsilon}$. By (b) and
the estimate (*) above, we have

$$p(f) \leq c(V) \|f\|_V$$

(a) $\Rightarrow$ (c). This implication is clear if we set $c(\varepsilon,V) = c(V)$.

(c) $\Rightarrow$ (b). For a given $\varepsilon > 0$ and a compact subset K of U, find
$r > 0$ and $\delta > 0$ such that $B_r(\xi) \subset U$ for all $\xi \in K$ and $2(r+\delta) < \varepsilon$.
Set $V = B_r(K)$ and apply the third inequality in 16.15 and the prop-
erty (c) to V and δ to obtain (b). $\square$

REMARK. In each estimate appearing in the above theorem, it is not

required that the right-hand side be finite. It is possible that
the right-hand side could be infinite. However, once p, K, and f
are given, we can find an open subset V of U containing K and ε > 0
such that

$$\sum_{m=0}^{\infty} \varepsilon^m \sup_{x \in V} \left\| \frac{1}{m!} \, \hat{d}^m f(x) \right\| < \infty \tag{**}$$

In fact, if f ∈ H(U;F) and K is a compact subset of U we can find
C ≥ 0, c ≥ 0 and an open subset V of U containing K such that

$$\sup_{x \in V} \left\| \frac{1}{m!} \, \hat{d}^m f(x) \right\| \leq Cc^m$$

for all m ∈ ℕ (if necessary, see either Corollary 2, 11.5 or 16.16).
Then choose ε > 0 such that cε < 1. Then the series (**) becomes
finite.

17.9 If U is ξ-balanced, then we have the following characterization
for compact-ported seminorms.

THEOREM. *If* U *is* ξ-*balanced and* K *is a* ξ-*balanced compact subset
of* U, *then a seminorm* p *on* H(U;F) *is ported by* K *if and only if
corresponding to every open subset* V *of* U *containing* K *there is a
real number* c(V) > 0 *such that for every* f *in* H(U;F) *we have*

$$p(f) \leq c(V) \sum_{m=0}^{\infty} \sup_{x \in V} \left\| \frac{1}{m!} \, \hat{d}^m f(\xi)(x-\xi) \right\|$$

Proof. The Taylor series of f at ξ converges to f uniformly on some
neighborhood of each compact subset of U (see Theorem 15.3). Therefore, the estimate in the theorem is true if p is ported by a compact
subset K regardless of whether K is ξ-balanced or not.

     Conversely, corresponding to every open subset V of U containing
K, let W be the balanced hull and r > 1 such that K ⊂ W ⊂ V and if

$z \in \mathbb{C}$, $|z| \leq r$, and $x \in W$, then $(1-z)\xi + zx \in V$ (see Lemma 15.3). It follows from the Cauchy integral formula 13.4 that

$$\frac{1}{m!} \hat{d}^m f(\xi)(x - \xi) = \frac{1}{2\pi i} \int_{|z|=r} \frac{f[\xi + z(x - \xi)]}{z^{m+1}} \, dz$$

Hence

$$\left\| \frac{1}{m!} \hat{d}^m f(\xi)(x - \xi) \right\| \leq r^{-m} \|f\|_V$$

for all $x \in W$. Let $c(V) = \dfrac{c(W)r}{r-1}$. Then we obtain

$$p(f) \leq c(V) \|f\|_V \qquad\qquad\qquad\qquad \Box$$

17.10 *Topology* τ_σ. So far the compact-ported seminorms are described in terms of a certain estimate; it is necessary to know whether there is an explicit formula describing τ_ω-continuous seminorms.

Corresponding to every compact subset K of U and a sequence $\alpha = (\alpha_m)$ of nonnegative real numbers such that $(\alpha_m)^{1/m} \to 0$ as $m \to \infty$, we define a seminorm $P_{\alpha,K}$ on $H(U;F)$ by

$$P_{\alpha,K}(f) = \sum_{m=0}^{\infty} \alpha_m \sup_{x \in K} \left\| \frac{1}{m!} \hat{d}^m f(x) \right\|$$

The topology τ_σ on $H(U;F)$ is defined by all seminorms of the form $P_{\alpha,K}$.

It is immediate that these seminorms are ported by K (see 17.8(b)), and

$$\tau_\infty \leq \tau_\sigma \leq \tau_\omega$$

It is not known yet if $\tau_\sigma = \tau_\omega$.

17.11   THEOREM. $\tau_\infty = \tau_\sigma$ *if and only if* dim E $< \infty$.

Proof.   If E is finite dimensional, then for every seminorm p ported by a compact subset K, we can find an open subset V of U containing K such that the closure $\overline{V}$ is compact and

$$p(f) \le c(V) \|f\|_{\overline{V}}$$

Thus p is τ_o-continuous, which shows that $\tau_0 = \tau_\omega$.

Conversely, suppose that $\tau_\infty = \tau_\sigma$. For simplicity we take U = E and F = $\mathbb{C}$. Let $\alpha = (\alpha_m)$ be a sequence of nonnegative numbers such that $(\alpha_m)^{1/m} \to 0$ as $m \to \infty$. Then the set

$$A = \{f \in H(E) : \sum_{m=0}^{\infty} \alpha_m \left\| \frac{1}{m!} \hat{d}^m f(0) \right\| < 1\}$$

is a τ_∞-neighborhood of 0. Therefore, there exist a compact subset K of E, r > 0, and m $\in \mathbb{N}$ such that B $\subset$ A, where

$$B = \{f \in H(E) : \sup_{x \in K} \left\| \frac{1}{k!} \hat{d}^k f(x) \right\| \le r, \ 0 \le k \le m\}$$

If we set

$$C = \{f \in H(E) : \left\| \frac{1}{(m+1)!} \hat{d}^{m+1} f(0) \right\| < \frac{1}{\alpha_{m+1}}\}$$

then B $\subset$ C, and hence C becomes a τ_m-neighborhood of 0; but this is absurd since the mapping defined by

$$q(f) = \left\| \hat{d}^{m+1} f(0) \right\|$$

is not continuous for the topology τ_m if dim E = ∞ (see Proof of 17.3).

Thus $\tau_\infty < \tau_\sigma$.   □

COROLLARY. $\tau_\infty = \tau_\omega$ *if and only if* dim E $< \infty$.

REMARK. In several complex variables, for E $= \mathbb{C}^n$, the fact that
$\tau_0 = \tau_\infty$ is relevant; but the topologies τ_ω and τ_σ, and the fact
that

$$\tau_0 = \tau_\infty = \tau_\sigma = \tau_\omega$$

are not mentioned since there seems to be no reason to introduce
τ_σ and τ_ω in finite dimensional spaces.

17.12 *Differential Operator* $\hat{d}^m$. We now show that the differential
operator

$$\hat{d}^m: H(U;F) \to H(U;P(^mE;F))$$

is continuous for the Nachbin topology τ_ω for all m.

THEOREM. *Let* τ *be a locally convex topology with* $\tau_\infty \le \tau \le \tau_\omega$. *Then
the differential operator*

$$f \in (H(U;F),\tau) \to \frac{1}{m!}\hat{d}^m f \in (H(U;P(^mE;F)),\tau)$$

is continuous.

Proof. It is enough to show for $\tau = \tau_\omega$ since the general case can
be proved similarly. Let p be ported by a compact subset K of U and
let $\varepsilon > 0$. Then there exists $c(\varepsilon) > 0$ such that for $g \in H(U;P(^mE;F))$
we have

$$p(g) \le c(\varepsilon) \sum_{k=0}^{\infty} \varepsilon^k \sup_{x \in K} \left\| \frac{1}{k!}\hat{d}^k g(x) \right\|$$

If $f \in H(U;F)$, then we have

$$\left\| \frac{1}{k!} \hat{d}^k \left[\frac{1}{m!} \hat{d}^m f \right] (\xi) \right\| = \left\| \frac{1}{m!} \hat{d}^m \left[\frac{1}{(k+m)!} \hat{d}^{k+m} f(\xi) \right] \right\| = \left\| \frac{1}{(k+m)!} \hat{d}^{k+m} f(\xi) \right\|$$

and so

$$P(\frac{1}{m!}\hat{d}^m f) \leq c(\varepsilon) \sum_{k=0}^{\infty} \varepsilon^k \sup_{x \in K} \left\| \frac{1}{(k+m)!}\hat{d}^{k+m} f(x) \right\|$$

$$\leq \frac{c(\varepsilon)}{\varepsilon^m} \sum_{k=0}^{\infty} \varepsilon^k \sup_{x \in K} \left\| \frac{1}{k!}\hat{d}^{k+m} f(x) \right\|$$

proving the continuity.   □

COROLLARY. *If* $\tau_\infty \leq \tau \leq \tau_\omega$, *then for each* $\xi \in U$ *the mapping*

$$f \in (H(U;F),\tau) \rightarrow \frac{1}{m!} \hat{d}^m f(\xi) \in P(^m E;F)$$

is continuous.

BOUNDED SUBSETS

17.13 τ_ω-*Bounded Subsets of* $H(U;F)$. The characterization of τ_0-bounded subsets of $H(U;F)$ in 16.16 shows that both topologies τ_0 and τ_ω share the same stock of bounded subsets. The reader can recognize that the definition of the Nachbin topology is also motivated by 16.16.

THEOREM. *A subset* X *of* $H(U;F)$ *is* τ_0-*bounded if and only if it is* τ_ω-*bounded.*

Proof. If X is τ_ω-bounded, then X is τ_0-bounded since $\tau_0 \leq \tau_\omega$. Conversely, if X is τ_0-bounded, then X satisfies the condition 16.16 (7): corresponding to every compact subset K of U there are $C \geq 0$

and c ≥ 0 and an open subset V of U containing K such that

$$\sup_{x \in V} \left\| \frac{1}{m!} \, \hat{d}^m f(x) \right\| \leq Cc^m$$

for all f $\in$ X and m = 0,1,... . If p is a seminorm of H(U;F) ported by K, then choose $\varepsilon > 0$ to satisfy $\varepsilon c < 1$. Then for some c(ε,V) > 0 we have

$$p(f) \leq c(\varepsilon,V) \sum_{m=0}^{\infty} \varepsilon^m \sup_{x \in V} \left\| \frac{1}{m!} \, \hat{d}^m f(x) \right\| \leq M < \infty$$

for all f $\in$ X where M = c(ε,V)C(1-εc)$^{-1}$. Thus X is τ_ω-bounded. $\square$

17.14 THEOREM. *If X is a* τ_ω-*bounded subset of* H(U;F), *then X is equicontinuous at every point of* U.

Proof. Let $\xi \in$ U, and C and c correspond to it by 16.16(4). For r > 0 and $B_r(\xi) \subseteq$ U, if x $\in B_r(\xi)$, we have

$$f(x) = \sum_{m=0}^{\infty} \frac{1}{m!} \, \hat{d}^m f(\xi)(x-\xi)$$

for all f $\in$ H(U;F). It follows that

$$\left\| f(x) - f(\xi) \right\| \leq \sum_{m=1}^{\infty} \left\| \frac{1}{m!} \, \hat{d}^m f(\xi) \right\| \|x - \xi\|^m \leq \sum_{m=1}^{\infty} Cc^m r^m = \frac{Ccr}{1-cr}$$

provided x $\in B_r(\xi)$ and cr < 1 and f $\in$ X, which shows that X is equicontinuous at ξ. $\square$

17.15 *Topology* $\tau_{m,D}$. Let D be a fixed subset of U meeting every connected component of U. For each m = 0,1,..., and x $\in$ D, define a seminorm $P_{m,x}$ on H(U;F) by

$$p_{m,x}(f) = \left\| \hat{d}^m f(x) \right\|$$

The topology $\tau_{m,D}$ is generated by all seminorms of the form $p_{k,x}$ where $k \leq m$ and $x \in D$. We also define $\tau_{\infty,D}$ similarly. Then we have

$$\tau_{0,D} \leq \tau_{1,D} \leq \dots \leq \tau_{\infty,D} \leq \tau_{\infty} \leq \tau_{\omega}$$

17.16   THEOREM. *Let* U *be connected and* x,y $\in$ U. *If* X *is a* τ_{ω}-*bounded subset of* H(U;F), *then*

$$\tau_{m,\{x\}} = \tau_{m,\{y\}}$$

on X *for any* m.

Proof. By translation (if necessary), we may assume that $0 \in X$. For a fixed m, let $k = 0,1,\dots,m$. Since X is bounded and

$$\hat{d}^k: (H(U;F),\tau_{\omega}) \rightarrow (H(U;P(^m E;F)),\tau_{\omega})$$

is continuous,

$$\hat{d}^k(X) = \{\hat{d}^k f: f \in X\}$$

is equicontinuous on U by 17.15.

For every $\xi \in U$, let V_{ξ} be a neighborhood of ξ such that if $x \in V_{\xi}$ then

$$\left\| \hat{d}^k f(x) - \hat{d}^k f(\xi) \right\| < \frac{1}{2}$$

for all $f \in X$. Thus

$$\left\| \hat{d}^k f(x) \right\| \leq \left\| \hat{d}^k f(\xi) \right\| + \frac{1}{2}$$

$$\left\| \hat{d}^k f(\xi) \right\| \leq \left\| \hat{d}^k f(x) \right\| + \frac{1}{2}$$

for any $x \in V$ and $f \in X$. Hence

$$\{ f \in X: \left\| \hat{d}^k f(x) \right\| < \frac{1}{2} \} \subset \{ f \in X: \left\| \hat{d}^k f(\xi) \right\| < 1 \}$$

$$\{ f \in X: \left\| \hat{d}^k f(\xi) \right\| < \frac{1}{2} \} \subset \{ f \in X: \left\| \hat{d}^k f(x) \right\| < 1 \}$$

for every $x \in V_\xi$, $f \in X$, and $k = 0, 1, \ldots, m$. This shows that $\tau_{m, \{x\}} = \tau_{m, \{\xi\}}$ on X for $x \in V_\xi$.

Let $A = \{ x \in U: \tau_{m, \{x\}} = \tau_{m, \{\xi\}} \}$. Then A is nonempty and it is open as we have seen in the preceding paragraph. We want to show that A is also closed; then $A = U$ since U is connected and the theorem is proved.

Let $y \in \bar{A} \cap U$. Then we can find a neighborhood V_y of y such that $\tau_{m, \{x\}} = \tau_{m, \{y\}}$ on X for any x in V_y. Choose x in $A \cap V_y$, $x \neq y$. Then $\tau_{m, \{x\}} = \tau_{m, \{\xi\}}$ on X. Thus, $\tau_{m, \{y\}} = \tau_{m, \{\xi\}}$ on X; i.e., $y \in A$. This shows that A is closed. □

17.17 THEOREM. *If* X *is a* τ_ω-*bounded subset of* $H(U; F)$ *then on* X *we have*

$$\tau_{\infty, D} = \tau_\infty = \tau_\omega$$

Proof. Since $\tau_{\infty, D} \leq \tau_\infty \leq \tau_\omega$, it suffices to show that $\tau_{\infty, D} \geq \tau_\infty \geq \tau_\omega$ on X. We assume that $0 \in X$. In light of Theorem 17.16, we can assume U is connected and contains the origin 0. Then we take $D = \{0\}$.

We first show that $\tau_\infty \geq \tau_\omega$ on X. Let p be a seminorm ported by a compact subset K. We claim that

$$\{ f \in X: p(f) < 1 \}$$

is a τ_ω-neighborhood of 0 in X. Since X is bounded, we can find C and c such that

$$\sup_{x \in K} \left\| \frac{1}{m!} \, \hat{d}^m f(x) \right\| \leq Cc^m$$

for all m and $f \in X$. Choose $\varepsilon > 0$ such that $\varepsilon c < 1$ and $c(\varepsilon) > 0$ by 17.8(b). Then for some $N > 1$ we have

$$Cc(\varepsilon) \sum_{m > N} (\varepsilon c)^m < \frac{1}{2}$$

Then the seminorm

$$q(f) = c(\varepsilon) \sum_{m=0}^{N} \varepsilon^m \sup_{x \in K} \left\| \frac{1}{m!} \, \hat{d}^m f(x) \right\|$$

is τ_∞-continuous, and if $q(f) < 1/2$, then $p(f) < 1$. This proves that $\tau_\infty \geq \tau_\omega$ on X.

We now show that $\tau_{\infty,D} \geq \tau_\infty$ on X. It suffices to show that for any compact set $K \subset U$ and m, the τ_∞-neighborhood

$$B_m = \{f \in X: \left\| \hat{d}^m f \right\|_K < 1\}$$

is a $\tau_{\infty,D}$-neighborhood. Since X is bounded, it is equicontinuous at each point of U. Hence for each $x \in K$ there exists a neighborhood V_x such that whenever $y \in V_x$,

$$\left\| \hat{d}^m f(x) - \hat{d}^m f(y) \right\| < \frac{1}{4}$$

for all $f \in X$. Hence we can find a finite number of open neighborhoods $V_1, \ldots V_m$ of points $x_1, \ldots, x_m$ in K (respectively) such that $K \subset V_1 \cup \ldots \cup V_m$. Then

$$\{f \in X: \|\hat{d}^m f(x_j)\| < \tfrac{1}{2}\}$$

are $\tau_{m,\{0\}}$-neighborhoods since U is connected; thus their intersection is a $\tau_{m,\{0\}}$-neighborhood of 0. Since $x_j \in K$, this intersection is contained in B_m. Hence B_m is a $\tau_{\infty,D}$-neighborhood, which shows that $\tau_{\infty,D} \geq \tau_\infty$ on D. $\quad\square$

COROLLARY. *If U is connected, then every* τ_ω*-bounded subset of* H(U;F) *is metrizable.*

Proof. $\tau_{\infty,\{0\}}$ is a metrizable topology. $\quad\square$

17.18 THEOREM. *Let* (f_n) *be a sequence in* H(U;F) *and* $f \in$ H(U;F). *Then the following are equivalent:*

(a) $f_n \to f$ *in the sense of* τ_ω;

(b) $\{f_n : n \in \mathbb{N}\}$ *is* τ_ω*-bounded and* $f_n \to f$ *in the sense of* $\tau_{\infty,D}$

Proof. The implication (a) $\Rightarrow$ (b) is clear. It remains to show (b) $\Rightarrow$ (a). Let X be the set of f and all f_n's. Then X is τ_ω-bounded. Thus on X, we have $\tau_{\infty,D} = \tau_\omega$ by the preceding theorem. Hence, $f_n \to f$ in the sense of τ_ω. $\quad\square$

COROLLARY. *Let U be* ξ*-balanced and* $f \in$ H(U;F). *Then the Taylor series of f about* ξ *converges to f in the sense of* τ_ω.

RELATIVELY COMPACT SETS IN H(U;F)

17.19 In the tradition of the classical analysis, we study the Montel theorem which states that if E and F are finite dimensional, then every bounded subset of H(U;F) is relatively compact for τ_o (hence for τ_ω). Although the theorem is true for $(H(U),\tau_o)$ when E is infinite dimensional (see 17.21), it turns out that in infinite dimensions, a subset X of H(U;F) is τ_ω-relatively compact if and only if X is not only τ_ω-bounded but X should satisfy an additional condition

of pointwise relative compactness of X respect to all differentials
$\hat{d}^m$ (see 17.22). Of course, any theorem of this type is related to
the classical Ascoli's theorem.

17.20 The proof of the following theorem is omitted since it is a
standard theorem in General Topology (see Kelley (1955), p. 234 or
Willard (1970), p. 287).

ASCOLI THEOREM. *Let X be a metric space and Y a Hausdorff locally
convex space. Then a subfamily F of* $C(X;Y)$ *is compact for the
topology* τ_o *of uniform convergence on compact sets if and only if*

   (a)   F is closed in $C(X;Y)$ *for* τ_o;

   (b)   the closure of $F[\xi]$ *is compact for each* $\xi \in X$ *where*

$F[\xi] = \{f(\xi): f \in F\}$

   (c)   F is equicontinuous at each point of X.

17.21   MONTEL THEOREM.   *If X is a* τ_o*-bounded subset of* $H(U;\mathbb{C})$, *then
X is relatively compact for* τ_o.

Proof. We use Ascoli's theorem in this proof. Let X be bounded for
τ_o. Then it is bounded for τ_ω; hence X is equicontinuous by 17.14.
Since the mapping $f \in H(U;\mathbb{C}) \to f(\xi) \in \mathbb{C}$ is continuous for each ξ in
U, $X[\xi]$ is bounded in $\mathbb{C}$, and hence it is relatively compact in $\mathbb{C}$.
It follows from Ascoli's theorem that X is relatively compact in
$C(U;F)$ for τ_o. Since $H(U;F)$ is closed in $C(U;F)$ for τ_o, X is rela-
tively compact in $H(U;F)$ for τ_o.   □

17.22   In order to show that the Montel theorem is not valid for
topologies τ, $\tau_\infty \leq \tau \leq \tau_\omega$, we need first the following theorem.

THEOREM. $P(^mE;F)$ *is a closed complementary subspace of* $H(U;F)$ *for
any topology* τ, $\tau_\infty \leq \tau \leq \tau_\omega$, *and on* $P(^mE;F)$ τ *induces the norm topology.*

Proof (Chae, 1971). It is easy to see that the norm topology on
$P(^mE;F)$ is finer than the topology τ, $\tau_\infty \leq \tau \leq \tau_\omega$. Let $\xi \in U$ be a

fixed point in U. Define a linear map T: H(U;F) → P(^{m}E;F) by

$$T(f) = \frac{1}{m!} \hat{d}^m f(\xi)$$

Then T is continuous for τ on H(U;F) and P(^{m}E;F) as a normed space
by 17.12, and the inclusion map P(^{m}E;F)↪ H(U;F) is trivially continuous. Therefore the composite map

$$P(^mE;F) \xhookrightarrow{\quad} H(U;F) \xrightarrow{\ T\ } P(^mE;F)$$

is continuous with respect to τ on the first two spaces and the norm
topology on the last. Since P = $\frac{1}{m!} \hat{d}^m P(\xi)$ for any P ∈ P(^{m}E;F), the
composite map is the identity map on P(^{m}E;F); hence τ is finer than
the norm topology on P(^{m}E;F). This shows that τ induces the norm
topology on P(^{m}E;F). Hence P(^{m}E;F) is closed in H(U;F) for τ. It
is easy to see that T^2 = T, which shows that P(^{m}E;F) is a complementary subspace of H(U;F).   □

COROLLARY. *The dual* E* *of* E *is a closed subspace of* H(U) *for* τ,
$τ_\infty$ ≤ τ ≤ $τ_\omega$ *and* τ *induces the norm topology on* E*.

17.23 THEOREM. *If* E *is infinite dimensional, a* τ-*bounded subset
of* H(U) *may not be relatively compact for* τ, $τ_\infty$ ≤ τ ≤ $τ_\omega$.

Proof. Let B* be the unit ball of E*. Then B* is a closed bounded
subset of H(U), but it cannot be compact since a compact set in an
infinite dimensional space is nowhere dense by Riesz's theorem 1.15. □
   This theorem shows that $τ_o$ and $τ_\omega$ do not share the same collection
of relatively compact subsets.

17.24 *Relatively Compact At A Point.* A subset X of H(U;F) is said
to be *relatively compact at a point* ξ ∈ U if, for every m ∈ ℕ , the
set

$$X[\xi,m] = \{\hat{d}^m f(\xi): f \in X\}$$

is relatively compact in $P(^mE;F)$.

If X is relatively compact for τ, $\tau_\infty \leq \tau \leq \tau_\omega$, then X is bounded for τ.   Moreover the mapping

$$f \in H(U;F) \rightarrow \hat{d}^m f(\xi) \in P(^mE;F)$$

is continuous for every $\xi \in U$ and $m \in \mathbb{N}$ with respect to τ and the norm topology.   Hence the image $X[\xi,m]$ is relatively compact in $P(^mE;F)$ for every $m \in \mathbb{N}$, that is X is relatively compact at every $\xi \in U$.

Let D be a subset of U meeting each connected component of U. Then we know that $\tau_{\infty,D} = \tau_\infty = \tau_\omega$ on each bounded subset of H(U;F). It follows that X is relatively compact for τ if and only if it is relatively compact for $\tau_{\infty,D}$.   In particular, if U is connected, then $X \subset H(U;F)$ is relatively compact for τ if and only if it is relatively compact for $\tau_{\infty,\{\xi\}}$ for some $\xi \in U$.

17.25 THEOREM (Nachbin, 1968).   A subset X of H(U;F) is relatively compact for τ, $\tau_\infty \leq \tau \leq \tau_\omega$, if and only if X is τ-bounded and relatively compact at every point of D.

Proof.   Assume that X is bounded for τ and relatively compact at each point of D.   Then X is relatively compact for τ if and only if it is relatively compact for $\tau_{\infty,X}$.   Now consider

$$S = \prod_{m=0}^{\infty} P(^mE;F)$$

the vector space of all formal power series from E to F endowed with the product topology.   We define a natural mapping Φ from H(U;F) into the vector space C(D;S) of all continuous S-valued functions on X by associating with every $f \in H(U;F)$ the function $\Phi(f)$ defined at every $x \in D$ by

$$\Phi(f)(x) = \left(\frac{1}{m!} \hat{d}^m f(x)\right)_{m \in \mathbb{N}} \in S$$

It is immediate that $\Phi(f) \in C(D;S)$ because each $\hat{d}^m f$ is holomorphic, hence continuous from D to $P(^m E;F)$. It is easy to see that Φ is linear, one-one, and open if we endow H(U;F) with $\tau_{\infty,D}$ and C(D;S) with the compact open topology. Hence it is sufficient to prove that $\Phi(X)$ is relatively compact in C(D;S). To do so, we use Ascoli's theorem. Since X is relatively compact at each point x of D, by our original assumption we see that the set

$$\Phi(X)[x] = \{\Phi(f)(x) : f \in X\}$$

is relatively compact in S. It remains to show that $\Phi(X)$ is equi-continuous at every point $x \in D$. But notice that

$$\{\hat{d}^m f : f \in X\}$$

is a bounded subset of $H(U;P(^m E;F))$ for τ; hence it is equicontinuous at each point $x \in D$. Thus $\Phi(X)$ is equicontinuous at each point $x \in D$, which shows that $\Phi(X)$ is relatively compact. $\square$

COROLLARY. *If* U *is connected, a bounded subset* X *of* H(U;F) *is relatively compact for* τ, $\tau_{\infty} \leq \tau \leq \tau_{\omega}$, *if and only if it is relatively compact at a single point of* U.

COMPLETENESS OF H(U;F)

17.26   Unlike the compact-open topology τ_o, proving H(U;F) is complete for τ_{ω} turns out to be much more complicated and difficult since the compact-ported seminorms of Nachbin are not given by explicit formulas. In this section we present an explicit formula for the τ_{ω}-continuous seminorms which define the topology τ_{ω} when U is a ξ-balanced open subset; and prove that $(H(U;F),\tau_{\omega})$ is complete.

17.27   THEOREM. *Let* U *be* ξ-*balanced. If* $f \in$ H(U;F), *then the Taylor series of* f *at* ξ *converges to* f *in the sense of* τ_{ω}.

Proof.   This is a consequence of 15.3.   □

17.28   *Characterization of Seminorms.*   We now show that the topology τ_ω can be defined by explicitly defined seminorms.

THEOREM (Dineen, 1970a).   *Let U be an open ξ-balanced subset of E and K a compact ξ-balanced subset of U.   Then the Nachbin topology τ_ω on H(U;F) is defined by seminorms of the form*

$$p_{\varepsilon,K}(f) = \sum_{m=0}^{\infty} \sup_{x \in B_{\varepsilon_m}(K)} \left\| \frac{1}{m!} \hat{d}^m f(\xi)(x - \xi) \right\|$$

where $\varepsilon = (\varepsilon_m) \in c_o^+$, where c_o^+ denotes the set of sequences of non-negative numbers tending to 0.

Proof.   For simplicity we take $\xi = 0$.   Let $f \in H(U;F)$ and K a balanced compact subset of U.   Then we can find $r > 0$ such that

$$\sum_{m=0}^{\infty} \left\| \frac{1}{m!} \hat{d}^m f(0) \right\|_{B_r(K)} < \infty \tag{*}$$

For $\varepsilon = (\varepsilon_m) \in c_o^+$, choose N such that $\varepsilon_m < r$ for any $m \geq N$.   Then

$$\left\| \frac{1}{m!} \hat{d}^m f(0) \right\|_{B_{\varepsilon_m}(K)} \leq \left\| \frac{1}{m!} \hat{d}^m f(0) \right\|_{B_r(K)} \tag{**}$$

It follows from (*) and (**) that $p_{\varepsilon,K}(f) < \infty$.   This shows that $p_{\varepsilon,K}$ is well-defined.   It is then obvious that $p_{\varepsilon,K}$ is a seminorm on H(U;F).

   We now show that $p_{\varepsilon,K}$ is τ_ω-continuous.   In fact, we show that $p_{\varepsilon,K}$ is ported by K.   Given $r > 0$ with $B_r(K) \subset U$ choose N such that $\varepsilon_m \leq r$ for $m \geq N$.   Set $V = B_r(K)$.   We then get

$$\sum_{m=N}^{\infty} \left\| \frac{1}{m!} \hat{d}^m f(0) \right\|_{B_{\varepsilon_m}(K)} \leq \sum_{m=N}^{\infty} \left\| \frac{1}{m!} \hat{d}^m f(0) \right\|_V$$

For $m = 0,1,\ldots,N-1$, there exists δ, $0 < \delta < 1$ such that

$$B_{\delta \varepsilon_m}(K) \subset B_r(K)$$

Then

$$\left\| \frac{1}{m!} \hat{d}^m f(0) \right\|_{B_{\varepsilon_m}(K)} \leq \delta^{-m} \left\| \frac{1}{m!} \hat{d}^m f(0) \right\|_V$$

and hence

$$p_{\varepsilon,K} \leq \delta^{-N+1} \sum_{m=0}^{\infty} \left\| \frac{1}{m!} \hat{d}^m f(0) \right\|_V$$

for all $f \in H(U;F)$, which shows that $p_{\varepsilon,K}$ is τ_ω-continuous.

Conversely, let p be a τ_ω-continuous seminorm ported by a compact balanced set K. Then for any $r > 0$ with $B_r(K) \subset U$ there exists $c(r) > 0$ such that

$$p(f) \leq c(r) \sum_{m=0}^{\infty} \left\| \frac{1}{m!} \hat{d}^m f(0) \right\|_{B_r(K)}$$

by 17.9. Since p is ported by K and $P(^mE;F) \subset H(U;F)$, let

$$K_m(r) = \inf \{c \colon p(P_m) \leq c \|P_m\|_{B_r(K)}, P_m \in P(^mE;F)\}$$

Then $K_m(r) \leq c(r)$ for all m; hence

$$\lim \sup \, [K_m(r)]^{1/m} \le 1$$

Thus we can choose a positive integer n_1 such that

$$[K_m(r)]^{1/m} \le 2 \qquad\qquad \text{for } m \ge n_1$$

Inductively, we choose $n_k > n_{k-1}$ and

$$[K_m(r/k)]^{1/m} \le 2 \qquad\qquad \text{for } m \ge n_k$$

Let

$$\varepsilon_m = \begin{cases} r & \text{for } m < n_2 \\ r/k & \text{for } n_k < m < n_{k+1} \end{cases}$$

Then $\varepsilon = (\varepsilon_m) \in c_o^+$ and $[K_m(\varepsilon_m)]^{1/m} \le 2$ for $m \ge n_1$.

Therefore, we have for some $C > 0$

$$K_m(\varepsilon_m) \le C2^m$$

for all m. Then

$$p(f) \le \sum_{m=0}^{\infty} p(\frac{1}{m!}\, \hat{d}^m f(0)) \le C \sum_{m-0}^{\infty} 2^m \|\frac{1}{m!}\, \hat{d}^m f(0)\|_{B_{\varepsilon_m}}(K)$$

or

$$p(f) \le C \sum_{m=0}^{\infty} \|\frac{1}{m!}\, \hat{d}^m f(0)\|_{B_{2\varepsilon_m}}(K) = Cp_{2\varepsilon,K}(f)$$

for all f ∈ H(U;F). This gives the required domination.   □

17.29 THEOREM. *If* U *is* ξ-*balanced, then* $(H(U;F),\tau_\omega)$ *is complete.*

Proof. We take ξ = 0 for simplicity. Let $(f_\alpha)_{\alpha \in A}$ be a Cauchy net
in $(H(U;F),\tau_\omega)$. Then for each m, $(\frac{1}{m!}\hat{d}^m f_\alpha(0))_{\alpha \in A}$ is a Cauchy net
in the Banach space $P(^mE;F)$; hence it converges to some P_m in $P(^mE;F)$.
For a compact balanced set K in U and $\varepsilon = (\varepsilon_m) \in c_o^+$, consider the
seminorm $p_{\varepsilon,K}$ defined in 17.28. For any given δ > 0 there exists
$\alpha_o \in A$ such that $\alpha_1, \alpha_2 \geq \alpha_o$ we have

$$\sum_{m=0}^{\infty} \left\| \frac{1}{m!} \hat{d}^m [f_{\alpha_1} - f_{\alpha_2}](0) \right\|_{B_{\varepsilon_m}(K)} \leq \delta$$

Hence for each n and $\alpha_1, \alpha_2 \geq \alpha_o$ we have

$$\sum_{m=0}^{n} \left\| \frac{1}{m!} \hat{d}^m [f_{\alpha_1} - f_{\alpha_2}](0) \right\|_{B_{\varepsilon_m}(K)} \leq \delta$$

Letting $\alpha_1 \to \infty$, we get

$$\sum_{m=0}^{n} \left\| P_m - \frac{1}{m!} \hat{d}f_{\alpha_2}(0) \right\|_{B_{\varepsilon_m}(K)} \leq \delta \qquad\qquad (*)$$

for all n and all $\alpha_2 \geq \alpha_o$. In particular, we have

$$\sum_{m=0}^{\infty} \left\| \frac{1}{m!} P_m \right\|_{B_{\varepsilon_m}(K)} \leq p_{\varepsilon,K}(f_{\alpha_o}) + \delta$$

Let $f = \Sigma P_m$. Then f ∈ H(U;F) by Theorem 14.23. Hence

$$p_{\varepsilon,K}(f - f_{\alpha_2}) \leq \delta$$

for all $\alpha_2 \geq \alpha$ by (*), which shows that $f_\alpha \to f$ in H(U;F) for τ_ω. □

17.30 The above proof of completeness of $(H(U;F),\tau_\omega)$ for balanced open sets U is due to Dineen (1970a). Chae (1971) has also proved that $(H(U;F),\tau_\omega)$ is complete for most open subsets including balanced sets using the space of holomorphic germs; later Mujica (1979) extended Chae's method and proved that $(H(U;F),\tau_\omega)$ is complete for every open subset U.

EXERCISES

17A *Nachbin Topology on* C(U;F). On the space C(U;F) of all continuous functions of U into F, $\tau_o = \tau_\infty = \tau_\omega$, where U is an open subset of a normed space E.

17B *Topology* $\tau_{m,X}$. Let X be a fixed subset of U meeting every connected component of U. A subset X of H(U;F) is τ_ω-bounded if and only if X is equicontinuous and X is bounded for $\tau_{m,X}$ for some m.

17C. Let U be an open subset of an infinite dimensional normed space E. For $\xi \in U$ and a sequence $(\alpha_n) \in c_o^+$, the seminorm

$$p(f) = \sum_{m=0}^{\infty} \left\| \frac{1}{m!} \hat{d}^m f(\xi) \right\|_{B_{\alpha_n}(0)}$$

is τ_ω-continuous but it is not τ_o-continuous.

17D. *Barrelledness of* $(H(U;F),\tau_m)$

(a) On E*, we have $\tau_o = \tau_m$ for all $m \in \mathbb{N}$

(b) $(H(U;F),\tau_m)$ is not barrelled for every m if E is infinite dimensional

(c) If $E = \ell^2$, $(H(U;F),\tau_\infty)$ is not barrelled (Dineen (1970a)).

17E *Relatively compact at a point*. Let $X \subset H(U;F)$ be such that

(1) X is relatively compact at some point $\xi \in U$;

(2) there exist $r > 0$ and $C \geq 0$ such that $B_r(\xi) \subset U$ and

$$\left\|\frac{1}{m!}\, \hat{d}^m f(\xi)\right\| < C(2r)^{-m}$$

for every $f \in X$ and $m \in \mathbb{N}$.

Then X is relatively compact at every point of $B_r(\xi)$.

17F *Entire Function*. Let $f = \Sigma\, P_m$ be a function on E with $P_m \in P(^m E;F)$ for all $m \in \mathbb{N}$. Then the following conditions are equivalent:

(a) $f \in H(E;F)$;

(b) For each compact balanced subset K and $(\alpha_m) \in c_o^+$,

$$\lim \|P_m\|_{B_{\alpha_m}(K)}^{1/m} = 0$$

(c) For each compact balanced set K and $(\alpha_m) \in c_o^+$,

$$\sum_{m=0}^{\infty} \|P_m\|_{B_{\alpha_m}(K)} < \infty$$

(d) For each compact convex subset K there exists $\delta > 0$ such that

$$\sum_{m=0}^{\infty} \|P_m\|_{B_r(K)} < \infty$$

17G. $(H(U;F),\tau_\omega)$ is quasi-complete; i.e., each bounded closed subset is complete.

17H. If E is separable, then every τ_ω-bounded subset of H(U;F) is metrizable.

18

The Bornological Topology on H(U;F)

18.1 This chapter is motivated by the topics we will discuss in
Chapter 19; some terminology which is defined in Chapter 19 is used
without specific definition in this introduction. In Chapter 19,
we will study the Cartan-Thullen theorem describing the phenomenon
of simultaneous holomorphic continuation of all mappings in an open
subset of E. (The term *analytic continuation* is interchangably used
with *holomorphic continuation*.) However, this will be too specific
for many purposes. For example, we may ask:

What is the largest domain E(U) into which all mappings in

H(U) can be holomorphically continued?

It is not practical to limit ourselves to a *schlicht* domain (that
is, a connected open subset in E). Analytic continuation of functions of one variable may lead to Riemann surfaces; similarly, we
may admit non-schlicht domains. Thus we can speak of a manifold
spread over a Banach space as a domain into which we can continue
holomorphically. This possibility does occur in finite dimensions
and the largest domain of simultaneous holomorphic continuation is
called the *envelope of holomorphy*.

Generalizing the concept of the envelope of holomorphy in infinite dimensions, Alexander (1968) recognized the necessity of a
barrelled topology on H(U) to show the uniqueness of such extensions
E(U) such that H(U) and H(E(U)) are homeomorphic and isomorphic
(see 19.27).

In 1970, Coeuré and Nachbin independently introduced such a

topology on H(U;F), which is the main theme of this chapter. This topology will be called, obviously, the Coeuré-Nachbin topology and denoted by τ_δ. Then τ_δ will be finer than τ_ω. In Coeuré (1970), E was assumed to be a separable space.

The question of the equality between the Nachbin topology and the Coeuré-Nachbin topology is quite important in establishing the Cartan-Thullen theorem. In fact, if U is holomorphically convex and these two topologies are identical on H(U), then U becomes a domain of holomorphy (see 19.17 and 19.18). This provides the motivation for categorizing those Banach spaces E for which these two topologies are identical on H(E). The concept of bounding sets becomes a natural tool in this study.

BORNOLOGICAL SPACES

18.2. Let E be a locally convex space, and let B denote the family of all bounded sets in E. Although the family B naturally depends on the topology of the space E, this does not in general define the topology uniquely as we have experienced in Chapter 17 for the space $(H(U;F), \tau_\omega)$ when U is an open subset of an infinite dimensional normed space E.

A space E is described as *bornological* if it is impossible to introduce on E a finer locally convex topology with the same family of bounded sets as there is in E. We have the following characterization of a bornological space:

A locally convex space E is bornological if and only if every balanced convex set absorbing each bounded set is a neighborhood of zero (Exercise 18A).

It is immediate that every normed space is bornological. More generally, every metrizable locally convex spcae is bornological.

18.3 *Inductive Limit.* A way of constructing a bornological space (respectively, a barrelled space) is to take the inductive limit of bornological spaces (respectively, barrelled spaces) in the following manner.

Let E be a vector space, $(E_\iota)_{\iota \in I}$ a family of locally convex spaces, and for each ι in I let f_ι be a linear map from E_ι into E. If E is the union of all $f_\iota(E_\iota)$, then the space E equipped with the finest locally convex topology for which all the maps f_ι are continuous is called the *inductive limit* of the family $(E_\iota)_{\iota \in I}$ with respect to the maps $(f_\iota)_{\iota \in I}$. We write this relation by

$$E = \varinjlim_{\iota \in I} E_\iota \qquad \text{or} \qquad E = \varinjlim E_\iota$$

In a simpler situation, each E_ι could be a vector subspace of E and each map f_ι the inclusion map $E_\iota \to E$.

18.4 We need the following theorem whose proof can be found in many standard textbooks, e.g. Horváth (1966).

THEOREM. *(1) The inductive limit of bornological spaces is bornological.*

   (2) The inductive limit of barrelled spaces is barrelled.

   (3) A sequentially complete bornological space is barrelled.

THE COEURÉ-NACHBIN TOPOLOGY

18.5. Let U be an open subset of a normed space E. For each countable open cover $I = (V_n)$ of U, let

$$H_I(U;F) = \{f \in H(U;F): \|f\|_{V_n} < \infty \text{ for all } n \in \mathbb{N}\}$$

Then $H_I(U;F)$ is a vector subspace of H(U;F). We endow $H_I(U;F)$ with the topology generated by the seminorms

$$f \in H_I(U;F) \to \|f\|_{V_n} \in \mathbb{R}^+$$

for all $V_n \in I$. It is easy to see that $H_I(U;F)$ is a metrizable locally convex space because its topology is defined by a countable number of seminorms. Furthermore, $H(U;F)$ is the union of all $H_I(U;F)$ where I ranges over all possible countable open covers of U. In fact, if $f \in H(U;F)$, let $I = (V_k)$ where

$$V_k = \{x \in U : \|f(x)\| < k\}$$

then f is clearly a member of $H_I(U;F)$.

The *Coeuré-Nachbin* topology on $H(U;F)$ is defined to be the inductive limit of the spaces $H_I(U;F)$ with respect to the inclusion maps $H_I(U;F) \rightarrow H(U;F)$, and it is denoted by τ_δ. By definition, then, the Coeuré-Nachbin topology is bornological because it is an inductive limit of metrizable spaces.

Notice that in this definition it is sufficient to use only those countable covers $I = (V_n)$ of U which are increasing; i.e., $V_n \subset V_m$ whenever $n \leq m$.

18.6 THEOREM. *A seminorm p on $H(U;F)$ is τ_δ continuous if and only if for each increasing countable open cover $I = (V_n)$ of U there exist $V_m \in I$ and $c > 0$ such that for any $f \in H(U;F)$*

$$p(f) \leq c\|f\|_{V_m} \tag{*}$$

Proof. The "if" part is clear, and we show the other part. Suppose that p is τ_δ continuous and the relation (*) does not hold. Then we can find a countable open cover $I = (V_n)$ of U and a sequence (f_n) in $H(U;F)$ such that

$$p(f_n) > n \|f_n\|_{V_n}$$

Set

$$g_n = f_n / \|f_n\|_{V_n}$$

provided the denominator is not equal to 0; otherwise, set $g_n = a_n f_n$

where $a_n > 0$ such that $p(g_n) > n$. It is clear that

$$\|g_n\|_{V_n} \leq 1$$

for all $n \in \mathbb{N}$. We claim that the sequence (g_n) is a bounded sequence

in $H_J(U;F)$ for some countable open cover J of U; hence it is bounded

for τ_δ although

$$p(g_n) > n$$

for all $n \in \mathbb{N}$, which shows that p can not be τ_δ continuous.

To show that (g_n) is bounded in $H(U;F)$, for each $n \in \mathbb{N}$, let

$$W_n = \{x \in U: \|g_m(x)\| \leq n \text{ for all } m\}$$

and let $J = (\text{Int } W_n)$, where Int W_n denotes the interior of W_n. Since

$$\|g_m\|_{V_n} \leq 1$$

for all $m \geq n$, it follows that J is an increasing sequence of open

sets covering U. Hence p is continuous on $H_J(U;F)$. By construction,

$$\|g_m\|_{W_n} \leq n$$

for every $m \in \mathbb{N}$; hence (g_n) is bounded in $H_J(U;F)$.   □

Intuitively, the theorem say that a variable $f \in H(U;F)$ tends to

a fixed $f_o \in H(U;F)$ with respect to τ_δ if f tends to f_o uniformly

on some neighborhood of every point of U.

18.7 THEOREM. *On* $H(U;F)$ *we have*

$$\tau_\omega \leq \tau_\delta$$

Proof. Let p be a τ_ω continuous seminorm on $H(U;F)$ ported by a compact set K. To show that p is τ_δ continuous, let $I = (V_n)$ be an increasing countable open cover of U. Then there exists $V_n \in I$ such that $K \subset V_n$. Since p is ported by the compact set K, we can find a real number $c > 0$ such that

$$p(f) \leq c\|f\|_{V_n}$$

for all $f \in H(U;F)$, which shows that p is τ_δ continuous. □

18.8 THEOREM. *The space* $(H(U;F),\tau_\delta)$ *is both bornological and barrelled.*

Proof. We know already that it is bornological. To show that it is barrelled, it suffices to prove the space $H_I(U;F)$ is barrelled (see 18.4(3)). Since $H_I(U;F)$ is bornological, to prove that $H_I(U;F)$ is barrelled it suffices to show $H_I(U;F)$ is sequentially complete. But this fact is easy to check because F is complete and locally bounded G-holomorphic mappings are holomorphic by 14.9. □

18.9 THEOREM. *For a subset X of $H(U;F)$, the following are equivalent.*

(a) *X is τ_o bounded;*

(b) *X is τ_δ bounded.*

Proof. Since $\tau_o \leq \tau_\delta$, the implication (b) → (a) is obvious. Conversely, suppose X is τ_o bounded. We want to show that there exists a countable open cover $I = (V_n)$ of U such that X as a subset of $H_I(U;F)$ is bounded; hence X is τ_δ bounded because the inclusion map $H_I(U;F) \to H(U;F)$ is continuous.

Choose an arbitrary τ_δ continuous seminorm p on $H(U;F)$. For each $n \in \mathbb{N}$, let

$$W_n = \{x \in U: \|f(x)\| \leq n \text{ for all } f \in X\}$$

and we let

$$V_n = \text{Int } W_n$$

Then $I = (V_n)$ becomes an increasing countable open cover of U. Since p is τ_δ continuous, by 18.6 there exists $c > 0$ such that

$$p(f) \leq c\|f\|_{V_N}$$

for all $f \in H(U;F)$. Hence

$$p(f) \leq cN < \infty$$

for all $f \in X$. This proves that X is τ_δ bounded.   □

18.10. The Coeuré-Nachbin topology τ_δ is, as we have seen above, the finest locally convex topology on H(U;F) with the same stock of bounded sets as the τ_o topology. Henceforth, we often call τ_δ the *bornological topology associated with* τ_o. This relation is indicated by

$$(\tau_o)_b = \tau_\delta$$

where the index 'b' stands for 'bornological.' Since

$$\tau_o \leq \tau_\omega \leq \tau_\delta = (\tau_o)_b$$

and τ_o bounded sets are τ_ω bounded, we also have

$$(\tau_\omega)_b = \tau_\delta$$

18.11 THEOREM. *The vector space* $P(^mE;F)$ *is a closed complementary subspace of* $(H(U;F),\tau_\delta)$ *for each* $m \in \mathbb{N}$. *In particular,* τ_ω *and* τ_δ *induce the same topology on* $P(^mE;F)$.

Proof. As in the proof of 16.20, consider the mapping $T: H(U;F) \to H(U;F)$ defined by

$$T(f) = \frac{1}{m!} \hat{d}^m f(0)$$

Then T is a projection, i.e., $T^2 = T$. We have to show that T is τ_δ continuous (compare with the proof of 16.20). In fact, let p be a τ_δ continuous seminorm on $H(U;F)$, and let V be a convex balanced neighborhood of zero. Applying Theorem 18.6 to the increasing countable open cover $(U \cap nV)_{n \in \mathbb{N}}$ of U, we can find $c > 0$ and $N \in \mathbb{N}$ such that

$$q(f) = p(\frac{1}{m!} \hat{d}^m f(0)) \leq c \left\| \frac{1}{m!} \hat{d}^m f(0) \right\|_{U \cap NV}$$

$$\leq cN^m \left\| \frac{1}{m!} \hat{d}^m f(0) \right\|_V \leq cN^m \|f\|_V$$

for all $f \in H(U;F)$. Hence T is τ_δ continuous. This completes the proof of the first part. Notice that q is a τ_ω continuous seminorm on $H(U;F)$ and $q = p$ on $P(^mE;F)$. Hence τ_δ and τ_ω induce the same topology on $P(^mE;F)$.   □

BOUNDING SETS

18.12. It has been shown that

$$\tau_o \leq \tau_\omega \leq \tau_\delta$$

if E is infinite dimensional. It may occur that $\tau_\omega = \tau_\delta$ on $H(E)$ for some interesting infinite dimensional space E. The notion of bounding set is closely related to finding those spaces E for which the Nachbin topology and the Coeuré-Nachbin topology are identical on $H(E)$. The investigation of bounding sets in Banach spaces was originally motivated by the study of analytic continuation in Banach spaces by H. Alexander (1968).

18.13 *Bounding Sets.* Let U be an open subset of a normed space E.
A subset A of U is said to be a *bounding subset* of U if every holomorphic map in H(U) is bounded on A.

   It is immediate from the definition of bounding sets that

   (a) Every compact subset of U is a bounding subset of U;

   (b) If A is a bounding subset of U, then its closure $cl_U A$ is
        also bounding;

   (c) If $f \in H(U;F)$ and A is a bounding subset of U, then f(A) is
        a bounding subset of F.

18.14 LEMMA. *If A is a bounding subset of U, then the mapping*

$$p_A(f) = \left\| f \right\|_A = \{ \left| f(x) \right| : x \in A \}$$

is a τ_δ *continuous seminorm on* H(U).

Proof. It suffices to show that the set

$$V = \{ f \in H(U) : p_A(f) \leq 1 \}$$

is a neighborhood of zero in $(H(U), \tau_\delta)$. In fact, V is clearly convex, balanced and absorbing. Furthermore, it is τ_δ closed since V
is the intersection of all sets of the form

$$\{ f \in H(U) : \left| f(x) \right| \leq 1 \}$$

where $x \in A$. In a barrelled space, we know that every closed convex,
balanced and absorbing set is a neighborhood of zero. Since $(H(U), \tau_\delta)$
is barrelled, V must be a neighborhood of zero; hence p_A is τ_δ continuous.   □

18.15 THEOREM. *If E is a normed space and contains a non-precompact
bounding subset, then* $\tau_\omega < \tau_\delta$ *on* H(E).

First recall that a nonempty subset A of a metric space M is called *precompact* or *totally bounded* if for $\varepsilon > 0$, A is contained in the union of a finite number of open balls of radius ε. It can be shown that a subset of a complete metric space is relatively compact if and only if it is precompact.

Proof. Suppose that A is a non-precompact bounding set in E. By Lemma 18.14, the seminorm $p_A(f) = \|f\|_A$ is τ_δ continuous on H(E). To show that $\tau_\omega < \tau_\delta$, it suffices to prove p_A is not τ_ω continuous. In fact, if p_A were τ_ω continuous and ported by a compact subset K, then for any neighborhood V of K there corresponds $c(V) > 0$ such that

$$\|f\|_A \leq c(V)\|f\|_V$$

for all $f \in H(E)$. It follows from this relation that

$$\|f\|_A \leq \|f\|_V$$

by replacing f with f^n, taking the nth root, and letting $n \to \infty$. Since V is an arbitrary open neighborhood of K, we obtain

$$\|f\|_A \leq \|f\|_K$$

Let L be the closed convex hull of K. Then L is compact by the Mazur theorem (see Dunford and Schwartz (1964), p. 416) and

$$\|f\|_A \leq \|f\|_L$$

Thus A is a subset of the compact set L by the Hahn-Banach separation theorem 16.7. This is absurd because A is not precompact. Hence, p_A cannot be τ_ω continuous. $\square$

COROLLARY 1. *If $\tau_\delta = \tau_o$ on H(E), then every bounding set in E is precompact.*

COROLLARY 2. *If a normed space* E *contains a non-precompact bounding subset, then* (H(E),τ_ω) *is not bornological.*

18.16 REMARKS. A remarkable example was given by Dineen (1971) showing that a bounding set in ℓ^∞ need not to be precompact. EXAMPLE. The non-compact set of unit vectors

$\quad$ e$_i$ = (0,...,0,1,0,...)

(where 1 is in the ith place) is a bounding subset of ℓ^∞. Thus the Coeuré-Nachbin topology is strictly finer than the Nachbin topology on H(ℓ^∞).

The proof of this result is very complicated and we refer the reader to Dineen (1981), Theorem 4.31. Motivated by the fact that if $\tau_\omega = \tau_\delta$ on H(U) and U is holomorphically convex, then U is a domain of holomorphy (see 19.17), Dineen (1972) proved the following important result.

THEOREM. *If* U *is a balanced subset of a Banach space with an uncon-ditional basis, then* (H(U),τ_ω) *is bornological.*

Since Dineen's proof for this theorem is very technical and complicated, we will not give the proof here. We only record here the definition of an unconditional basis for a Banach space for the reader's convenience. A Schauder basis (e$_n$) for a Banach space E (see 20.8 for the definition of a Schauder basis) is said to be *an unconditional basis* if for any x $\in$ E with

$$x = \sum_{n=1}^\infty x_n e_n$$

and for any r > 0 there exists a finite subset J$_r$ of $\mathbb{N}$ such that for any finite subset J of $\mathbb{N}$ which contains J$_r$ we have

$$\left\| x - \sum_{n \in J} x_n e_n \right\| \leq r$$

The spaces ℓ^p, $1 \le p < \infty$, $L^p[0,1]$, $p > 1$ and c_o all have unconditional bases and the finite product of spaces with an unconditional basis also has an unconditional basis. An example of Banach spaces without any unconditional basis is the Lebesgue space $L^1[0,1]$.

18.17 Aside from the question of the equality $\tau_\omega = \tau_\delta$, classifying those Banach spaces in which every bounding subset is relatively compact is very important in analytic continuation. In particular, it is easy to establish the following equivalent statements:

(a) Every bounding set is relatively compact.

(b) For every sequence (x_n) in E without accumulation point in E, there exists $f \in H(E)$ such that

$$\sup \{|f(x_n)| : n \in \mathbb{N}\} = \infty$$

The statement (b) appears frequently in the next two chapters.

THEOREM (Dineen, 1971). *In a separable Banach space, every bounding set is relatively compact.*

Proof. We adapt here the proof given by Schottenloher (1976). Let A be a bounding subset of E and let (x_n) be a dense sequence in E. For any real number $r > 0$ and $m \in \mathbb{N}$, let

$$V_m = B_r(x_1, \ldots, x_m)$$

Then (V_m) is an increasing countable open cover of E. Since

$$P_A(f) = \|f\|_A$$

is a τ_δ continuous seminorm on H(E), it follows from 18.6 that for some $m \in \mathbb{N}$ and $c > 0$ we have

$$\|f\|_A \le c\|f\|_{V_m} \qquad (*)$$

for all $f \in H(E)$. As in the proof of 18.15, replacing f with f^n in
(*), taking the nth root, and letting $n \to \infty$, we obtain

$$\|f\|_A \leq \|f\|_{V_m}$$

for all $f \in H(E)$. It follows from the Hahn-Banach separation
theorem 16.7 that A must be contained in the closed convex hull of
V_m. Thus A is a precompact subset of E. Since E is complete, A
must be relatively compact, and this completes the proof.   □

18.18. Using the above result we further classify those spaces for
which the same result holds.

THEOREM (Dineen, 1971; Hirschowitz, 1970). *Let* E *be a Banach space*
which is isomorphic to a subspace of the Banach space C(T) *of con-*
tinuous functions on a sequentially compact Hausdorff space T *with*
the sup norm, then the bounding subsets of E *are relatively compact.*

Proof (Schottenloher, 1976). Every bounded nonrelatively compact
set contains a bounded sequence without any convergent subsequences.
Let (x_n) be a bounded sequence without any convergent subsequences.
We may assume that $x_n \neq x_m$ for $n \neq m$.
    We now claim that the set $\{x_n : n \in \mathbb{N}\}$ is not a bounding set in E.
For $r > 0$, find $t_{nm} \in T$ such that

$$|x_n(t_{nm}) - x_m(t_{nm})| = \|x_n - x_m\| \geq r$$

for $m \neq n$. Since T is sequentially compact, we can find a convergent
subsequence $(t_k)_{k \in \mathbb{N}}$ and a subsequence (y_n) of (x_n) such that

$$\|y_n(t_k) - y_m(t_k)\| \geq r$$

whenever $m, n > k$. Now consider the compact subset S which consists

of the sequence (t_k) and its limit and the Banach space $C(S)$. From
our assumptions about the sequence (x_n), we note that the restric-
tions $y_k|_S$ have no convergent subsequences in $C(S)$. By Theorem
18.17, we can find $h \in H(C(S))$ with

$$\sup \{ |f(y_k|_S)| : k \in \mathbb{N} \} = \infty$$

Since the restriction map $r: C(T) \to C(S)$, $x \to x|_S$ is linear and
continuous, the function $f = h \circ r$ satisfies $f \in H(C(T))$ and

$$\sup \{ |f(x_n)| : n \in \mathbb{N} \} = \infty$$

This completes the proof.   □

18.20.   The class of Banach spaces which can be embedded into $C(T)$,
where T is a sequentially compact Hausdorff space, includes reflexive
spaces E since the closed unit ball U^o of E^* is weak* sequentially
compact and E is canonically embedded into $C(U^o)$.

   There is no complete characterization of those Banach spaces in
which bounding subsets agree with relatively compact subsets.

18.21.   We conclude this chapter with the following theorem.

THEOREM.   *A bounding subset of an infinite dimensional Banach space*
is nowhere dense.

Proof.   We have shown in 15.11 that for any infinite dimensional
Banach space E there exists an entire function f of unbounded types;
i.e., f is unbounded on a bounded neighborhood of each point of E.
Hence, no ball in E is a bounding subset.   □

EXERCISES

18A *Bornological Property.*   A locally convex space E is bornological
if and only if every balanced convex set absorbing each bounded set
is a neighborhood of zero.

18B. If E is finite dimensional, then $\tau_0 = \tau_\delta$ on H(E).

18C *Bounding Sets.* Let U be an open subset of a normed space E.
Then

    (a) Every compact subset of U is a bounding subset of U.

    (b) If A is a bounding subset of U, then its closure $cl_U A$ is
       also a bounding subset of U.

    (c) If A is a bounding set of U and $f \in H(U;F)$, then $f(A)$ is a
       bounding subset of F.

18D. The following are equivalent in a Banach space E.

    (a) Every bounding subset is relatively compact.

    (b) For every sequence (x_n) in E without any accumulation point
       in E there exists $f \in H(E)$ such that f is unbounded on the
       sequence.

18E. Not every holomorphic mapping $f \in H(c_0)$ can be continued
      holomorphically to ℓ^∞. (Hint: 18.16.)

18F. Let $g \in H(U)$. Then the multiplication map

$$f \in H(U) \rightarrow f \cdot g \in H(U)$$

      is continuous with respect to the topology τ_δ.

19

Domains of Holomorphy

19.1. This chapter is devoted to the study of domains of holomorphy
in infinite dimensional spaces. We discuss the basic properties of
domains of holomorphy in normed spaces, and their relationship with
various convexity properties of domains arising in finite or infinite
dimensional spaces. In particular, we study the Cartan-Thullen
theorem in this chapter and the Levi problem in the next chapter.

HOLOMORPHIC CONVEXITY

19.2. Let U be an open subset of a normed space. We recall that U
is *convex* if for every x, y in U the closed line segment [x,y] is in
U. This is the classical definition of geometric convexity of a set.
However, it is of little use in several complex variables since
geometric convexity is not preserved under biholomorphic mappings,
which is a trivial consequence of the Riemann mapping theorem for
the complex plane. However, it is a well-known fact that the Riemann
mapping theorem fails to hold for $\mathbb{C}^n$. Thus we explore some other
concepts of convexity: they are holomorphic convexity, polynomial
convexity, plurisubharmonic convexity, and pseudoconvexity. These
various convexities will be shown to be equivalent to domain of
holomorphy for a Banach space with a Schauder basis.

19.3 *Convex Hull.* We first recall the classical definition of convex
hull. Given a subset X of a normed space, the *convex hull* co(X) of
X is the smallest convex set containing X and *closed convex hull*

299

$\overline{co}(X)$ of X is then the smallest closed convex set containing X.

For a compact set K, we have the following characterization of the closed convex hull of K.

THEOREM. *Let A(E) be the family of all affine linear functionals on* E. *If K is compact, then the closed convex hull of K is given by*

$$\overline{co}\ K = \{x \in E: |f(x)| \le \|f\|_K \text{ for all } f \in A(E)\} \qquad (*)$$

Proof. For simplicity, we denote the set in (*) by $K_{A(E)}$, and for each f in E*, we set

$$K_f = \{x \in E: |f(x)| \le \|f\|_K\}$$

Then K_f is a closed and convex set containing K. It is clear that $K_{A(E)}$ is the intersection of K_f for all $f \in A(E)$. Hence $K_{A(E)}$ is also a closed convex set containing K. Thus

$$\overline{co}\ K \subset K_{A(E)}$$

On the other hand, let $y \in E \setminus \overline{co}\ K$. We use the Hahn-Banach separation theorem (see Dunford and Schwartz (1964), p. 417) to find a real linear functional $p \in (E_{\mathbb{R}})^*$ such that

$$p(y) > \sup \{p(x): x \in \overline{co}\ K\}$$

Denote the right side of the preceding inequality by L for the sake of simplicity. Then the function defined by

$$q(x) = p(x) - ip(ix)$$

is a continuous complex linear functional in E* (see Lemma 2.9). Let

$$M = \{z \in \mathbb{C}: Re(z) \le L\}$$

If $x \in \overline{co}$ K, then $Re(q(x)) = p(x) \leq L$; hence $q(\overline{co}$ K$) \subset$ M. Since $Re(q(y)) = p(y) > L$, $q(y)$ does not belong to M. By the Mazur theorem stating that the closed convex hull of a compact set in a normed space E is precompact (compact if E is a Banach space) (see Dunford and Schwartz (1964), p. 416), we can find a complex number a and $r > 0$ such that

$$q(\overline{co} \ K) \subset \overline{B}_r(a) \qquad\qquad q(y) \in E \setminus \overline{B}_r(a) \qquad\qquad (**)$$

Let $f(x) = q(x) - a$. Then $f \in A(E)$ and it follows from (**) that

$$\left| f(y) \right| = \left| q(y) - a \right| > r \geq \left| q(x) - a \right| = \left| f(x) \right|$$

for all $x \in K$, which shows that

$$\left| f(y) \right| > \left\| f \right\|_K$$

Hence $y \in E \setminus K_{A(E)}$, which proves the set equality.   □

19.4 *Functional Convex Hull.* Motivated by the preceding characterization of the closed convex hull of a compact set, we now introduce the notion of convex hull with respect to a given family of holomorphic mappings. Let U be an open subset of a normed space E.

Given a subset X of U and a subfamily Λ of H(U), we define the *Λ-convex hull* of X by

$$\hat{X}_\Lambda = \{x \in U: \left| f(x) \right| \leq \left\| f \right\|_X \text{ for all } f \in \Lambda\}$$

By Theorem 19.3, the closed convex hull of a compact set K is then the A(E)-convex hull of K.

If no confusion can arise, we omit the subscript Λ and write

$$\hat{X} \qquad\qquad \text{instead of} \qquad\qquad \hat{X}_\Lambda$$

It is immediate that for any subset X the following properties
hold true:

(1) $X \subset \hat{X}$

(2) $\hat{X}$ is closed in U.

(3) $\hat{\phi} = \phi$ and $\hat{U} = U$

(4)   If $X \subset Y \subset U$, then $\hat{X} \subset \hat{Y}$.

(5)   For any $f \in \Lambda$ and $X \subset U$, we have

$$\| f \|_{\hat{X}} = \| f \|_{X}$$

(6) $\hat{\bar{X}} = X$ (where $\bar{X}$ denotes the closure of X in U).

(7)   If $\Lambda \subset \Gamma \subset H(U)$, then

$$\hat{X}_{\Gamma} \subset \hat{X}_{\Lambda}$$

19.5.   Before introducing the concept of Λ-convexity, we study the
geometric convexity of an open set. First we need a definition.

A set B is said to be U-*bounded* if B is a bounded subset of U
and there is a neighborhood V of zero such that

B + V ⊂ U

It is clear that a bounded set B is U-bounded if and only if

$d(B, CU) > 0$

where $CU = E \setminus U$.

THEOREM. *For an open subset of* E, *the following statements are
equivalent.*

(a)   U *is convex.*

(b)   *For each compact subset* K *of* U, *the set* $\hat{K}_{A(E)}$ *is* U-*bounded.*

Proof. (a) → (b). If K is a compact subset of U, then we can find a convex balanced neighborhood V of zero such that K + V ⊂ U. Since U is convex, it is clear that co K + V ⊂ U. Hence we have

$$\overline{\text{co}} \ K + \tfrac{1}{2}V \subset \text{co} \ K + V \subset U$$

But the A(E)-convex hull of K is the closed convex hull of K, which was shown in 19.3. Hence the A(E)-convex hull is U-bounded.

(b) → (a). Let a, b ∈ U. We claim that the compact line segment [a,b] is contained in U. For K = {a,b}, the A(E)-convex hull of K is the line segment [a,b]. Hence [a,b] is U-bounded, which proves that U is convex. □

19.6 *Λ-Convexity*. We can use the condition (b) of Theorem 19.5 as definition of convexity. Motivated by this fact, we introduce the following general concept of convexity.

Let U be an open subset of E and Λ ⊂ H(U). U is said to be *Λ-convex* if for each compact subset K of U, the Λ-convex hull $\hat{K}_{\Lambda}$ is U-bounded.

In particular, we have the following specific definition when Λ is specified.

An open set U is said to be *holomorphically convex* if U is H(U)-convex. We will call the H(U)-convex hull the *holomorphic convex hull*. Since A(E) ⊂ H(U), for any subset X of U, we have

$$\hat{X}_{H(U)} \subset \hat{X}_{A(E)}$$

Thus if K is a compact subset of U, then the holomorphic convex hull of K is precompact (compact if E is a Banach space) and contained in the closed convex hull of K.

19.7 EXAMPLE. In ℂ, every connected open set is holomorphically convex.

In fact, if K is a compact subset of an open connected subset U

of $\mathbb{C}$, then the holomorphic convex hull of K, denoted by $\hat{K}$ in this
proof, is compact. We now show that $\hat{K}$ is U-bounded. Let $r = d(K,CU)$
and suppose that w in U satisfies $d(w,CU) < r$. Choose u in CU such
that $|w - u| = d(w,CU)$. Then the function $f_w(z) = 1/(z - u)$ is
holomorphic on U and $\|f\|_K < |f_w(w)|$, and hence w does not belong to
$\hat{K}$. Therefore, $\hat{K}$ is U-bounded.   □

REMARK. The reader should consider carefully why the argument in the
above example breaks down in $\mathbb{C}^n$, $n \geq 2$. It is the fact that a
holomorphic function in $\mathbb{C}$ has isolated zeros that makes us consider
the function f_w in the argument. In higher dimensions, zeros of a
holomorphic mapping are never isolated (see Nachbin (1970), Chap. 2).

19.8 THEOREM. *Every convex open set in a normed space is holomor-*
phically convex.

Proof. Since every convex open set is A(E)-convex and the holomorphic convex hull of a compact set K is contained in the closed convex
hull of K, it is clear that every convex open set is holomorphically
convex.   □

DOMAINS OF HOLOMORPHY

19.9 *Domains of Holomorphy.* An open subset U of a normed space E
is called a *domain of holomorphy* if there do not exist two nonempty
connected open subsets U_1 and U_2 in E such that $U_2 \subset U \cap U_1$, $U_1 \not\subset U$ and
for every $f \in H(U)$ there exists a holomorphic mapping f_1 on U_1 such
that $f = f_1$ on U_2.

  The definition of domain of holomorphy is complicated because it
must take into account the possibility that the boundary of U could
intersect itself. Setting aside this technical condition, it is
easy to see that an open set U is not a domain of holomorphy if
every holomorphic mapping f on U can be continued holomorphically
to some slightly larger open set and this open set does not depend
on f.

Intuitively, U is a domain of holomorphy if there is no nonempty part of the boundary of U beyond which every holomorphic map on U can be continued holomorphically. Suppose that a mapping $f \in H(U)$ admits a holomorphic continuation beyond the boundary of U. It is easy to see from the principle of analytic continuation (12.9) that we can find two nonempty connected open sets U_1 and U_2 and a mapping $f_1 \in H(U_1)$ such that $U_2 \subset U \cap U_1$, $U_1 \not\subset U$ and $f = f_1$ on U_2. Thus a domain of holomorphy would not allow simultaneous holomorphic continuation of holomorphic mappings beyond its boundary.

19.10. The unit disk D in $\mathbb{C}$ is a domain of holomorphy since the function defined by

$$f(z) = \sum_{n=o}^{\infty} 2^{-n} z^{(2^n)} \qquad\qquad (*)$$

is continuous on $\bar{D}$ and holomorphic on D. However, on the boundary $z = e^{i\theta}$, the function is nowhere differentiable since it is the classical Weierstrass nowhere differentiable function (see Chae, 1980, p. 189). This proves that D is a domain of holomorphy.

We can further prove that every connected open subset of $\mathbb{C}$ is a domain of holomorphy. In fact, let U be a connected open set in $\mathbb{C}$. For any connected open set V containing U properly, let w be a boundary point of U which is in V. Then the mapping $f(z) = 1/(z - w)$ is holomorphic in U but cannot be continued holomorphically to V.

The situation in several complex variables is completely different. Some open sets are trivially domains of holomorphy, however, not all open sets in $\mathbb{C}^n$ are domains of holomorphy. Of course, D^2 is a domain of holomorphy in $\mathbb{C}^2$. In fact, the function

$$g(x,y) = f(x) f(y)$$

cannot be continued holomorphically to a larger open set containing D^2, where f is the function defined in $(*)$.

19.11 *Domains of Existence.* An open subset U of E is said to be
the *domain of existence* of a holomorphic mapping f on U if there do
not exist two nonempty connected open sets U_1 and U_2 in E and f_1
$\in H(U_1)$ such that $U_2 \subset U \cap U_1$, $U_1 \not\subset U$ and $f_1 = f$ on U_2.

The following theorem follows immediately from definitions of
domains of holomorphy and domains of existence.

THEOREM. *If U is a domain of existence of a holomorphic mapping,
then U is a domain of holomorphy.*

19.12. We now give some sufficient conditions for an open set U to
be a domain of holomorphy or a domain of existence. The following
lemma is repeatedly used for this purpose.

LEMMA. *For an open set U, let U_1 and U_2 be nonempty connected open
sets with $U_2 \subset U \cap U_1$ and $U_1 \not\subset U$. If V_2 is the connected component
of $U \cap U_1$ containing U_2, then there exists a point $\xi \in \delta U \cap U_1 \cap \delta V_2$
which is a limit of a sequence (x_n) in V_2 (δU denotes the boundary
of U).*

Proof. Let $\gamma: [0,1] \to U_1$ be an arbitrary path joining a point in
V_2 to a point in $U_1 \setminus U$, and let

$$t_o = \inf \{t: \gamma(t) \in U_1 \setminus V_2\}$$

$$t_1 = \inf \{t: \gamma(t) \in U_1 \setminus U\}$$

Then it is clear that $\gamma(t_o) \in U_1 \cap \delta U_2$, $\gamma(t_1) \in U_1 \cap \delta U$, and $t_o \leq t_1$.
We show that $t_o = t_1$; hence $\gamma(t_o) \in U_1 \cap \delta U \cap \delta U_2$. In fact, if t_o
< t_1, then $\gamma: [0,t_o] \to U \cap U_1$ becomes a path which connects a point.
in V_2 to a point outside of V_2, which is impossible because V_2 is a
connected component of $U \cap U_1$. Let $\xi = \gamma(t_o)$ and $x_n = (t_o - 1/n)$ for
all n > N for some suitable N. Then $x_n \in V_2$ and $x_n \to \xi$.   □

19.13 THEOREM. *Let U be an open subset of E. If for every sequence
(x_n) in U which converges to a boundary point $\xi \in \delta U$ there corresponds*

a mapping $f \in H(U)$ *such that*

$$\sup_n \left| f(x_n) \right| = \infty$$

then U *is a domain of holomorphy.*

Proof. Suppose the contrary and find two nonempty connected open sets U_1 and U_2 with $U_2 \subset U \cap U_1$ and $U_1 \not\subset U$ such that for every holomorphic mapping f on U there corresponds a mapping $f_1 \in H(U_1)$ with $f = f_1$ on U_2. We can assume without loss of generality that U_2 is a connected component of $U \cap U_1$. It follows then from Lemma 19.12 that there exists a sequence (x_n) in U_2 converging to a point ξ in $\delta U \cap U_1 \cap \delta U_2$. By the hypothesis, f is unbounded on the sequence (x_n), however, $f_1(x_n) \to f_1(\xi)$. This is absurd since $f = f_1$ on U_2. $\square$

19.14 THEOREM. *Let* $f \in H(U)$. *If for every convex open set* V *of* E *meeting* δU, *f is unbounded on each connected component of* $U \cap V$, *then* U *is a domain of existence of* f.

Proof. Suppose the contrary and find two nonempty connected open sets U_1 and U_2 with $U_2 \subset U \cap U_1$ and $U_1 \not\subset U$ and find a mapping f in $H(U_1)$ such that $f_1 = f$ on U_2. We may assume that U_2 is a connected component of $U \cap U_1$. It follows from Lemma 19.12 that there exists ξ in $\delta U \cap U_1 \cap \delta U_2$ which is the limit of a sequence (x_n) in U_2. Since f_1 is locally bounded, we can find a convex neighborhood of zero such that $\xi + V \subset U_1$ and f_1 is bounded on $\xi + V$. Then a tail of the sequence (x_n) will eventually be in $\xi + V$. Thus $(\xi + V) \cap U_2$ is nonempty. Let $p \in (\xi + V) \cap U_2$ and W a connected component of $U \cap (\xi + V)$. Since U_2 is a connected component of $U \cap U_1$, we have $W \subset U_2$. Then it follows from the hypothesis of the theorem, f is unbounded on W; this contradicts the fact that $f = f_1$ on W, and f_1 is bounded on W. $\square$

19.15 THEOREM. *Every domain of holomorphy is holomorphically convex.*

Proof (Noverraz, 1973). Let U be a domain of holomorphy, and suppose that U is not holomorphically convex. Then there is a compact set K in U such that

$$d(\hat{K}, CU) = 0 \qquad\qquad (*)$$

where $\hat{K}$ denotes the holomorphic convex hull of K. Let $r > 0$ be such that $d(K, CU) = r$. By (*), we can find $\xi \in \hat{K}$ satisfying

$$d(\xi, CU) < r/2$$

Fix y in E with $\|y\| < r$. Choose $s > 1$ to satisfy

$$d(K_y, CU) > 0$$

where K_y denotes $K + \{zy: z \in \mathbb{C}, |z| \leq s\}$. If $f \in H(U)$, we can find a convex balanced neighborhood W of zero such that $K_y + sW \subset U$ and $|f|$ is bounded by a constant $M > 0$ on $K_y + sW$.

For each $w \in W$, $a \in \mathbb{C}$, $|a| \leq 1$, and $m \in \mathbb{N}$, consider the following holomorphic mapping

$$x \in U \rightarrow \frac{1}{m!} \hat{d}^m f(x)(ay + w) \in \mathbb{C}$$

Since $\xi \in \hat{K}$, we have

$$\left| \frac{1}{m!} \hat{d}^m(\xi)(ay + w) \right| \leq \sup_{x \in K} \left| \frac{1}{m!} \hat{d}^m f(x)(ay + w) \right|$$

$$= \sup \left| \frac{1}{2\pi i} \int_{|z|=s} \frac{f[x + z(ay+w)]}{z^{m+1}} dz \right| \leq \frac{M}{s^m}$$

by the Cauchy integral formula 13.4 since $x + z(ay + w) \in K_y + sW$.

The above inequality shows that the power series

$$\sum_{m=0}^{\infty} \frac{1}{m!} \hat{d}^m f(\xi) (x-\xi)$$

converges normally for $x = \xi + ay + w$, for any $a \in \mathbb{C}$, for any $y \in E$ with $\|y\| < r$, and for any $w \in W$. Hence the power series determines a holomorphic mapping, say f', on $B_r(\xi)$. Notice that $B_r(\xi)$ is not contained in U since $d(\xi, \mathbb{C}U) < r/2$, but $f = f'$ on a neighborhood of ξ contained in $B_r(\xi) \cap U$. This is absurd since U is a domain of holomorphy. From this contradiction we conclude that there is a neighborhood V of zero such that $\hat{K} + V \subset U$; hence U must be holo-morphically convex.   □

19.16. We now ask when the converse of Theorem 19.15 is true. In the case where $E = \mathbb{C}^n$, the Cartan-Thullen Theorem (see Cartan-Thullen, 1932) asserts the following.

THEOREM. *For an open set* U *in* $\mathbb{C}^n$, *the following are equivalent.*

   (a) U *is a domain of existence.*

   (b) U *is a domain of holomorphy.*

   (c) U *is holomorphically convex.*

Although these statements are equivalent for the finite dimen-sional cases, it is not obvious that they are equivalent for infi-nite dimensions. As a matter of fact, it is not true in general. Hirschowitz (1969) and Josefson (1974) produced counter-examples showing the fallacy of the reverse implications. Their examples are drawn from non-separable Banach spaces. Thus to establish the theorem for infinite dimensional spaces, we need some restrictions on the spaces of holomorphic mappings as shown in the following theorem.

19.17 THEOREM. *If* U *is holomorphically convex and the topologies* τ_ω *and* τ_δ *are identical on* H(U), *for every sequence* (x_n) *in* U *con-verging to a boundary point* $\xi \in \delta U$ *there exists a mapping* $f \in H(U)$

such that

$$\sup \{|f(x_n)| : n \in \mathbb{N}\} = \infty$$

Proof. Assume $\tau_\omega = \tau_\delta$ on $H(U)$. Suppose that there is a sequence (x_n) in U converging to a boundary point $\xi \in \delta U$ such that every holomorphic mapping on U is bounded. Then the set $B = \{x_n : n \in \mathbb{N}\}$ is a bounding subset of U. Thus the seminorm defined by

$$p(f) = \|f\|_B$$

is τ_δ continuous; hence it is τ_ω continuous on $H(U)$. Thus there exists a compact set K of U such that p is ported by K; i.e., for every neighborhood V of K with $V \subset U$ there corresponds a constant $c > 0$ such that

$$p(f) \leq c\|f\|_V$$

for all $f \in H(U)$. Replacing f with f^n, taking the nth root, and letting $n \to \infty$, we have

$$p(f) \leq \|f\|_V$$

for all $f \in H(U)$. Since this holds for every V containing K, we have

$$p(f) \leq \|f\|_K$$

for all $f \in H(U)$; it means that for every $n \in \mathbb{N}$ we have

$$|f(x_n)| \leq \|f\|_K$$

for all $f \in H(U)$; i.e.,

$B \subset \hat{K}$

Because $\hat{K}$ is closed in U, we have $\xi \in \delta U \cap \delta \hat{K}$ and

$d(\hat{K}, CU) = 0$

This contradicts the fact that U is holomorphically convex and

$d(\hat{K}, CU) > 0$

which completes the proof. □

COROLLARY. *If U is holomorphically convex and the topologies* τ_ω *and* τ_δ *are identical on H(U), then U is a domain of holomorphy.*

We summarize previous results in the following theorem.

19.18 CARTAN-THULLEN THEOREM. *Let U be an open subset of a normed space E. Consider the following statements:*

(a) *For every sequence* (x_n) *in U converging to a boundary point* $\xi \in \delta U$, *there exists a mapping* $f \in H(U)$ *such that*

$\sup\{|f(x_n)| : n \in \mathbb{N}\} = \infty$

(b) *U is a domain of existence of a holomorphic mapping;*

(c) *U is a domain of holomorphy;*

(d) *U is holomorphically convex.*

Then the following implications hold:

(a) ⇒ (c)   *and*   (b) ⇒ (c) ⇒ (d)

If the topologies τ_ω *and* τ_δ *are identical on H(U), the statements (a), (c) and (d) are equivalent.*

REMARK. As remarked before, in the above theorem, the reverse implications are not valid in general. Hirschowitz (1969) shows that the unit ball of $C([0,\Omega])$, where Ω is the first uncountable ordinal, is not the domain of existence of a holomorphic mapping:

hence (c) → (b) is not true. Josefson (1973) shows an example of a
set in $c_o(\Gamma)$ showing (d) → (c) is not true, where Γ is an uncountable
set (see 20.19).

B-DOMAINS OF HOLOMORPHY

19.20 *Bounded Type.* Recognizing the difficulties in establishing
the reverse implications in the Cartan-Thullen theorem for infinite
dimensional Banach spaces, Dineen (1970) introduced b-domains of
holomorphy, domains of holomorphy of bounded type, by replacing the
space H(U) of holomorphic mappings with the subspace $H_b(U)$ of H(U)
consisting of all holomorphic mappings of bounded type, and successfully formulated a Cartan-Thullen theorem for Banach spaces.

A holomorphic mapping f $\in$ H(U) is said to be of *bounded type* if
f is bounded on every U-bounded subset of U.

In 15.11, it was shown that every infinite dimensional Banach
space E admits an entire function of unbounded type. This shows
that the space $H_b(E)$ is a proper subspace of H(E). This is also
true for any open subset U of E; $H_b(U)$ is a proper subspace of H(U)
if E is infinite dimensional.

We endow the space $H_b(U)$ with the topology τ_b of uniform convergence on U-bounded sets; i.e., the topology generated by the seminorms of the form:

$$p_B(f) = \|f\|_B$$

where B is a U-bounded subset of U. Then it is easy to show that
$(H_b(U), \tau_b)$ is complete by noticing $\tau_o \leq \tau_b$ on $H_b(U)$.

Furthermore, $(H_b(U), \tau_b)$ is metrizable since U is the union of
the increasing sequence (U_n) of U-bounded sets such that every U-
bounded subset of U is contained in some U_m. Hence $(H_b(U), \tau_b)$ is
a Fréchet space.

19.21 THEOREM. *Let U be an open subset of a Banach space. Then
the following are equivalent:*

(a) E *is infinite dimensional.*

(b) $H_b(U)$ *is a proper subspace of* $H(U)$.

Proof. We prove (a) $\rightarrow$ (b). By 15.11, there exists an entire function f of unbounded type. This shows that for some bounded set X
f is unbounded on X. Let

$$M = \sup \{\|x\|: x \in X\}$$

We may assume that U contains the zero vector O. Find r > 0 and
s > 0 such that d(O,CU) > r, M < s and

$$B_r(O) \subset U \qquad \text{and} \qquad X \subset B_s(O)$$

Let X' = (r/s)X. Then X' is also bounded and $X' \subset B_r(O)$; i.e, X'
is a U-bounded set. Let

$$g(x) = f[(s/r)x]$$

Then g is clearly holomorphic on U and g is unbounded on X'.   □

19.22 *B-Domain of Holomorphy and B-Holomorphic Convexity.* In the
definition of a domain of holomorphy (19.9), if we replace the space
H(U) with the space $H_b(U)$, we have a definition of a *b-domain of
holomorphy*. In the definition of a holomorphically convex set, replacing H(U) with $H_b(U)$ and compact sets with U-bounded sets, we have
the notion of a *b-holomorphically convex set.*

Notice that a b-domain of holomorphy is a domain of holomorphy
if E is a finite dimensional space because of 19.21.

19.23 CARTAN-THULLEN THEOREM. *Let U be an open subset of a normed
space E. Then the following are equivalent:*

(a) *For every sequence* (x_n) *in U which converges to a boundary
point* $\xi \in \delta U$ *there exists* $f \in H_b(U)$ *such that*

$$\sup \, \left| f(x_n) \right| \, = \, \infty$$

(b)   U *is a b-domain of holomorphy;*

(c)   U *is b-holomorphically convex.*

Proof.   The proof that (a) → (b) is identical to that of 19.13.

(b) → (c).   This is again a modification of the proof in 19.15 using the result of Exercise 19D which can be proved easily by use of the Cauchy inequalities and the Cauchy-Hadamard formula.

Let U be a b-domain of holomorphy and suppose that U is not b-holomorphically convex.   Then for some U-bounded set B in U, we have

$$d(\hat{B},CU) = 0 \tag{1}$$

where $\hat{B}$ denotes the $H_b(U)$-convex hull of B.   Let $r > 0$ be such that $d(B,CU) = r$.   By (1), there exists $\xi \in \hat{B}$ such that

$$d(\xi,CU) < r/2$$

We claim that for any $f \in H_b(U)$ there is a mapping $g \in H_b(B_r(\xi))$ such that $f = g$ on a neighborhood of ξ contained in $B_r(\xi) \cap U$.

Let $f \in H_b(U)$ and let $s > 0$ be the radius of convergence of the Taylor series of f at ξ.   It follows from Exercise 19D(c) that $s \geq r$.   Hence we can define a holomorphic mapping $g\colon B_r(\xi) \to \mathbb{C}$ by

$$g(x) = \sum_{m=1}^{\infty} \frac{1}{m!} \, \hat{d}^m f(\xi)(x-\xi)$$

By the Cauchy inequalities, we can see that $g \in H_b(B_r(\xi))$.   Since $f \in H_b(U)$, it follows that $f = g$ on $B_{r/2}(\xi)$, a connected open subset of U.   Thus U cannot be a b-domain of holomorphy.

(c) → (a).   Suppose that (a) does not hold.   Then there exists a sequence (x_n) in U converging to a boundary point $\xi \in \delta U$ on which every mapping in $H_b(U)$ is bounded. Define a seminorm p on $H_b(U)$ by

$$p(f) = \sup \{|f(x_n)| : n \in \mathbb{N}\}$$

Then p is τ_b-continuous. In fact, if $B_p = \{f: p(f) \leq 1\}$, then B_p is the intersection of the closed and convex sets $\{f: |f(x_n)| \leq 1\}$ for all n since $f \to |f(x_n)|$ is continuous for τ_b. It is clear that B_p is absorbing by hypothesis. Hence B_p is a barrel. We know that every Fréchet space is barrelled, and hence B_p is a neighborhood of zero in $H_b(U)$.

It follows that there exist a U-bounded subset B in U and $c > 0$ such that

$$p(f) \leq c\|f\|_B \tag{2}$$

for all $f \in H_b(U)$. By the familiar method, in (2), replacing f with f^n, taking the nth roots, and letting $n \to \infty$, we obtain

$$p(f) \leq \|f\|_B$$

for all $f \in H_b(U)$. This shows that for any $f \in H_b(U)$,

$$|f(x_n)| \leq \|f\|_B$$

Hence $x_n \in \hat{B}$, where $\hat{B}$ is the $H_b(U)$-convex hull of B. Because $\hat{B}$ is closed in U, we have $\xi \in \hat{B}$ since $x_n \to \xi$. Thus $d(\hat{B}, CU) = 0$, which is a contradiction since B is U-bounded and U is b-holomorphically convex. □

19.24 REMARK. Inspired by Dineen's result, Matos (1972) proved a similar Cartan-Thullen theorem for various subalgebras of $H(U)$. Independently of Dineen, Schottenloher (1974) considered a much more general situation by defining regular classes and admissible coverings for a Riemann domain spread over a Banach space. For each regular class, he showed a Cartan-Thullen theorem.

ENVELOPES OF HOLOMORPHY

19.25 *Envelope of Holomorphy.* For every open connected set U there
exists a maximal domain to which every holomorphic mapping in U can
be continued holomorphically. Clearly, a domain of holomorphy is
the envelope of holomorphy of itself. For an open set U which is
not a domain of holomorphy, there is such a maximal domain only if
we can admit domains spread over a normed space. The subject of
domains (connected manifolds) spread over a normed space has been
studied actively and constructions of the envelope of holomorphy
for an open set U are presented in several different ways: some
using sheaves of holomorphic mappings and others using continuous
homomorphisms on the space H(U). These topics would occupy several
chapters, and presenting them in this book would be over-ambitious;
hence we refrain from doing so. We refer the interested reader to
Hirschowitz (1972), Novarraz (1973), Coeuré (1974), Schottenloher
(1974a), and Alexander (1968).

19.26. In passing we present one theorem which relates to the
opening remark in 18.1. Let E(U) be the envelope holomorphy. Then
E(U) has the same class of holomorphic mappings as U. In contrast
to the finite dimensional case, the extension mapping

$$f \in H(U) \rightarrow \hat{f} \in H(E(U))$$

need not be an open map with respect to the compact open topology
τ_o or the Nachbin topology τ_ω although it is continuous (see Josefson,
1974). Thus H(U) and H(E(U)) are not homeomorphic under the exten-
sion map $f \rightarrow \hat{f}$, where $\hat{f}$ denotes the holomorphic continuation of f.

19.27 THEOREM. *Let U and V be connected open subsets of a normed
space E with U $\subseteq$ V such that every holomorphic mapping f $\in$ H(U) has
a (unique) continuation $\hat{f} \in$ H(V). Then the extension map*

$$f \in H(U) \rightarrow \hat{f} \in H(V)$$

is a homeomorphism with respect to the topology τ_δ.

Proof. Recall the definition of the topology τ_δ given in 18.5:
the topology τ_δ on $H(U)$ is the inductive limit of the spaces $H_I(U)$,
where

$$H_I(U) = \{f \in H(U) : \|f\|_{V_n} < \infty \text{ for all } n\}$$

for any countable open cover $I = (V_n)$ of U. The topology on $H_I(U)$
is generated by seminorms

$$p_n(f) = \|f\|_{V_n}$$

and $H_I(U)$ is a Fréchet space under this topology. Notice that we
can take $H(U)$ as the inductive limit of the space $H_I(U)$ without
sacrificing the topology τ_δ, where each cover $I = (V_n)$, not necessarily open, satisfies the following condition

$$V_n + \delta_n B_1(0) \subset V_{n+1} \tag{*}$$

for some $\delta_n > 0$.

It is clear that the extension map is surjective and its inverse
is continuous. Hence it remains to show the mapping

$$f \in H(U) \rightarrow \hat{f} \in H(V)$$

is continuous with respect to the topology τ_δ. It suffices to show
that for any countable open cover I for U with the condition (*),
there exists a countable open cover J for V such that the mapping

$$f \in H_I(U) \rightarrow \hat{f} \in H_J(V)$$

is continuous; this follows from the following diagram

$$f \in H(U) \rightarrow \hat{f} \in H(V)$$
$$\uparrow \qquad\qquad \uparrow$$
$$f \in H_I(U) \rightarrow \hat{f} \in H_J(V)$$

in which the continuity in the lower direction implies the continuity in the upper direction. The upward arrow in the diagram denotes the inclusion map.

For this purpose, for each $y \in V$ and I, we consider the mapping

$$f \in H_I(U) \rightarrow \hat{f}(y) \in \mathbb{C}$$

and show that it is continuous on $H_I(U)$. (This shows that each $y \in V$ is considered as a continuous homomorphism on $H_I(U)$.) In this proof we need the uniform boundedness principle or the Banach-Steinhaus theorem for barrelled spaces. These were presented in 2.17 and Ex. 2K for Banach spaces. We record the following theorem whose proof is found in Horváth (1966).

THEOREM. *(1) Every pointwisely bounded family of continuous linear maps from a barrelled space into a locally convex space is equicontinuous.*

(2) If (A_n) is a sequence of continuous linear maps from a barrelled space into a Hausdorff locally convex space which converges pointwisely to a map A, then A is a continuous linear map.

We now return to our proof. Consider the following set

$$S = \{y \in V: f \in H_I(U) \rightarrow \hat{f}(y) \text{ is continuous}\}$$

Then it is not empty since $U \subseteq S$. We claim that S is both open and closed; hence $S = V$.

Let $\xi \in S$. Since $f \rightarrow \hat{f}(\xi)$ is continuous, there exist $V_n \in I$ and $c > 0$ such that for all $f \in H_I(U)$ we have

$$\left| \hat{f}(\xi) \right| \le c \|f\|_{V_n}$$

Replacing f with f^m, taking the mth root, and letting $m \to \infty$, we obtain

$$\left| \hat{f}(\xi) \right| \le \|f\|_{V_n} \tag{1}$$

for all $f \in H_I(U)$.

We claim that if $x \in V$ such that $\|x-\xi\| < \delta_n$ then $x \in S$; hence S is open. Recall that $V_n + \delta_n B_1(0) \subseteq V_{n+1}$. Apply the Cauchy inequalities 13.5 to the Taylor series of $\hat{f}$ at ξ to get

$$\left\| \hat{f}(x) \right\| \le \sum_{m=0}^{\infty} \left\| \frac{1}{m!} \hat{d}^m \hat{f}(\xi)(x-\xi) \right\| \le \sum_{m=0}^{\infty} \left\| \frac{1}{m!} \hat{d}^m \hat{f} \right\|_{V_n} \|x-\xi\|^m$$

$$\le \|f\|_{V_k} \sum_{m=0}^{\infty} \left(\|x-\xi\|/\delta_n \right)^m \le c \|f\|_{V_k} \tag{2}$$

where $k \ge n+1$, $\|x-\xi\| < \delta_n$, and c denotes the infinite sum in (2); thus $x \in S$. This proves that S is open.

We now show that S is closed. Let (x_n) be a sequence in S which converges to a point x in V. We claim that $x \in S$. Since each $\hat{f}$ is continuous on V, we have $\hat{f}(x_n) \to \hat{f}(x)$. Considering (x_n) as a sequence of continuous linear maps from $H_I(U)$ into $\mathbb{C}$, we know that $x \in S$ by statement (2) in the theorem quoted above. This proves that S is closed, and hence $S = V$.

We now construct a countable open cover J for V corresponding to each countable open cover I satisfying (*). For each n, let

$$W_n = \{ y \in V: \left| \hat{f}(y) \right| \le \|f\|_{V_n} \text{ for all } f \in H_I(U) \}$$

It is clear that $J = (W_n)$ is a countable cover for V. The mapping

$$f \in H_I(U) \rightarrow \hat{f} \in H_J(V) \tag{3}$$

is well-defined. Furthermore, for each n, the set

$$\{f \in H_I(U) : \left\| \hat{f} \right\|_{W_n} \leq 1\}$$

is a barrel in the barrelled space $H_I(U)$; hence it is a neighborhood
of zero. Hence the mapping in (3) is continuous. This completes
the proof that the extension map is a homeomorphism.   □

19.28 REMARK. The proof of Theorem 19.27 given above indicates
that the envelope of holomorphy of a connected open subset U of a
Banach space E can be constructed using the spectrum of $(H(U), \tau_\delta)$
in the following way. First the set of all continuous homomorphisms
on $H_I(U)$ is endowed with a canonical structure of a complex manifold
spread over E. Take   a suitable component $E_I(U)$ of the manifold
which is the maximal domain of holomorphic continuations of all
mappings in $H_I(U)$. Then the envelope of holomorphy is the intersection of all $E_I(U)$ for all countable covers I of U determining
the topology τ_δ. See Schottenloher (1974a).

EXERCISES

19A. Let U be a connected open set of E. Then U is C(U)-convex.

19B. *Bounded Type.* Let U be an open subset of a normed space E.

  (a) If $f \in H_b(U)$, then for each $m \in \mathbb{N}$, $\hat{d}^m f$ is bounded on any
       U-bounded subset of U.

  (b) If $\Sigma\, P_m(x - \xi)$ is a power series with radius of convergence
       $r > 0$, then it represents a mapping $f \in H_b(B_r(\xi))$.

  (c) If B is a U-bounded subset of U and $r = d(B, CU)$, then for
       any $f \in H_b(U)$ and $\xi \in \hat{B}_{H_b(U)}$ the radius of convergence of
       the Taylor series of f at ξ is at least r.

19C. For an open subset U of a normed space E and a Banach space F,

(a) Define the notions of domains of holomorphy and holomorphic convexity with respect to F-valued functions.

(b) Establish a Cartan-Thullen theorem for F-valued mappings.

(c) Establish Theorem 19.27 for F-valued mappings.

(d) Study the relationship between domains of holomorphy with respect to ℂ-valued and F-valued mappings.

20

The Levi Problem in Banach Spaces

20.1 *Levi Problem*. We have introduced the concept of pseudoconvex domains in 15.28. Recall that an open subset U of a normed space E is said to be *pseudoconvex*, if the function $-\log d_U$ is plurisubharmonic on U, where d_U is the boundary distance function defined by

$$d_U(x) = d(x, \complement U)$$

In the preceding chapter we descirbed domains in terms of holomorphic mappings, but pseudoconvexity does not use any holomorphic mappings directly in its definition. However, holomorphic mappings and plush (short for plurisubharmonic) mappings are related intimately through domains of holomorphy and pseudoconvexity. In 1911, E. E. Levi asked whether every pseudoconvex domain in $\mathbb{C}^2$ is a domain of holomorphy. This became the celebrated problem known as the *Levi problem* and solved by K. Oka in 1942 and extended to domains in $\mathbb{C}^n$ by several mathematicians, K. Oka in 1953, F. Norguet in 1954, and H. Bremerman in 1954. It is easy to show the converse of the Levi problem is true (see 20.6). Thus in $\mathbb{C}^n$, an open set U is a domain of holomorphy if and only if U is pseudoconvex. See Hörmander (1967) and Gunning and Rossi (1965). This is a very deep result in several complex variables to which many major branches of mathematics applied; and it is considered as the highlight of several complex variables.

In this chapter we extend the solution of the Levi problem for $\mathbb{C}^n$ to a Banach space with a Schauder basis, which is due to L. Gruman and C. O. Kiselman (1972). The Levi problem for a general separable Banach space still remains unsolved.

Unlike the previous chapters, this chapter is not self-contained. We use some finite dimensional results on the Levi problem to solve the infinite dimensional case by approximating infinite dimensional sets with finite dimensional ones. Our main reference on the finite dimensional Levi problem will be Hörmander (1966) and references for the infinite dimensional case are Noverraz (1973) and (1979).

PSEUDOCONVEXITY

20.2 *Plurisubharmonic Convexity.* Let U be an open subset of E and denote Ps(U) the space of all plush functions on U. For a compact subset K of U, we define the *P-convex hull* (or *plurisubharmonic convex hull*) of K in U by

$$\hat{K}_{Ps(U)} = \{x \in U: f(x) \le \sup_{t \in K} f(t) \text{ for all } f \in Ps(U)\}$$

An open subset U is called *P-convex* or *plurisubharmonic convex* if for every compact set K in U, its P-convex hull is U-bounded.

20.3 LEMMA. *The P-convex hull of a compact set is always a subset of the holomorphic convex hull; the P-convex hull of a compact set is precompact.*

Proof. Let K be a compact subset of an open set U. Since the holomorphic convex hull is precompact, it suffices to show that

$$\hat{K}_{Ps(U)} = \hat{K}_{H(U)}$$

Let $x \in \hat{K}_{Ps(U)}$. If $f \in H(U)$, then $\log |f| \in Ps(U)$ (see 15.25). Thus

$$\log |f(x)| \le \sup_{t \in K} \log |f(t)|$$

from which we conclude that

$$|f(x)| \leq \|f\|_K$$

for all $f \in H(U)$. Hence $x \in \hat{K}_{H(U)}$.   □

20.4 THEOREM. *Every holomorphically convex open set is P-convex.*

Proof. Let U be a holomorphically convex open subset of a normed
space E and let $f \in H(U)$. Then $|f|$ is plush on U by Example 15.25;
hence $|f| \in Ps(U)$. Let K be a compact subset of U. Then every
point in the P-convex hull of K belongs to the holomorphic convex
hull; hence the P-convex hull of K is U-bounded.   □

20.5 THEOREM. *Let U be an open subset of a normed space* E. *The*
following statements are equivalent:

  (a)   U is pseudoconvex.

  (b)   U is finitely pseudoconvex; i.e., $U \cap F$ *is pseudoconvex in*
         F for any finite dimensional subspace F of E.

  (c)   U is P-convex.

Proof. It is immediate that the statements (a) and (b) are equivalent
from the definition of pseudoconvexity and the definition of plush
functions.

       (a) → (c). Let K be a compact subset of U. Then $-\log d_U$ is
plush on U. Hence if x is a member of the P-convex hull $\hat{K}_{Ps(U)}$ of
K, then

$$-\log d(x,CU) \leq \sup_{t \in K} -\log d(t,CU)$$

Applying the exponential function to this relation, we get

$$d(x,CU) \geq d(K,CU)$$

Since $d(K,CU) > 0$, we have $d(\hat{K}_{Ps(U)},CU) > 0$. Therefore $\hat{K}_{Ps(U)}$ is U-bounded.

(c) → (b). In this proof we use the fact that in $\mathbb{C}^n$ an open set is pseudoconvex if and only if it is P-convex (see Hörmander, 1966, p. 46). Let F be a finite dimensional subspace of E. We want to show that $U \cap F$ is P-convex.

Let K be a compact subset of $U \cap F$. We need to show that the $Ps(U \cap F)$-convex hull of K is $U \cap F$-bounded. Since K is also a compact subset of U, $\hat{K}_{Ps(U)}$ is U-bounded and

$$\hat{K}_{Ps(U)} \subset \overline{co}\ K \subset F$$

Hence $\hat{K}_{Ps(U)}$ is precompact in $U \cap F$ and $U \cap F$-bounded. Thus it is sufficient to show that

$$\hat{K}_{Ps(U \cap F)} \subset \hat{K}_{Ps(U)}$$

Let $\xi \in U \cap F \setminus \hat{K}_{Ps(U)}$. Then there exists $v \in Ps(U)$ such that

$$u(\xi) = v(\xi) > \sup_{t \in K} v(t) = \sup_{t \in K} u(t)$$

where u is the restriction of v to $U \cap F$. This proves that

$$\xi \in U \cap F \setminus \hat{K}_{Ps(U \cap F)} \qquad \square$$

20.6 COROLLARY. *Every holomorphically convex open set is pseudoconvex; in particular, every convex open set is pseudoconvex.*

20.7 THEOREM. *If U is a pseudoconvex open set and v is a plush function on U, then for any $\beta \in \mathbb{R}$,*

$$U_\beta = \{x \in U: v(x) < \beta\}$$

is pseudoconvex.

Proof. Since v is upper semicontinuous, U_β is open. We show that U_β is finitely pseudoconvex. Let F be a finite dimensional subspace of E, and let K be a compact subset of $U_\beta \cap F$. Find $r \in \mathbb{R}$, such that

$$\sup_{x \in K} v(x) \leq r < \beta$$

Then for any point $\xi \in \hat{K}_{Ps(U \cap F)}$, we have

$$v(\xi) \leq \sup_{x \in K} v(x) \leq r < \beta$$

Thus $\xi \in U_\beta \cap F$, i.e., $\hat{K}_{Ps(U \cap F)} \subset U_\beta \cap F$. Notice that we can consider $Ps(U \cap F)$ as a subspace of $Ps(U_\beta \cap F)$ by restricting each mapping in $Ps(U \cap F)$ to the subset $U_\beta \cap F$. Hence

$$\hat{K}_{Ps(U_\beta \cap F)} \subset \hat{K}_{Ps(U \cap F)} \subset U_\beta \cap F$$

Since U is pseudoconvex, $\hat{K}_{Ps(U \cap F)}$ is compact in $U \cap F$, and hence it is compact in $U_\beta \cap F$. This shows that $\hat{K}_{Ps(U_\beta \cap F)}$ is compact in $U_\beta \cap F$. Therefore, $U_\beta \cap F$ is a pseudoconvex open set in F.   □

THE LEVI-OKA THEOREM FOR BANACH SPACES

20.8 *Schauder Basis.* A sequence (e_n) in a Banach space E is called a *Schauder basis* if for each x in E there exists a unique sequence (x_n) of scalars such that

$$\lim_{m \to \infty} \left\| x - \sum_{n=1}^{m} x_n e_n \right\| = 0$$

i.e.,

$$x = \sum_{n=1}^{\infty} x_n e_n$$

We write each member $x \in E$ with respect to the basis (e_n) by

$$x = (x_1, x_2, \ldots, x_m, \ldots)$$

if there can be no misunderstanding.

Banach proved that any basis is a Schauder basis. See McArthur (1972) for the development of the theory of basis through 1972. It is not hard to show that if a Banach space has a Schauder basis, then the space is separable. The converse, known as the *basis problem* and posed by S. Banach as early as 1932, remained an unsolved problem in functional analysis until 1972. In that year P. Enflo (1973) announced that there exists a separable Banach space without a basis.

The standard unit vectors $e_1, e_2, \ldots$ in the Banach space ℓ^p form a Schauder basis. For other examples and a study of this topic, see Lindenstrauss and Tzafriri (1977) and Singer (1970).

20.9. We present the major theorem of the chapter, the Levi-Oka theorem for Banach spaces with basis.

THE LEVI-OKA THEOREM (Gruman and Kiselman, 1972). *Let E be a Banach space with a Schauder basis and let U be a pseudoconvex open subset of E. Then U is a domain of existence of a holomorphic mapping.*

L. Gruman (1974) was the first to give a complete solution to the Levi problem in an infinite dimensional space in 1972. He applied the L^2 estimates and existence theorems for the Cauchy-Riemann equations in pseudoconvex domains in $\mathbb{C}^n$ to an inductive construction of a holomorphic mapping f to show that a pseudoconvex domain in a separable Hilbert space is a domain of existence of the holomorphic mapping f. L. Gruman and C. O. Kiselman (1972) immediately extended the Gruman method to a Banach space with a Schauder basis.

Our plan for the proof is as follows. Since E is separable, we can find a sequence (b_n) which is dense in the boundary δU. Corresponding to this sequence there exists a sequence (a_n) in U such that every b_k is a limit of some subsequence of (a_n). Then we construct a holomorphic function f which is unbounded on every subsequence of (a_n), and show that U is the domain of existence of f.

20.10. We now give a proof showing the existence of the sequence mentioned above through a series of lemmas. We use frequently some finite dimensional results from Hormander (1967).

Let E be a Banach space with a Schauder basis (e_n), and let E_m be the vector subspace of E generated by $\{e_1, \ldots e_m\}$. Then

$$E_\infty = \cup\{E_m : m \in \mathbb{N}\}$$

is dense in E. Moreover, the projection $P_m : E \to E_m$ defined by

$$(x_1, x_2, \ldots, x_m \ldots) \to (x_1, x_2, \ldots, x_m, 0, 0, \ldots)$$

is a continuous linear map in L(E;E) for each m. Since the sequence (P_m) converges pointwisely to the identity map i_E, for each $x \in E$, there exists $M_x > 0$ such that $\|P_m(x)\| \leq M_x$ for all m. By the Banach-Steinhaus theorem 2.17, we conclude that the sequence (P_m) is bounded in L(E;E) and equicontinuous on E. We also notice that

$$P_m \circ P_n = P_{\min\{m,n\}}$$

for any m and n in $\mathbb{N}$.

For an open pseudoconvex subset U of E and $m \in \mathbb{N}$, let

$$U_m = U \cap E_m$$

$$K_m = \{x \in U : \|x\| \le m, \ d(x, CU) \ge m^{-1}\}$$

$$L_m = \{x \in U : \|P_m(x) - x\| \le \tfrac{1}{2}d(x, CU)\}$$

Then (K_m) is an increasing sequence of closed bounded subsets of U such that its union is U, and L_m is closed in E and $P_m(L_m) \subset U$ for any $m \in \mathbb{N}$.

20.11 LEMMA. *Let* $T = U_m \cap K_n \cap L_p$ *for any* $m, n, p \in \mathbb{N}$. *Then* T *is compact and the holomorphic convex hull of* T *satisfies*

$$\hat{T}_{H(U_m)} = T$$

A compact set satisfying the lemma is called *Runge compact*.

Proof. Let $\hat{T}$ denote the holomorphic convex hull of T with respect to $H(U_m)$ and T_{Ps} the P-convex hull of T with respect to $Ps(U_m)$ for simplicity of notation.

Since T is closed and bounded in U_m, T is compact in a finite dimensional space E_m by the Heine-Borel theorem. For any $x \in \hat{T}$, we want to show that $x \in T$. Because U is pseudoconvex, U_m is a pseudocompact open subset of E_m.

In Hörmander (1966), Theorem 4.3.4 states that if T is a compact subset in a pseudoconvex open set U in $\mathbb{C}^n$, then the holomorphic convex hull and the plurisubharmonic convex hull of T in U are identical (a consequence of Levi-Oka Theorem in $\mathbb{C}^n$). Hence we have $\hat{T} = T_{Ps}$ and $x \in T_{Ps}$.

To show $x \in T$, it is enough to show $x \in K_n$ and $x \in L_p$. First we prove that $x \in K_n$. Since U is pseudoconvex, the mapping

$$x \in U_m \to -\log d(x, CU)$$

is plush on U_m. It follows from $x \in T_{Ps}$ that

$$-\log \, d(x,CU) \leq \sup_{t \in T} -\log \, d(t,CU) \leq -\log \, n^{-1}$$

because $T \subset K_n$. Thus $\dot{d}(x,CU) \geq n^{-1}$. On the other hand, the mapping

$$x \in U_m \rightarrow \log \, \|x\|$$

is plush on U_m by 15.25, hence

$$\log \, \|x\| \leq \sup \, \log \, \|t\| \leq \log \, n$$

because $T \subset K_n$. This shows that $\|x\| \leq n$; hence $x \in K_n$.

With a similar argument, we can show that $x \in L_p$. In fact, notice that the mapping

$$x \in U_m \rightarrow \log \, \|P_p(x)-x\|$$

is plush on U_m (see 15.25). Hence a parallel argument to the above case shows $x \in L_p$. Therefore $x \in T$.   ▽

The symbol ▽ is used not only to indicate the end of the proof of the lemma but also to suggest that the main proof is continuing.

20.12 LEMMA. *For any compact subset K of U there exists* $q = q(K) \in \mathbb{N}$ *such that for any* $m \geq q$,

$$P_m(K) \subset K_q$$

Proof. For $x \in K$, $y \in E$ and $m \in \mathbb{N}$, we have

$$\|P_m(x)-y\| \geq \|x-y\|-\|x-P_m(x)\|$$

Taking the infimum for $y \in CU$, we obtain

$$d(P_m(x), CU) \geq d(x, CU) - \sup_{t \in K} \|t - P_m(t)\|$$

Hence

$$d(P_m(x), CU) \geq d(K, CU) - \sup_{t \in K} \|t - P_m(t)\| \qquad (1)$$

Since $P_m \to i_E$ uniformly on K, we can find $n_1 = n_1(K) \in \mathbb{N}$ such that $n_1 \geq 2/d(K, CU)$ and

$$\sup \|t - P_m(t)\| \leq \tfrac{1}{2} d(K, CU)$$

for all $m \geq n_1$. Thus we have from (1) that

$$d(P_m(x), CU) \geq \tfrac{1}{2} d(K, CU)$$

for all $m > n_1$ and $x \in K$. This shows that

$$d(P_m(x), CU) \geq 1/n_1$$

The sequence (P_m) being uniformly bounded on K, for some $r > 0$ we have $\|P_m(x)\| < r$ for all $x \in K$ and $m \in \mathbb{N}$. Choose $p = p(K) \in \mathbb{N}$ to satisfy $p > r$. Then

$$P_m(K) \subset B_r(0) \subset B_p(0)$$

If we let $q = \max \{n_1, p\}$, for $x \in K$ and $m \geq q$ we have

$$d(P_m(x), CU) \geq 1/q \qquad \|P_m(x)\| \leq q$$

which shows that for $m \geq q$

$$P_m(K) \subset K_q \qquad\qquad \triangledown$$

20.13 LEMMA. *For any compact subset* K *of* U *there exists* $q = q(K)$
$\in \mathbb{N}$ *such that if* $m > n > q$, *then*

$$P_m(K) \subset L_n$$

Proof. By the preceding lemma, we can find $p = p(K)$ in $\mathbb{N}$ such that
for any $m \geq p$, $P_m(K) \subset K_p \subset U$. To prove that $P_m(K) \subset L_n$, we should
check if

$$\left\| P_n(P_m(x)) - P_m(x) \right\| \leq \tfrac{1}{2} d(P_m(x), CU)$$

or, equivalently if $n < m$ and $x \in K$

$$\left\| P_n(x) - P_m(x) \right\| \leq \tfrac{1}{2} d(P_m(x), CU) \tag{1}$$

Since $P_m \to i_E$ uniformly on K, there exists $n_1 = n_1(K)$ in $\mathbb{N}$ such
that for $m \geq n_1$ and $x \in K$

$$\left\| P_m(x) - x \right\| \leq (1/5) d(K, CU) \leq (1/5) d(x, CU)$$

Hence we have for all $x \in K$ and $m > n > n_1$

$$\left\| P_n(x) - P_m(x) \right\| \leq \left\| P_n(x) - x \right\| + \left\| P_m(x) - x \right\| \leq (2/5) d(x, CU) \tag{2}$$

We also have that for $x \in K$ and $m \geq n_1$

$$d(P_m(x),CU) \geq d(x,CU) - \|P_m(x)-x\| \geq d(x,CU) - (1/5)d(x,CU)$$

$$= (4/5)d(x,CU) \tag{3}$$

Suppose that the relation (1) is false. Then there exists $\xi \in K$ such that

$$\|P_n(x)-P_m(x)\| > \tfrac{1}{2}d(P_m(\xi),CU)$$

Then from (3) we have

$$\|P_n(x)-P_m(x)\| > (2/5)d(\xi,CU)$$

which contradicts the relation (2); hence (1) holds. Therefore, if we take $q = \max\{p,n_1\}$. Then for $m > n > q$, we obtain the desired relation

$$P_m(K) \subset L_n$$

This completes the proof. ◁

20.14 LEMMA. *Let* $B = \{b_n : n \in \mathbb{N}\}$ *be a countable dense set in* δU. *Then there exists a sequence* (a_n) *in* $U \cap E_\infty$ *and a strictly increasing sequence* (n_k) *in* $\mathbb{N}$ *such that*

(1) $a_k \in (K_{n_k} \setminus K_{n_{k-1}}) \cap E_{n_k}$ $\quad (k \geq 2)$

(2) *Each point of* B *is the limit of a subsequence of* (a_n).

Proof. Let $\sigma: \mathbb{N} \to \mathbb{N}$ be a surjection such that for each $\sigma^{-1}(i)$ is an infinite set. For $b_{\sigma(1)} \in B$, let $(b_{m,\sigma(1)})$ be a sequence in $U \cap E_\infty$ converging to $b_{\sigma(1)}$ such that

$$\|b_{\sigma(1)} - b_{m,\sigma(1)}\| < 1/m$$

Then $b_{m,\sigma(1)}$ is not in K_m for any $m \in \mathbb{N}$. Since (K_m) is an increasing sequence of bounded closed sets such that its union is U, we can find $n_2 > n_1$ with $n_1 = 1$ such that

$$b_{n_1,\sigma(1)} \in (K_{n_2} \setminus K_{n_1}) \cap E_{n_2}$$

Inductively, in general, we choose $n_{i+1} > n_i$ to satisfy

$$b_{n_i,\sigma(i)} \in (K_{n_{i+1}} \setminus K_{n_i}) \cap E_{n_{i+1}}$$

We then define the desired sequence (a_n) by letting

$$a_{i+1} = b_{n_i,\sigma(i)}$$

Now we verify that the statement (2) holds. Let $n \in \mathbb{N}$. Since the $\sigma^{-1}(n)$ is infinite, we can find a strictly increasing sequence $\{i_1, i_2, \ldots, i_k, \ldots\}$ in $\sigma^{-1}(n)$. Then the subsequence

$$a_{i_k+1} = b_{n_{i_k},\sigma(i_k)} = b_{n_{i_k},n} \to b_n$$

as $k \to \infty$. $\qquad\qquad\qquad\qquad\qquad\qquad\qquad\qquad \triangledown$

We shall replace the sequence (n_i) in Lemma 20.14 with $\mathbb{N}$ in the subsequent lemmas without losing the generality because $n_i \to \infty$ as $i \to \infty$. In this way, we can avoid the cumbersome sub-subscripts. We now restate the lemma with this symbolic adjustment.

LEMMA. *Let* $B = \{b_n : n \in \mathbb{N}\}$ *be a countable dense set in* δU. *Then there exists a sequence* (a_n) *in* $U \cap E_\infty$ *such that*

(a) $a_n \in (K_n \setminus K_{n-1}) \cap E_n$ $\qquad (n \geq 2)$

(b) *Each point of* B *is the limit of a subsequence of* (a_n).

20.15 LEMMA. *Let* $T_n = U_n \cap K_{n-1} \cap L_{n-1}$ *for each* $n \in \mathbb{N}$. *Then there exists a sequence* (f_n) *such that for each* n

(a) $f_n \in H(U_n)$

(b) $\left\| f_n - f_{n-1} \circ P_{n-1} \right\|_{T_n} \leq 2^{-n}$

(c) $\left| f_n(a_n) \right| \geq n$

Proof. We use the following approximation theorem from Hörmander (1966), Theorem 4.3.2.

THEOREM. *Let* U *be a pseudoconvex open set in* $\mathbb{C}^n$ *and* K *a compact subset of* U *such that* $\hat{K}_{Ps(U)} = K$. *Then every holomorphic function which is holomorphic in a neighborhood of* K *can be approximated uniformly on* K *by functions in* $H(U)$.

We use induction. It is easy to find a function $f_1 \in H(U_1)$ such that $\left| f_1(a_1) \right| \geq 1$. Suppose that there exist mappings $f_1, \ldots, f_{i-1}$ satisfying the conditions (a), (b), and (c). We want to find f_i in $H(U_i)$ meeting the same conditions. Notice that T_i is a compact subset of U_i by Lemma 20.11 and

$$P_{i-1}: U_i \rightarrow U_{i-1}$$

$$f_{i-1}: U_{i-1} \rightarrow \mathbb{C}$$

Hence, if we set

$$V_i = [P_{i-1}]^{-1}(U_{i-1})$$

we have

$$T_i \subset V_i \subset H_i$$

$$f_{i-1} \circ P_{i-1} \in H(V_i)$$

Since U is pseudoconvex, U_i is a pseudoconvex open set in the finite dimensional space E_i. By the approximation theorem quoted above, there exists a mapping $h_i \in H(U_i)$ such that

$$\|h_i - f_{i-1} \circ P_{i-1}\|_{T_i} \leq 2^{-(i+1)}$$

Recall that T_i is a Runge compact set by Lemma 20.11. Since a_i is not in K_{i-1}, a_i is not in T_i. Hence a_i is not in $(\hat{T_i})_{H(U_i)} = T_i$. It means that there is a mapping $g_i \in H(U_i)$ such that for some $r_i \in \mathbb{R}$

$$|g_i(a_i)| > r_i > \|g_i\|_{T_i}$$

We assume $r_i = 1$ (if necessary, by dividing the inequalities by r_i). Hence we can find a number $m(i) \in \mathbb{N}$ such that

$$\left| [g_i]^{m(i)}(a_i) \right| > i + \left| h_i(a_i) \right|$$

$$\left\| [g_i]^{m(i)} \right\|_{T_i} < 2^{-(i+1)}$$

Denote

$$G_i = [g_i]^{m(i)}$$

$$f_i = G_i + h_i$$

Then we have

$$f_i \in H(U_i)$$

We now show that f_i satisfies the conditions (b) and (c). In fact,

$$\left\| f_i - f_{i-1} \circ P_{i-1} \right\|_{T_i} = \left\| G_i + h_i - f_{i-1} \circ P_{i-1} \right\|_{T_i}$$

$$\leq \left\| G_i \right\|_{T_i} + \left\| h_i - f_{i-1} \circ P_{i-1} \right\|_{T_i} \leq 2^{-(i+1)} + 2^{-(i+1)} = 2^{-i}$$

which shows the condition (b) holds. Furthermore,

$$\left| f_i(a_i) \right| = \left| G_i(a_i) + h_i(a_i) \right| \geq \left| G_i(a_i) \right| - \left| h_i(a_i) \right|$$

$$> i + \left| h_i(a_i) \right| - \left| h_i(a_i) \right| = i$$

Hence the lemma is proved. ▽

20.16 LEMMA. *The sequence* $(f_n \circ P_n)$ *converges to a holomorphic mapping* $f \in H(U)$ *such that*

$$\lim_{n \to \infty} \left| f(a_n) \right| = \infty$$

and U is the domain of existence of f.

Proof. Consider the sequence of functions

$$F_n = f_n \circ P_n$$

Then F_n is defined on U for all $n \in \mathbb{N}$. We claim that the sequence (F_n) is uniformly convergent on each compact subset of U; hence its limit, say f, is holomorphic on U since $(H(U), \tau_0)$ is complete (see 16.13).

Let K be a compact subset of U. Choose $q \in \mathbb{N}$ be such that for every $m \geq q$, $P_m(K) \subset K_q$ by Lemma 20.12; for any m and n, $m > n > q$, $P_m(K) \subset L_n$ by Lemma 20.13. Then for any $m > q+1$, we have

$$P_m(K) \subset T_m$$

Moreover,

$$\left\| F_{m+p} - F_m \right\|_K \leq \sum_{i=1}^{p} \left\| F_{m+i} - F_{m+i-1} \right\|_K$$

$$= \sum_{i=1}^{p} \left\| f_{m+1} \circ P_{m+i} - f_{m+i-1} \circ P_{m+i-1} \right\|_K$$

$$= \sum_{i=1}^{p} \left\| f_{m+i} \circ P_{m+i} - f_{m+i-1} \circ P_{m+i-1} \circ P_{m+i} \right\|_K$$

$$= \sum_{i=1}^{p} \left\| f_{m+i} - f_{m+i-1} \circ P_{m+i-1} \right\|_{P_{m+i}(K)}$$

$$\leq \sum_{i=1}^{p} \left\| f_{m+i} - f_{m+i} \circ P_{m+i-1} \right\|_{T_{m+1}} \leq \sum_{i=1}^{p} 2^{-(m+i)} < 2^{-m} \qquad (1)$$

The last inequality holds by Lemma 20.15. Therefore, the convergence of (F_n) on K is uniform. Thus we have a mapping $f \in H(U)$ defined by

$$f = \lim F_n$$

We claim now that f is unbounded on the sequence (a_n). For each $m \in \mathbb{N}$, notice that from Lemma in 20.14 we have

$$a_m \in U_m \subset U_{m+i}$$

$$a_m \in K_m \subset K_{m+i}$$

Since $\| P_{m+i-1}(a_m) - a_m \| = 0 \; (a_m \in U_m)$, we have

$$a_m \in L_{m+i-1}$$

Hence, for any $i \in \mathbb{N}$ we obtain

$$a_m \in T_{m+i}$$

It follows from (1) that

$$\left| F_{m+p}(a_m) - F_m(a_m) \right| \le 2^{-m}$$

Letting $p \to \infty$, we get

$$\left| f(a_m) - f_m(a_m) \right| \le 2^{-m}$$

Thus

$$\left| f(a_m) \right| \ge \left| f_m(a_m) \right| - 2^{-m} \ge m - 2^{-m}$$

Thus f is unbounded on (a_n).

Suppose that U is not a domain of existence for f. Then there exist a $\in$ U, s > r = d(a,CU) and g $\in$ H(B_s(a)) such that f = g on B_r(a). Then the restriction on F_n to the set $E_n \cap B_r$(a) can be continued holomorphically to $E_n \cap B_s$(a). Since U_n is pseudoconvex in E_n, it is a domain of holomorphy in E_n by the Levi-Oka theorem for $\mathbb{C}^n$. Hence $E_n \cap B_s$(a) must be contained in U_n. Thus f = g on $E_n \cap B_s$(a). Choose b $\in$ B_s(a) $\cap$ δU $\cap$ δV, where V is the connected component of U $\cap$ B_s(a) containing B_r(a), and find $b_m \in$ B (B is as in Lemma 20.14), and t > 0 such that $B_t(b_m) \subset B_s$(a). Then there exists a subsequence (a_{n_j}) of (a_n) such that $a_{n_j} \in E_{n_j}$ and $a_{n_j} \to$ b as j $\to \infty$. Then f = g on a tail of this subsequence. But f is unbounded on this subsequence, which is a contradiction since g is locally bounded at b_m. Therefore, U is the domain of existence of f.

This completes the proof of the Levi-Oka theorem.   □

20.17. We summarize in the following theorem the main results of Chapter 19 and Chapter 20.

THEOREM. *Let* U *be an open subset of a normed space* E. *Consider the following conditions:*

    (a) U *is a domain of existence of a holomorphic mapping.*

    (b) U *is a domain of holomorphy.*

    (c) U *is holomorphically convex.*

    (d) U *is pseudoconvex.*

Then (a) $\to$ (b) $\to$ (c) $\to$ (d). *If* E *is a Banach space with a Schauder basis, then all of these conditions are equivalent.*

20.18 REMARK. It can be shown that the Levi-Oka theorem holds for separable Banach spaces with the Banach approximation property (see Noverraz (1973, 98-104). The class of Banach spaces with the Banach approximation property includes

    (a) Banach spaces with Schauder bases;

    (b) The space $L^p(X,\mu)$, $1 \le p < \infty$, where X is a locally compact space and μ is a positive Radon measure;

(c)   The space C(K), where K is a compact metric space.

In order to prove the Levi-Oka theorem in this setting, we apply the Pełczynski theorem stating that a separable Banach space with the Banach approximation property is a complementary subspace of a Banach space with a Schauder basis (Pełczynski, 1971).

20.19.   We do not know yet if the Levi problem is solvable for all separable spaces.   Although there are nonseparable spaces on which the Levi problem has no solution, it is still possible that the Levi problem could be solved for any separable Banach space.

Josefson (1974) produced an open pseudoconvex set in a nonseparable Banach space which is not a domain of holomorphy.   In fact, for an uncountable index set A, let $c_0(A)$ be the space of complex valued functions f on A which are arbitrarily small off finite sets in the sense that for any r > 0 the set $\{\nu \in A: |f(\nu)| \geq r\}$ is finite. Then $c_0(A)$ becomes a Banach space under the supremum norm.   The unit ball of this space is a pseudoconvex open set which is not a domain of holomorphy.

20.20.   The Levi problem has been generalized to various infinite dimensional spaces having a Schauder basis or the Banach approximation property.   See Noverraz (1973) and (1975), Pomes (1974), Dineen (1975), Schottenloher (1976), Dineen-Noverraz-Schottenloher (1976).

EXERCISES

20A.   Let U be an open subset of a normed space E.   Define a mapping $\delta_U: U \times E \to \mathbb{C}$ by

$$\delta_U(x,a) = \sup \{r: r \in \mathbb{R}^+, x+rza \in U \text{ for all } z \in \mathbb{C}, |z| \leq 1\}$$

Then

   (a) $\delta_U(x,a) = \inf \{|z|: z \in \mathbb{C}, x+za \in E \backslash U\}$
   (b) $d_U(x,CU) = \inf \{\delta_U(x,a): \|a\| = 1\}$
   (c) δ_U is lower semicontinuous on $U \times E$.

(d)   U is pseudoconvex if and only if $-\log \delta_U(x,a)$ is plush on
U × E.

20B.   If U is a pseudoconvex open subset of E and $f \in H(U)$, then for
any $r > 0$, $\{x \in U: |f(x)| < r\}$ is a pseudoconvex open subset.

20C.   An open subset U is pseudoconvex if and only if U is the union
of open sets U_n satisfying cl $U_n \subset U_{n+1}$ and

$$U_n = \{x \in U: v_n(x) < 0\}$$

where v_n is a plush function on a neighborhood of U_n (this is true
for finite dimensions).

20D.   Let Psc(U) be the space of all continuous plush functions on
U.   Then U is Psc(U)-convex if and only if U is Ps(U)-convex.

20E.   Let E and F be normed spaces and f: E → F an open surjective
continuous linear map.   Let V be a nonempty open subset of F and
U an open pseudoconvex subset of E containing $f^{-1}(V)$.   Then

(a) $U = f^{-1} \circ f(U)$

(b)   f(U) is pseudoconvex in F.

21

Spaces of Holomorphic Germs

21.1. Spaces of germs of holomorphic mappings on compact subsets of
Banach spaces were studied by Chae (1970, 1971) in connection with
the Nachbin topology τ_ω on the spaces $H_\theta(U;F)$ of all holomorphic
mappings on open subsets U of E for any holomorphic type θ. As a
special case, when we take θ as the current type, we obtain $H_\theta(U;F)$
= H(U;F). For finite dimensional spaces, spaces of holomorphic germs
were studied by Köther, Grothendieck, Dias, Martineau, and many
others. The crux of this study is that through the interplay between
the space H(U;F) and the spaces H(K;F) for all compact subsets K of
U, we can obtain information about the Nachbin topology on H(U;F);
for example, the completeness of H(U;F). The question of completeness of the topology τ_ω on H(U;F) was investigated by Dineen, Chae,
and Aron with some partial affirmative answers. It was a refinement
of Chae's idea to use spaces of holomorphic germs on compact subsets
of U that eventually led Mujica (1979) to the general completeness
theorem for $(H(U;F), \tau_\omega)$.

TOPOLOGY ON THE SPACES H(K;F)

21.2 *The Spaces* H(K;F). Let K be a compact subset of a normed space
and let F be a Banach space. Let h(K;F) be the set of all F-valued
mappings which are defined and holomorphic on some open subset of E
containing K; i.e.,

h(K;F) = $\cup$ {H(U;F): U is an open set containing K}

Two mappings f_1 and f_2 in h(K;F) defined on open subsets U_1 and U_2, respectively, are said to be *equivalent modulo* K if there is an open set U of E with $K \subset U \subset U_1 \cap U_2$ such that $f_1 = f_2$ on U. This relation is indeed an equivalence relation on h(K;F), and we call each equivalence class in h(K;F) a *holomorphic germ* or a *germ of holomorphic mappings*. We denote by [f] the equivalence class modulo K determined by f.

The quotient space of h(K;F) with respect to this relation will be denoted by H(K;F). The space H(K;F) becomes a vector space over $\mathbb{C}$ if we define vector operations in the following manner:

$$[f + g] = [f] + [g]$$

$$[zf] = z[f]$$

for every f and g in h(K;F) and $z \in \mathbb{C}$.

If there is no confusion, we write each germ determined by a holomorphic mapping f by f.

21.3 *The Space* $H^\infty(U;F)$. For an open subset U of E, the vector subspace of H(U;F) consisting of all bounded holomorphic mappings on U is denoted by $H^\infty(U;F)$. The space $H^\infty(U;F)$ becomes a normed space with respect to the norm

$$f \in H^\infty(U;F) \rightarrow \|f\|_U$$

THEOREM. $H^\infty(U;F)$ *is a Banach space.*

Proof. Let (f_n) be a Cauchy sequence in $H^\infty(U;F)$. Define f on U as the pointwise limit of f_n. Then f is holomorphic and bounded since H(U;F) is complete with respect to the compact open topology τ_0. Clearly, the convergence is uniform on U.   □

21.4 *Topology on* H(K;F). For a fixed compact set K and any open neighborhood U of K, there exists a natural linear mapping

$$T_U : \overset{\infty}{H}(U;F) \to H(K;F)$$

assigning every $f \in \overset{\infty}{H}(U;F)$ to its equivalence class $[f]$ modulo K.

The natural topology on H(K;F) is defined as the inductive limit of the spaces $\overset{\infty}{H}(U;F)$, for all open subsets U of E containing K; i.e., the natural topology on H(K;F) is the finest locally convex topology such that the linear mapping T_U is continuous for each open subset U containing K. Hence we can write

$$H(K;F) = \lim_{\longrightarrow} \overset{\infty}{H}(U;F)$$

It is clear that a seminorm p on H(K;F) is continuous if and only if p o T_U is continuous on $\overset{\infty}{H}(U;F)$ for every open subset U of E containing K.

It is convenient to identify $\overset{\infty}{H}(U;F)$ with the subspace $T_U \overset{\infty}{H}(U;F)$ of H(K;F). Hence we can consider H(K;F) as the inductive limit of subspaces $\overset{\infty}{H}(U;F)$ for all open subsets U containing K.

We exhibit some classes of continuous seminorms on H(K;F) which are easily recognizable in the following two theorems.

21.5   THEOREM.   *For each* m $\in \mathbb{N}$ *and a bounded subset B in E the seminorm defined by*

$$p_{m,B}([f]) = \sup_{\substack{\xi \in K \\ x \in B}} \left\| \frac{1}{m!} \hat{d}^m f(\xi)(x) \right\|$$

is continuous on H(K;F) *where f is a representative of* $[f]$.

Proof.   The mapping is well defined since every $f \in [f]$ will take the same value on K. We show that p o T_U is a continuous seminorm on $\overset{\infty}{H}(U;F)$ for any open set U containing K. Let $r > 0$ be such that $B_r(K) \subset U$. It follows from the Cauchy inequalities 13.6 that for any $f \in \overset{\infty}{H}(U;F)$ we obtain

$$P_{m,B}(f) \leq C^m \left\| \tfrac{1}{m!} \hat{d}^m f \right\|_K \leq C^m (1/s^m) \|f\|_U$$

where $C = \sup \{ \|x\| : x \in B \}$.   □

21.6   THEOREM.   *Let K be a compact subset of E and let* $a = (a_n)$ *be a sequence of positive numbers such that*

$$\lim [a_n]^{1/n} = 0$$

Then the mapping defined by

$$P_a([f]) = \sum_{n=0}^{\infty} a_n \left\| \tfrac{1}{m!} \hat{d}^m f \right\|_K$$

is a continuous seminorm on H(K;F) where $f \in [f]$.

Proof.   For every open subset U containing K, $p_a \circ T_U$ is continuous on $\overset{\infty}{H}(U;F)$ by the Cauchy inequalities.   Therefore, p_a is continuous on $H(K;F)$.   □

21.7   THEOREM. $H(K;F)$ *is a bornological and barrelled space.*

Proof.   The inductive limit of bornological (respectively, barrelled) spaces is also bornological (respectively, barrelled) (see 18.2). Therefore, $H(K;F)$ is both bornological and barrelled.   □

21.8.   We now discuss an alternative method of obtaining the space $H(K;F)$.   For an open set U containing K, there corresponds a natural linear mapping

$$f \in H(U;F) \rightarrow [f] \in H(K;F)$$

assigning each $f \in H(U;F)$ to its equivalence class $[f]$.   We assume that $H(U;F)$ are endowed with the Nachbin topology τ_ω if there is no specific mention. Hence we can consider the inductive limit of the spaces $(H(U;F),\tau_\omega)$ for all open subsets U of E containing K. We also write

$$\varinjlim_{K \subset U} (H(U;F),\tau_\omega) \qquad \text{or} \qquad \varinjlim H(U;F)$$

We now show that these two inductive limits are identical.

21.9 THEOREM. $H(K;F) = \varinjlim H^\infty(U;F) = \varinjlim (H(U;F),\tau_\omega)$

Proof. Let τ and τ' be the topologies on H(K;F) defined by the inductive limits $\varinjlim H^\infty(U;F)$ and $\varinjlim H(U;F)$ respectively. Since the inclusion mapping $H^\infty(U;F) \to H(U;F)$ is continuous, the identity mapping

$$(H(K;F),\tau) \to (H(K;F),\tau')$$

is continuous. Hence $\tau' \leq \tau$.

On the other hand, let p be a seminorm on H(K;F) which is τ-continuous. Then, corresponding to every open subset U of E containing K, there exists a real number c > 0 such that

$$p \circ T_U(f) \leq c\|f\|_U$$

for all $f \in H^\infty(U;F)$. This implies that corresponding to every open subset V of U (U fixed) containing K, there is c > 0 such that

$$p \circ T_U(f) \leq c\|f\|_V$$

for all $f \in H(U;F)$. Thus $p \circ T_U$ is a τ_ω-continuous seminorm on H(U;F). Therefore p is τ'-continuous; and hence $\tau \leq \tau'$. □

21.10 THEOREM. H(K;F) *is a Hausdorff locally convex space.*

Proof. Notice that a locally convex space E is Hausdorff if and
only if for any $x \in E$, $x \neq 0$, there exists a continuous seminorm p
on E such that $p(x) \neq 0$, or equivalently, if $p(x) = 0$ for every
continuous seminorm p on E, then $x = 0$.

Let $f \in H(K;F)$ be such that $p(f) = 0$ for every continuous seminorm on $H(K;F)$. In particular, the following map is a continuous
seminorm on $H(K;F)$:

$$p_m(f) = \left\| \frac{1}{m!} \hat{d}^m f \right\|_K$$

for all $m \in \mathbb{N}$. Thus $p_m(f) = 0$ for each $m \in \mathbb{N}$ implies that $\hat{d}^m f(\xi) = 0$
for all $\xi \in K$, from which we conclude that $f = 0$ on an open neigh-
borhood of K.   □

21.11 THEOREM. H(K;F) *is an inductive limit of an increasing*
sequence of Banach spaces.

Proof. Let (U_m) be the sequence of open neighborhoods of K defined
by

$$U_m = \{x \in E: d(x,K) < 1/m\}$$

This sequence will be called the fundamental sequence of open neighborhoods of K. Then

$$H(K;F) = \varinjlim H^{\infty}(U_m;F)$$ □

21.12 (DF)-*Spaces.* Grothendieck (1954) introduced an important
class of locally convex spaces called (DF)-spaces. A locally convex space E is said to be a (DF)-space if it satisfies

   (1) E has a fundamental sequence of bounded sets;

(2) Every strongly bounded countable union of equicontinuous
     subsets of E is equicontinuous.

In this chapter we need the following three facts about (DF)-spaces.

THEOREM (Grothendieck, 1954). (a) *Every normed space is a* (DF)-
  space.

  (b) *An inductive limit* E *of an increasing sequence* (E_n) *of* (DF)-
       spaces is a (DF)-*space, and every bounded subset of* E *is
       contained in the* E-*closure of a bounded subset of some* E_m.

  (c) *A* (DF)-*space is complete if it is quasicomplete; i.e., every
       bounded closed set is complete.*

For a proof of these properties, see Grothendieck (1954) or (1958);
in English (1973).

COROLLARY. H(K;F) *is a* (DF)-*space.*

BOUNDED SETS IN H(K;F)

21.13 *Regularity.* In the preceding section we showed that H(K;F)
has some good topological properties since it is an inductive limit
of an increasing sequence of Banach spaces. We now characterize
bounded sets in H(K;F). We first introduce the following terminology.

    Let E = $\varinjlim E_l$ be an inductive limit of locally convex spaces.
The inductive limit is said to be *regular* if each bounded subset of
E is contained and bounded in some E_l.

    We show in this section that H(K;F) is a regular inductive limit.

21.13 THEOREM. *For a locally connected compact subset* K *of* E *and a
subset* X *of* H(K;F), *the following statements are equivalent.*

  (1) X *is a bounded subset of* H(K;F).

  (2) *There exist* C > 0 *and* c > 0 *such that*

$$\sup_{x \in K} \left\| \frac{1}{m!}\, \hat{d}^m f(x) \right\| \leq Cc^m$$

for all f ∈ X *and* m ∈ ℕ .

(3) *There exists an open set* U *containing* K *such that* X *is contained° and bounded in* H$^{\infty}$(U;F).

Proof. (1) → (2). Since X is bounded, it follows from 21.6 that the seminorms p_a are bounded on X for any sequence a = (a_n) such that $[a_n]^{1/n}$ → 0 as n → ∞. It is a classical result that if $s_{m,\nu}$ ≥ 0 are real numbers for m ∈ ℕ and ν ∈ Γ, where Γ is a set, then

$$\sup_{\gamma \in \Gamma} \sum_{m=0}^{\infty} a_m s_{m,\nu} < \infty$$

holds for every sequence a = (a_n) of positive numbers such that $[a_m]^{1/m}$ → 0 as m → ∞ if and only if there are C > 0 and c > 0 such that

$$s_{m,\nu} \leq Cc^m$$

for every m ∈ ℕ and ν ∈ Γ. Therefore, the fact that every seminorm of the form p_a is bounded on X implies (2).

(2) → (3). Assume (2) is true. We choose a real number r > 0 such that 2rc < 1. Since K is compact and locally connected, we may cover K with a finite number of open balls $B_r(\xi_1),\ldots,B_r(\xi_n)$ all centered in K such that K ∩ $B_r(\xi_j)$ is connected for all j, 1 ≤ j ≤ n. Let s = r/4. If ξ ∈ K, then $B_{2s}(\xi)$ is contained in some $B_r(\xi_j)$.

For each f ∈ X and ξ ∈ K, consider

$$f_{\xi}(x) = \sum_{m=0}^{\infty} \frac{1}{m!} \hat{d}^m f(\xi)(x-\xi)$$

Then f_{ξ} is holomorphic and bounded by C/(1-2cr) on $B_{2r}(\xi)$ since (2) holds.

We claim that if ξ, η ∈ K and x ∈ $B_s(\xi)$ ∩ $B_s(\eta)$, then

$$f_\xi(x) = f_\eta(x)$$

In fact, we have $\|\eta - \xi\| < 2s$, and hence both ξ and η belong to the connected set $K \cap B_r(\xi_j)$, for some $j \le n$, with $B_{2s}(\xi) \subset B_r(\xi_j)$. Then

$$f_\xi(x) = f_{\xi_j}(x) \tag{1}$$

for $x \in B_{2s}(\xi)$ by 11.11. Since $\eta \in B_{2s}(\xi)$, there exists $\delta > 0$ with $\delta \le 2s - \|\eta - \xi\|$ such that

$$f_\eta(x) = f_\xi(x) \tag{2}$$

for all $x \in B_\delta(\eta) \subset B_{2s}(\xi) \cap B_s(\eta)$ by 11.11. Since ξ and η belong to the same connected set $K \cap B_r(\xi_j)$, it follows from (1) and (2) above and the principle of analytic continuation 12.9 that

$$f_\xi(x) = f_\eta(x)$$

for all $x \in B_s(\xi) \cap B_s(\eta)$.

Let $U = B_s(K)$ and define $g: U \to F$ by

$$g(x) = f_\xi(x)$$

if $x \in B_s(\xi)$. Then g is holomorphic on U such that $g \in [f]$ and bounded by $C/(1-2cr)$, which shows that (3) holds.

The implication (3) → (1) is obvious.   □

21.14   REMARK.   The implication (2) → (3) is not true if K is not locally connected as shown by the following example.

EXAMPLE (R. Aron).   Let $E = \mathbb{C}$.   For each $n \in \mathbb{N}$, let

$$V_n = \{z \in \mathbb{C} : \text{Re } z < 2/(2n+1)\}$$

$$W_n = \{z \in \mathbb{C} : \text{Re } z > 2/(2n+1)\}$$

$$U_n = V_n \cup W_n$$

Let $f_n : U_n \to \mathbb{C}$ be defined to be 1 on V_n and 0 on W_n. Then f_n is holomorphic on U_n. If $K = \{0,1,1/2,\ldots,1/n,\ldots\}$, then K is compact and $f_n \in H(K)$ for all n. The sequence (f_n) satisfies the condition (2) of the theorem but it does not satisfy (3).

However, for any compact K, (1) and (2) are equivalent. There is a direct proof of this fact which shows some complicated structures of $H(K;F)$. We present here a proof using a Grothendieck's theorem 21.12(b). Each proof has its own advantage; one showing the intricacy of the structure, the other as a part of a big smooth machine. We give here the other proof due to Mujica (1979).

21.15 THEOREM. *Let K be a compact subset of a normed space. Then*

$$H(K;F) = \lim_{\longrightarrow} H^{\infty}(U;F)$$

is regular.

We first prove the following lemma.

21.16 LEMMA. *Let K be a compact subset of E and let $U_r = B_r(K)$ where $r > 0$. Then the closed unit ball of the normed space $H^{\infty}(U_r;F)$ is a closed subset of $H(K;F)$.*

Proof. Let B be the closed unit ball of $H^{\infty}(U_r;F)$. Since $H(K;F)$ may not be metrizable, we can not use a sequence argument to show that B is closed in $H(K;F)$, and hence we use nets instead of sequences. Let f be a limit of a Cauchy net $(f_\gamma)_{\gamma \in \Gamma}$ in B for the induced topology from $H(K;F)$. We claim that $f \in B$.

We first show that the net is a Cauchy net in $H(U_r;F)$ for the compact open topology τ_o. Let L be a compact subset U_r. Then for a positive real number s < r we have from the Cauchy inequalities that

$$\sup_{\xi \in K} \left\| \frac{1}{m!} \, \hat{d}^m f(\xi) \right\| \leq 1/s^m$$

for any f ∈ B and m ∈ ℕ. For β, γ ∈ Γ we have

$$\left\| f_\beta - f_\gamma \right\|_L \leq \sum_{m=0}^{\infty} \sup_{\substack{\xi \in K \\ x \in L}} \left\| \frac{1}{m!} \, \hat{d}^m [f_\beta - f_\gamma](\xi)(x-\xi) \right\|$$

$$\leq \sum_{m=0}^{N} \sup_{\substack{\xi \in K \\ x \in L}} \left\| \frac{1}{m!} \, \hat{d}^m [f_\beta - f_\gamma](\xi)(x-\xi) \right\| + \sum_{m=N+1}^{\infty} 2 \left(\frac{r}{s} \right)^m$$

Notice that each term of the first summation above is a continuous seminorm $P_{m,L-K}$ in 21.5, thus the first summation is a continuous seminorm on H(K;F). Since r/s < 1, for ε > 0 we can find N such that

$$\sum_{m=N+1}^{\infty} 2[r/s]^m \leq \varepsilon/2$$

Since the net is Cauchy in H(K;F), we can find α ∈ Γ such that for β , γ ≥ α we obtain

$$\left\| f_\beta - f_\gamma \right\|_L \leq \varepsilon$$

This shows that the Cauchy net is a τ_o-Cauchy net in H(U$_r$;F). Hence it converges to a holomorphic mapping g ∈ H(U$_r$;F). It is clear that g ∈ B.

It remains to show that f ∈ [g]. Since for any m and ξ ∈ K,

$$\hat{d}^m f_\gamma(\xi)(x) \to \hat{d}^m f(\xi)(x)$$

for any $x \in E$ by 21.5, it suffices to show that

$$\hat{d}^m f_\gamma (\xi)(x) \to \hat{d}^m g(\xi)(x)$$

For $\xi \in K$, $m \in \mathbb{N}$, $x \in E$ and $s < r$, consider the compact subset

$$L = \{\xi + zx: z \in \mathbb{C}, |z| \leq s\}$$

Then

$$\left\| \hat{d}^m [f_\gamma - g](\xi)(x) \right\| \leq [m!/s^m] \| f_\gamma - g \|_L$$

Since $f_\gamma \to g$ in $(H(U_r;F), \tau_0)$, we obtain from the preceding inequality

$$\hat{d}^m f_\gamma (\xi)(x) \to \hat{d}^m g(\xi)(x)$$

This completes the proof that B is a closed subset of $H(K;F)$.   ▽

Proof of Theorem. Let X be a bounded subset of $H(K;F)$ and let (U_m) be the fundamental sequence of open neighborhoods of K. Then by Grothendieck's theorem 21.12(b), there is a bounded subset Y in some $H^\infty(U_m;F)$ such that

$$X \subset cl_{H(K;F)} Y$$

Since Y is bounded in $H^\infty(U_m;F)$, there is $\rho > 0$ such that

$$Y \subset \rho B$$

where B is the closed unit ball of $H^\infty(U_m;F)$. By the preceding lemma, B is a closed subset of $H(K;F)$, and hence

$X \subset \rho B$

This shows that X is contained and bounded in $H^\infty(U_m;F)$.   □

COROLLARY. $H(K;F) = \varinjlim H(U;F)$ *is a regular inductive limit.*

Proof. It follows from the fact that the inclusion map

$$H^\infty(U;F) \to H(U;F)$$

is continuous.   □

We mentioned that H(K;F) might not be metrizable. Indeed, H(K;F) is not metrizable for any compact set K. To show this fact we need a preliminary lemma.

21.17 LEMMA. *Let* (U_m) *be the fundamental sequence of open neighborhoods of* K. *Then*

$$H^\infty(U_m;F) \neq H^\infty(U_n;F) \qquad if \qquad m \neq n$$

Proof. Assume that m < n. We prove that there is

$$f \in H^\infty(U_n;F) \setminus H^\infty(U_m;F)$$

It is sufficient to consider for the case $F = \mathbb{C}$. We apply the Hahn-Banach separation theorem 16.7. Let V_n be the closed convex balanced hull of U_n; i.e., V_n is the smallest convex balanced closed set containing U_n. Then U_m cannot be completely contained in V_n. In fact, let $y \in K$ be such that

$$\|y\| = \sup \{\|x\| : x \in K\}$$

For t, $1/n < t < 1/m$, we have $\xi = (1+t/\|y\|)y \in U_m$. We claim that ξ cannot be in V_n. If x is in the convex balanced hull of U_n, then

$$x = a_1 x_1 + \ldots + a_k x_k$$

where $x_i \in U_n$ and $a_i \geq 0$ and $a_1 + \ldots + a_k \leq 1$. Hence $\|x\| < \|y\| + 1/n$. Thus if ξ is in V_n, $\|\xi\| \leq \|y\| + 1/n$; but $\|\xi\| > \|y\| + 1/n$.

Thus we have a setting for the Hahn-Banach separation theorem. Since ξ is not in the closed convex balanced set V_n, there is $A \in E^*$ such that

$$A(\xi) = 1 \qquad \text{and} \qquad \sup \{|A(x)| : x \in V_n\} < 1$$

Define

$$f(x) = \sum_{m=0}^{\infty} A^m(x)$$

Then $f \in H(U_n)$, but f cannot be continued holomorphically to U_m since $\xi \in U_m$.   □

21.18 THEOREM. *H(K;F) is a non-metrizable space.*

Proof. If M is a metrizable locally convex space and (x_n) is a sequence in M, it is easy to prove that there is a sequence (a_n) of positive real numbers such that $\{a_n x_n : n \in \mathbb{N}\}$ is a bounded subset in M. This is known as the *Mackey countability condition* (see Horváth, 1966, p. 116). We use this property to prove that H(K;F) is not metrizable.

Suppose that H(K;F) is metrizable. Choose $f_n \in H^{\infty}(U_n;F)$ which is not in $H^{\infty}(U_{n-1};F)$, where (U_n) is the fundamental sequence of open neighborhoods of K. By the Mackey countability condition, there exists a sequence (a_n) of positive numbers such that

$$X = \{a_n f_n : n \in \mathbb{N}\}$$

is bounded in H(K;F). Thus X is contained and bounded in some
$H^\infty_{\bullet}(U_m;F)$. This is a contradiction because $a_m f_m$ does not belong to
$H^\infty(U_n;F)$ for any n < m.   □

21.19. Let H*(K;F) be the dual of H(K;F) consisting of all continuous linear functionals on H(K;F). The *strong topology* on H*(K;F) is
defined by all seminorms of the form p_B:

$$p_B(\phi) = \sup \{|\phi(f)| : f \in B\}$$

where B is a bounded subset of H(K;F). Although H(K;F) is not
metrizable, we show that the dual H*(K;F) is metrizable.

THEOREM. H*(K;F) *is a Fréchet space.*

Proof. Let B_m be the closed unit ball of $H^\infty(U_m;F)$ for each m, where
(U_m) is the fundamental sequence of open neighborhoods of K. Then
the family of seminorms p_m on H*(K;F) defined by

$$p_m(\phi) = \sup \{|\phi(f)| : f \in B_m\}$$

generates the strong topology on H*(K;F) because every bounded set
in H(K;F) is contained in ρB_m for some m and $\rho > 0$. Hence H*(K;F)
is metrizable since its topology is defined by a countable family
of seminorms.

Let (ϕ_m) be a Cauchy sequence in H*(K;F). For each $f \in$ H(K;F),
the sequence $(\phi_m(f))$ is a Cauchy sequence in $\mathbb{C}$. Let ϕ be the pointwise limit of (ϕ_m). Then ϕ is continuous and linear. Thus H*(K;F)
is complete.   □

COMPLETENESS OF H(K;F)

21.20 THEOREM. *Let* (U_n) *be the fundamental sequence of open neigh-*
borhoods of K and let X be a bounded subset of $H^\infty(U_k;F)$ *for some k.*

If $(f_\gamma)_{\gamma \in \Gamma}$ is a $H(K;F)$-Cauchy net in X, then X is a Cauchy net in the Banach space $\overset{\infty}{H}(U_n;F)$ for every $n > k$.

Proof. X being bounded in $\overset{\infty}{H}(U_k;F)$, there exists $C > 0$ such that

$$\|f\|_{U_k} \leq C$$

for all $f \in X$. It follows from the Cauchy inequalities that

$$\left\|\frac{1}{m!} \hat{d}^m f_\gamma\right\|_K \leq Ck^m$$

for all $\gamma \in \Gamma$ and for all $m \in \mathbb{N}$. Applying the Cauchy inequalities again if $n > k$ and $x \in U_n$ then

$$\left\|[f_\beta - f_\gamma](x)\right\| \leq \sum_{m=0}^\infty \left\|\frac{1}{m!} \hat{d}^m[f_\beta - f_\gamma]\right\|_K (1/n)^m$$

$$\leq \sum_{m=0}^N \left\|\frac{1}{m!} \hat{d}^m[f_\beta - f_\gamma]\right\|_K (1/n)^m + (2C) \sum_{m=N+1}^\infty (k/n)^m$$

Since each term of the first sum is a continuous seminorm on $H(K;F)$, $(f_\gamma)_{\gamma \in \Gamma}$ is a Cauchy net in $H(K;F)$, and $k/n < 1$, given a real number $\varepsilon > 0$ we can find $N \in \mathbb{N}$ and $\alpha \in \Gamma$ such that for any $\beta, \gamma \geq \alpha$

$$\|f_\beta - f_\gamma\|_{U_n} \leq \varepsilon$$

Thus the Cauchy net is a Cauchy net in the Banach space $\overset{\infty}{H}(U_n;F)$. $\square$

Combining Theorem 21.15 and Theorem 21.21 we obtain the following result.

21.21 THEOREM. *Let X be a bounded subset of $H(K;F)$. Then there is $m \in \mathbb{N}$ such that*

(a)   X *is contained and bounded in* $H^\infty(U_m;F)$.

(b)   *A net* (f_γ) *in* X *is* $H(K;F)$-*Cauchy if and only if it is* $H^\infty(U_m;F)$-*Cauchy.*

(c)   *On* X *both* $H(K;F)$ *and* $H^\infty(U_m;F)$ *induce the identical topology.*

21.22   COROLLARY. *Every bounded subset of* $H(K;F)$ *is metrizable.*

21.23   THEOREM. $H(K;F)$ *is a complete locally convex space.*

Proof. By Theorem 21.21, every closed bounded set in $H(K;F)$ is complete; i.e., $H(K;F)$ is quasicomplete. Since $H(K;F)$ is a (DF)-space, $H(K;F)$ must be complete by Grothendieck's theorem 21.12(c).   □

TOPOLOGY τ_π *ON* $H(U;F)$

21.24   *Projective Limit.* Let Γ be an index set. For each $\gamma \in \Gamma$, let $\pi_\gamma: E \to E_\gamma$ be a linear mapping of a vector space E into a locally convex space E_γ. The *projective topology* on E is the smallest locally convex topology on E for which each mapping π_γ is continuous.

Let Γ be a directed set and assume that for any $\alpha, \beta \in \Gamma$, $\alpha \leq \beta$ there exists a continuous linear map $\pi_{\alpha\beta}: E_\beta \to E_\alpha$ such that

(a) $\pi_{\alpha\alpha}$ is the identity map;

(b) $\pi_{\alpha\beta} \circ \pi_{\beta\gamma} = \pi_{\alpha\gamma}$ whenever $\alpha \leq \beta \leq \gamma$.

Then the family of locally convex spaces and linear maps $\{E_\alpha, \pi_{\alpha\beta}\}$ is called a *projective system*. Let X denote the Cartesian product of the spaces E , and denote

$$E = \{(x_\alpha) \in X: \pi_{\alpha\beta}(x_\beta) = x_\alpha \text{ for } \alpha \leq \beta\}$$

with the induced topology. Then E is called the *projective limit* of $\{E_\alpha, \pi_{\alpha\beta}\}$ and is denoted by

$$E = \varprojlim E_\alpha$$

It is easy to see that if π_γ is the restriction to E of the canonical projection $X \to E_\gamma$ then E has the projective topology.

We have the following theorem from Horváth (1966), p. 153 or Schaefer (1971), Prop. 5.3.

THEOREM. *The projective limit of a family of complete locally convex spaces is complete.*

21.25 *Projective Limit* $H(U;F)$. Let U be an open subset of a normed space. Order the family of all compact subsets of U by set inclusion. For any compact subsets K and J of U with $K \subset J$, the natural inclusion

$$H(J;F) \to H(K;F)$$

is a continuous linear map satisfying the conditions (a) and (b) of a projective system.

For each compact subset K of U, consider the canonical map

$$\pi_K: f \in H(U;F) \to [f]_K \in H(K;F)$$

where $[f]_K$ denotes the holomorphic germ on K determined by f.

THEOREM. *Let U be an open subset of a normed space E. Then*

$$H(U;F) = \varprojlim_{K \subset U} H(K;F)$$

where K ranges over all compact subsets of U.

Proof. Let $(f_K) \in \varprojlim H(K;F)$. Since the family of compact subsets of U is directed by inclusion, if we define

$$f(x) = f_{\{x\}}(x)$$

for each x ∈ U, then f is well defined. We claim that f ∈ H(U;F)
and $[f]_K = f_K$; hence we have algebraically

$$H(U;F) = \varprojlim_{K \subset U} H(K;F)$$

First notice that if a ∈ K, then

$$\hat{d}^m f_{\{a\}}(a) = \hat{d}^m f_K(a)$$

for all m ∈ ℕ. If a ∈ U and V = $B_r(a) \subset U$ for some r > 0, then
$f_{\{a\}} \in H^\infty(V;F)$. For x ∈ V, consider the compact line segment L =
[a,x] in V. Then

$$f(x) = f_{\{x\}}(x) = f_L(x) = \sum_{m=0}^{\infty} \frac{1}{m!} \hat{d}^m f_L(a)(x-a)$$

$$= \sum_{m=0}^{\infty} \frac{1}{m!} \hat{d}^m f_{\{a\}}(a)(x-a)$$

This shows that f ∈ H(U;F). Furthermore, if K is a compact subset
of U and a ∈ K, then for any m ∈ ℕ

$$\hat{d}^m [f]_K(a) = \hat{d}^m f_{\{a\}}(a) = \hat{d}^m f_K(a)$$

Hence $[f]_K = f_K$.   □

21.26 *Topology* τ_π *on* H(U;F). The topology τ_π on H(U;F) is the
projective topology for which all maps π_K: H(U;F) → H(K;F) are continuous for all compact subsets K of U.

THEOREM. *On* H(U;F) *we have*

$$\tau_\sigma \leq \tau_\pi \leq \tau_\omega$$

Proof. It is clear that $\tau_\pi \leq \tau_\omega$ since π_K is continuous for τ_ω. For the remaining inequality, recall that the topology τ_σ introduced in 17.10 is generated by the seminorms of the form in 21.6. Hence every τ_σ open subset of $H(U;F)$ is τ_π open.   □

COROLLARY. *The topologies* τ_ω, τ_π, *and* τ_σ *have the same family of bounded sets and on each bounded set, they are equivalent.*

21.27 THEOREM. $(H(U;F),\tau_\pi)$ *is complete.*

Proof. This is a consequence of Theorem 21.24.   □

COROLLARY. *The topologies* τ_ω *and* τ_σ *on* $H(U;F)$ *are quasicomplete.*

21.28. The following theorem describes a situation in which $\tau_\pi = \tau_\omega$. This is a special case of a much more general result (see 21.32).

THEOREM. *If* U *is a balanced open subset of a normed space, then*

$$\tau_\pi = \tau_\omega \text{ on } H(U;F)$$

Proof. Since 21.26 holds, it suffices to show that every seminorm ported by a compact subset of U is τ_π continuous. Since U is balanced, we only need to consider seminorms ported by balanced compact sets. Let p be a seminorm on $H(U;F)$ ported by a balanced compact set K in U. We claim that there is a continuous seminorm q on $H(K;F)$ such that $p = q \circ \pi_K$.

Since the Taylor series at O of a holomorphic mapping on a balanced open set converges for τ_ω (see 17.27), we may assume that

$$p(f) = p\left(\sum_{m=0}^{\infty} \frac{1}{m!} \hat{d}^m f(0) \right) = \sum_{m=0}^{\infty} \frac{1}{m!} p(\hat{d}^m f(0))$$

for $f \in H(U;F)$. Define a seminorm q on $H(K;F)$

$$q(f) = \sum_{m=0}^{\infty} \frac{1}{m!} p(\hat{d}^m f(0))$$

for $f \in H(K;F)$. If V is an open set containing K, then there exists $c(V) > 0$ such that

$$p(f) \leq c(V) \|f\|_V$$

for all $f \in H(U;F)$ since p is ported by K. Hence for $f \in H(K;F)$, we obtain

$$q(f) \leq c(V) \sum_{m=0}^{\infty} \frac{1}{m!} \|\hat{d}^m f(0)\|_V$$

This estimate shows that q is continuous by 17.9. It is clear that

$$p = q \circ \pi_K$$

Thus p is τ_π continuous.   □

COROLLARY. *If* U *is a balanced open subset of a normed space, then* $(H(U;F),\tau_\omega)$ *is complete.*

THE RUNGE PROPERTY

21.29 U-*Runge Compact Sets.* In the proof of Theorem 21.28 stating that if U is a balanced open set then $\tau_\omega = \tau_\pi$, we used the fact that the Taylor series of a holomorphic mapping f on U around O converges to f for the topology τ_ω; i.e., every holomorphic mapping on a balanced open set can be approximated uniformly on a neighborhood of a compact set by polynomials. This is a result of the classical Runge approximation theorem (see Hörmander (1967), pp. 6-9). Motivated by the

Runge approximation theorem, we introduce the following concept which
helps us to attain the equality $\tau_\omega = \tau_\pi$ on some $H(U;F)$.

Let U be an open subset of a normed space E. A compact subset K
of U is said to be *U-Runge* if for every $f \in H(K;F)$ there is a
sequence (f_n) in $H(U;F)$ such that $[f_n]_K \to f$ in $H(K;F)$.

If every compact subset of U is U-Runge, then U is said to have
the *Runge property*.

Notice that the definition of a U-Runge compact set is independent
of the choice of Banach spaces F since F can be equivalently replaced
with the complex field $\mathbb{C}$.

21.30 THEOREM. *A compact subset K of U is U-Runge if and only if
for every open subset V of U containing K, there exists an open
subset W of V containing K such that given any $f \in H^\infty(V;F)$ there
exists a sequence (f_n) in $H(U;F) \cap H^\infty(W;F)$ converging to f in the
sence of $H^\infty(W;F)$.*

Proof. The "only if" part is clear. Suppose that K is a U-Runge
compact subset of U. If V is an open subset of U containing K and
$f \in H^\infty(V;F)$, then $[f]_K \in H(K;F)$. Thus there exists a sequence (f_n)
in $H(U;F)$ such that $[f_n]_K \to [f]_K$. Since $H(K;F)$ is a regular
inductive limit (see 21.15), there exists an open subset W of V
satisfying the requirement.   □

21.31 EXAMPLES. (a) If U is a ξ-balanced open subset of a normed
               space E, then U has the Runge property.

       (b) A biholomorphic image of an open set with the Runge property
            also has the Runge property.

Proof. (a) If K is a ξ-balanced compact subset of U and $f \in H(K;F)$,
then there exists a ξ-balanced open set V, $K \subset V \subset U$ such that $f \in
H^\infty(V)$. Hence the Taylor series of f at ξ converges to f in $H^\infty(V;F)$;
thus in $H(K;F)$.

       (b) Let U and V be open subsets of normed spaces E and F respec-
tively, and assume that they are holomorphically equivalent (see
13.14). We claim that if U has the Runge property then V has the

Runge property. Let $\Psi: U \to V$ be a biholomorphic map. If K is a compact subset of U, we claim that $L = \Psi(K)$ is V-Runge. But this can be proved by observing that H(K) and H(L) are isomorphic as topological vector spaces. $\qquad\qquad\qquad\qquad\qquad\qquad\square$

21.32 THEOREM (Chae, 1970). *If* U *satisfies the Runge property, then*

$$\tau_\omega = \tau_\pi \text{ on } H(U;F)$$

Proof. We need to show that $\tau_\omega \leq \tau_\pi$ on H(U;F). Since U has the Runge property, the topology τ_ω is determined by the family of all seminorms ported by U-Runge compact sets.

Let K be a U-Runge compact subset of U and $[f] \in H(K;F)$. We can consider that H(K;F) is the inductive limit of Banach spaces $H^\infty(V;F)$ for all open subsets V with $K \subset V \subset U$. Hence there is an open subset V of U containing K such that $f \in H^\infty(V;F)$ with $[f]_K = [f]$. By 21.30, there exists an open subset W of V containing K and a sequence (f_n) in $H(U;F) \cap H^\infty(W;F)$ such that $f_n \to f$ in $H^\infty(W;F)$. Let p be a τ_ω continuous seminorm on H(U;F) which is ported by K. It follows that there is a $c(W) > 0$ such that

$$\left|p(f_m) - p(f_n)\right| \leq p(f_m - f_n) \leq c(W)\|f_m - f_n\|_W$$

Therefore, $\lim p(f_m)$ exists as $m \to \infty$.

Define a seminorm q on H(K;F) by

$$q([f]) = \lim_{m \to \infty} p(f_m)$$

if $[f] \in H(K;F)$ is such that a representative of $[f]$ is the limit of a sequence (f_m) in $H(U;F) \cap H^\infty(W;F)$ in the sense of $H^\infty(W;F)$. It is easy to check that q is a well-defined seminorm on H(K;F).

To show that the seminorm p is continuous on H(U;F) for the topology τ_π, it suffices to prove that q is continuous on H(K;F) since

$$p(f) = q(\pi_K(f))$$

for $f \in H(U;F)$.

Since $H(K;F)$ is the inductive limit of Banach spaces, it is sufficient to show that for any sequence (f_m) in $H^\infty(V;F)$ converging to 0 in $H^\infty(V;F)$, where $K \subset V \subset U$, $q([f_m]) \to 0$ as $m \to \infty$.

Since K is U-Runge, there exists an open subset W of V containing K such that for each m there corresponds a sequence $(f_{m,n})_{n \in \mathbb{N}}$ in $H(U;F) \cap H^\infty(W;F)$ converging to f_m uniformly on W. For each m, choose an integer $m(n) > m$ satisfying the following two conditions:

$$\left\| f_m - f_{m,m(n)} \right\|_W < 1/m \tag{1}$$

$$\left| q([f_m]) - p(f_{m,m(n)}) \right| < 1/m \tag{2}$$

Since (f_m) converges to 0 in $H^\infty(W;F)$, it follows from (1) that the sequence $(f_{m,m(n)})$ converges to 0 in $H^\infty(W;F)$. This shows that

$$p(f_{m,m(n)}) \to 0$$

as $m \to \infty$. Hence we conclude from (2) that

$$q([f_m]) \to 0$$

as $m \to \infty$.   □

The preceding theorem has the following application.

21.33 THEOREM. *If U is an open subset of a normed space E satisfying the Runge property, then $(H(U;F), \tau_\omega)$ is complete.*

21.34 REMARK. We do not know whether every open subset of a Banach space E satisfies the Runge property. It is not known whether or not every open subset in $\mathbb{C}^n$ satisfies the Runge property. Furthermore, we do not know if $\tau_\omega = \tau_\pi$ in general.

COMPLETENESS OF $(H(U;F),\tau_\omega)$

21.35. In this section we prove that $(H(U;F),\tau_\omega)$ is complete for any open subset of a normed space by refining the method presented in the preceding sections. This work is due to Mujica (1979). We first need to study some properties of inductive limits.

21.36 *Bounded Retraction.* The inductive limit

$$(E,\tau) = \varinjlim (E_\gamma, \tau_\gamma)$$

is said to be *boundedly retractive* if every bounded subset X of E is contained in some E_γ and $\tau = \tau_\gamma$ on X.

It is obvious that a boundedly retractive inductive limit is regular. The space $H(K;F)$ is a good example of a countable inductive limit which is boundedly retractive (see 21.21).

21.36 THEOREM. *A bounded retractive inductive limit of Banach spaces is complete.*

Proof. This is an easy exercise. $\square$

21.37 THEOREM. *Let*

$$(E,\tau) = \varinjlim (E_n, \tau_n)$$

be a boundedly retractive inductive limit of an increasing sequence of Banach spaces (E_n,τ_n) *and let F be a subspace of E. Then*

$$(F,\tau) = \varinjlim (E_n \cap F, \tau_n)$$

is also boundedly retractive. If we denote

$$F_n = cl_{\tau_n} (E_n \cap F)$$

the τ_n-closure of $E_n \cap F$, then the inductive limit

$$(F',\tau') = \lim_{\longrightarrow} (F_n,\tau_n)$$

is boundedly retractive and F' is the completion of F.

Proof. Since E is boundedly retractive, it is clear that F is also boundedly retractive.

We now prove the second part. Since F_n is a closed subspace of E_n, it is a Banach space. Hence F' is an inductive limit of an increasing sequence of Banach spaces; consequently, F' is a (DF)-space.

Consider the following diagram

$$
\begin{array}{ccccc}
F & \longrightarrow & F' & \to & E \\
\uparrow & & \uparrow & & \uparrow \\
E_n \cap F & \to & F_n & \longrightarrow & E_n
\end{array}
$$

where each arrow indicates the natural inclusion map which is continuous. Hence we have

$$\tau' \geq \tau$$

Let B be a bounded subset of F'. We claim that there exists m such that B is contained in $E_m \cap F'$ and $\tau' = \tau_m$. For each k let B_k be the closed unit ball of E_k. Since F' is a (DF)-space, by 21.12(b) there exist a positive integer n and $r > 0$ such that B is contained in the τ'-closure of $r(B_n \cap F)$. Denote

$$Y = cl_{\tau'}(B_n \cap F) \qquad \text{and} \qquad X = cl_{\tau}(B_n \cap F)$$

Since $Y \subset X$ and E is a boundedly retractive inductive limit, we can find m and s > 0 such that

$$X \subset sB_m$$

and we have $\tau_m = \tau$ on X. Hence on the set rY we have

$$\tau_m \geq \tau' \geq \tau = \tau_m$$

from which we have $\tau' = \tau_m$ on rY; thus on B.

It remains to show that Y is contained and bounded in F_m. In fact, let $x \in Y$ and let (x_γ) be a τ'-Cauchy net in $B_n \cap F$ converging to x. Then (x_γ) is a Cauchy net in E_m since $Y \subset X \subset sB_m$. E_m being complete, (x_γ) converges to a point y in E_m. Since E is boundedly retractive, $x \in E$ and E is Hausdorff, we conclude that x = y. This proves that

$$Y \subset cl_{\tau_m} (B_n \cap F) \subset cl_{\tau_m} (B_m \cap F) \subset F_m$$

Hence

$$B \subset rY \subset F_m$$

and $\tau' = \tau_m$ on B. Thus F' is boundedly retractive. By 21.26, F' is complete and it is clear from the above argument that F' is the completion of F.   □

21.38. We now apply the above theorem to the space H(K;F) since H(K;F) is a boundedly retractive inductive limit of an increasing sequence of Banach spaces.

For an open subset U of a normed space E and compact subset K of U, consider the following inductive limits:

$$H_U(K;F) = \varinjlim_{\substack{K\subset V \\ V\subset U}} H^\infty(V;F) \cap H(U;F)$$

$$\hat{H}_U(K;F) = \varinjlim_{\substack{K\subset V \\ V\subset U}} \overline{H^\infty(V;F)} \cap H(U;F)$$

where the horizontal line denotes the closure in the Banach space $H^\infty(V;F)$, and both inductive limits are with respect to the induced norm topology of $H^\infty(V;F)$. Identifying $H(U;F)$ with a subspace of $H(K;F)$, we have the following conclusion.

THEOREM. $\hat{H}_U(V;F)$ and $H_U(K;F)$ are *boundedly retractive inductive limits and* $\hat{H}_U(K;F)$ *is the completion of* $H_U(K;F)$.

21.39 THEOREM (Mujica, 1979). *If* U *is an open subset of a normed space* E, *then*

$$(H(U;F),\tau_\omega) = \varprojlim_{K\subset U} H_U(K;F) = \varprojlim_{K\subset U} \hat{H}_U(K;F)$$

Proof. Algebraically, both equalities follow from Theorem 21.25. Let τ be the projective topology on

$$H(U;F) = \varprojlim H_U(K;F)$$

First we show that the canonical mapping

$$\pi_K: f \in H(U;F) \to [f]_K \in H_U(K;F)$$

is τ_ω-continuous. In fact, if p_K is a continuous seminorm on $H_U(K;F)$ then for every open subset V of U containing K, there exists $c(V) > 0$ such that

$$p_K([f]) \leq c(V)\|f\|_V$$

where $f \in [f]$, and $[f] \in H_U(K;F)$. Hence if $f \in H(U;F)$ then

$$p_K \circ \pi_K(f) \leq c(V)\|f\|_V$$

so $p_K \circ \pi_K$ is a τ_ω-continuous seminorm on $H(U;F)$; thus $\tau \leq \tau_\omega$.

To show that $\tau \geq \tau_\omega$, let p be a seminorm on $H(U;F)$ ported by a compact subset K of U. Define p_K to satisfy

$$p = p_K \circ \pi_K$$

Then p_K is a well-defined seminorm on $H_U(K;F)$. In fact, if f and g are in $H(U;F)$ such that $\pi_K(f) = \pi_K(g)$, then there exists an open set V, $K \subset V \subset U$, such that f, g $\in H^\infty(V;F) \cap H(U;F)$ and f = g on V. Thus

$$p(f - g) \leq c(V)\|f - g\|_V = 0$$

Hence p_K is well-defined. It suffices to show that p_K is continuous on $H_U(K;F)$.

Since p is ported by K, for any open subset V of U containing K, p_K is continuous on $H^\infty(V;F) \cap H(U;F)$; thus p_K is continuous on $H_U(K;F)$. This proves that $\tau_\omega \leq \tau$. Therefore $\tau = \tau_\omega$.

It remains to show that the second equality is also valid. For a seminorm p on $H(U;F)$ ported by a compact subset K, we want to show that there exists a continuous seminorm p_K on $H_U(K;F)$ such that

$$p = p_K \circ \pi_K$$

Since we know such a seminorm exists for $H_U(K;F)$, it suffices to prove that every continuous seminorm on $H_U(K;F)$ admits a continuous extension of $\hat{H}_U(K;F)$. This is obvious since $\hat{H}_U(K;F)$ is the completion of $H_U(K;F)$. This completes the proof.   □

COROLLARY. *For any open subset of a normed space* E, *the space*
$(H(U;F), \tau_\omega)$ *is complete.*

EXERCISES

21A. The Banach space $P(^mE;F)$ is a complementary subspace of $H(K;F)$
for each $m \in \mathbb{N}$.

21B. Let X be a subset of $H(K;F)$. Then the following are equivalent.
  (a) X is compact in $H(K;F)$.
  (b) X is contained and compact in some $H^\infty(U;F)$ where U is an
       open neighborhood of K.
  (c) X is contained and compact in some $(H(U;F), \tau_\omega)$ where U is
       an open neighborhood of K.

21C. If (f_n) be a sequence in $H(K;F)$ converging to 0, then there
exists a sequence (a_n) of positive numbers with $a_n \to \infty$ such that
$(a_n f_n)$ still converges to 0.

21D. If K is a compact subset of a normed space E, then

$$H(K;F) = \varinjlim (H(U;F), \tau_\pi)$$

21E. A locally convex space E is said to satisfy the *Mackey conver-*
gent condition if for any sequence (x_n) in E converging to 0 there
exists a sequence (a_n) of positive numbers with $a_n \to \infty$ such that
$a_n x_n \to 0$. If K is a compact subset of a normed space E, then $H(K;F)$
satisfies the Mackey condition.

APPENDIX

Historical Notes on Analyticity as a Concept in Functional Analysis

by Angus E. Taylor*

INTRODUCTION

This is an essay--one of a projected series--on certain aspects of
the history of functional analysis. The emphasis of this essay is
on the way in which the classical theory of analytic functions of a
complex variable was extended, generalized, and came to play a significant role in functional analysis. One can perceive two lines of development: (i) the extension of the classical theory to cases in which
the function of a complex variable has its values in a function space
or in an abstract space, and (ii) the development of a theory of analytic functions from one general space to another (where by a "general
space" we mean a function space or an abstract space). The essay
also seeks to place the history of these developments in proper per-
spective in the narrative of the development of functional analysis
as a whole.

Before dealing explicitly with the main subject of the essay it
is desirable to sketch some relevant details of the history of functional analysis. The explicit emergence of the subject as a distinct
and separate branch of mathematics may perhaps be considered as
beginning with a series of five notes published by Vito Volterra in

*Reprinted from *Problems in Analysis: A Symposium in Honor of
Salomon Bochner,* R. C. Gunning (editor), Princeton University Press,
1970. We wish to acknowledge Professor A. E. Taylor and the
Princeton University Press for allowing us to reproduce the article
for this book.

1887 (see Volterra [55], [56], [57], [58] and [59]). The ideas
underlying functional analysis were of course much older. At the
1928 International Congress in Bolgona Jacques Hadamard mentioned
Jean Bernoulli's problem of the curve of quickest descent and the
pioneering work of the Bernoullis and Euler in the calculus of variations as the true and definitive foundation of "le calcul fonctionnel" (see Hadamard [24]). Volterra took the significant step
of focussing attention on functions for which the independent
variable was a function or a curve. This is not to say that Volterra
was the first to study operations by which functions are transformed
into other functions (or other entities). He was not. The "functional
operations" considered by mathematicians prior to the work of Volterra
were studied, however, from a somewhat different point of view, more
algebraic than analytic. Mathematicians examined the manipulations
which could be performed with functional operations, viewed as
symbols. (See Pincherle [2] and references cited therein.) In
particular, Volterra seemed to be making a new venture in subjecting
his "functions of lines" to analysis comparable to that applied in
calculus and the classical theory of functions.

In our current terminology, Volterra considered the domain of one
of his functions to be a class of functions (in the classical sense)
or a class of curves or surfaces. Volterra's functions were numerically valued; thus the ranges were sets of numbers. Volterra drew
his examples from boundary-value problems for partial differential
equations and from the calculus of variations.

In Volterra's time the name "functional" was still in the future.
In the first of his 1887 notes, he referred to "functions which
depend on other functions." The title of one of the later notes in
this series was "Sopra le funzioni dipendenti da linee." This gave
rise to the term "fonctions de lignes" which persisted in its Italian,
French, and English forms for a considerable period. It is asserted
by Maurice Fréchet and Paul Lévy that the noun "functional" (fonc-
tionelle) originated with Hadamard (see Fréchet [14], p. 2, and Lévy
[32], p. 8). Since Hadamard's pioneering work [22] on a general

method of representing linear functionals, the term "functional" has rather generally been reserved for a numerically valued function whose argument varies over a class of functions or some abstract set. However, such names as "operazioni funzionali" and "le calcul fonctionnel" were being used before 1900 in connection with operations which map functions into other functions (see Pincherle [41] and [42]). In [41], Pincherle has the following to say with reference to "le calcul fontionnel": "On reunirait sous ce titre les chapitres de l'analyse ou l'élément variable n'est plus le nombre, mais la fonction considerée en elle-meme." The name "analyse fontionnelle" was introduced by Lévy, according to Fréchet (see [14], p. 3).

In the early work on functional analysis certain algebraic and analytical operations were available in the nature of the explicit situation, there being no "abstract space" under consideration. It was possible to imitate, to a degree, the formulation of concepts such as continuity of a functional and uniform convergence of a sequence of functionals without a fully explicit treatment of metric and topological notions in the underlying class of functions. It was even possible to calculate such things as "variations" or differentials and functional derivatives without a fully explicit truly general definition. Most commonly these things were done merely by using absolute values and uniformity ideas, assuming the members of the underlying class of functions to be bounded and continuous. It was the work of Fréchet which led to abstraction and the explicit introduction of metrical and topological concepts into the general setting. With the greater generality and abstraction introduced by Fréchet and by E. H. Moore the name "general analysis" gained some currency. Many writers have tried to maintain a restriction on terminology, using "functional" always, as Hadamard had done, for a numerically valued function. In spite of this, "functional analysis" has come to mean not only functional calculus in the sense of Hadamard (l'étude des fonctionnelles), but also general analysis in the sense of Fréchet and Moore--that is, the study of functions (transformations) of a very general character, mapping one set onto another. The sets

may be abstract or they may be composed of mathematical objects
having a certain amount of structure: continuous functions, linear
operators, matrices, measurable sets, and so forth.

These historical notes present some of the results of an attempt
to search out the pioneering work on analytic functions in general
analysis, that is, of functions from one set to another which are
analytic in a suitable sense as an extension of the concept of complex analytic functions of a complex variable.

The essay is divided into four parts, dealing respectively with
generalizations of the concept of a polynomial, analytic functions
of a complex variable with values in a general space, analytic
functions from one general space to another, and the role in spectral
theory of abstractly valued analytic functions of a complex variable.

POLYNOMIAL OPERATIONS

One line of development of the generalization of the notion of
analyticity follows the Weierstrassian point of view, which places
the power series at the center of the theory. For this it is necessary
to have a functional analysis counterpart of the monomial function
defined by the expression $a_n z^n$ (a_n and z complex, n a natural number).

The initial steps in generalizing the concept of a polynomial
suitably for functional analysis seem to have been taken by Fréchet.
His first application was not to a generalization of analytic functions, but to a generalization of the Weierstrass theorem on approximation of continuous functions by polynomials. In a paper published
in 1909 [8] he considers how ordinary real-valued polynomials of one
real variable may be characterized as continuous functions such that
the application of certain differencing operations to these functions
leads to an identically vanishing result. The starting point is the
observation that f (real and continuous) is a first degree polynomial
in one real variable if

$$\Delta_2 f = f(x + y) - f(x) - f(y) + f(0) \equiv 0$$

For a polynomial of degree n the corresponding identity is $\Delta_{n+1}f \equiv 0$, where $\Delta_{n+1}f$ is a certain sum involving terms $\pm f(x_{i_1} + \ldots + x_{i_k})$ and $\pm f(0)$, where $(i_1, \ldots, i_k)$ is a combination of integers chosen from the aggregate $(1, 2, \ldots, n+1)$ and $k = 1, 2, \ldots, n+1$. For example,

$$\Delta_3 f = f(x_1 + x_2 + x_3) - f(x_2 + x_3) - f(x_3 + x_1) - f(x_1 + x_2)$$

$$+ f(x_1) + f(x_2) + f(x_3) - f(0)$$

In this 1909 paper Fréchet extends the considerations to real functions of several real variables and then to real functions (i.e., functionals) of an infinite sequence of real variables. For this latter case the nature of continuity and of the domain of definition of the function f are not discussed very clearly or generally. Here a continuous functional for which $\Delta_{n+1}f \equiv 0$ for some n is called a "fonctionnelle d'ordre entier n."

In the following year (1910) Fréchet published a paper [9] on continuous (real-valued) functionals defined on the class of real continuous functions which we now denote by C[a,b], where [a,b] is a finite real interval. Here he carries over to this different setting the concept of a "fonctionnelle d'ordre entier n" from the 1909 paper. The definition of such a functional U calls for U to be continuous and for $\Delta_{n+1}U$ to vanish identically. Fréchet generalizes Hadamard's 1903 theorem on the representation of linear functionals by showing that if U is a continuous functional on C[a,b] it can be represented in the form

$$U(f) = \lim_{n \to \infty} [u_n^{(0)} + U_n^{(1)}(f) + \ldots + U_n^{(r_n)}(f)]$$

where

$$U_n^{(r_k)}(f) = \int_a^b \ldots \int_a^b u_n^{(r_k)}(x_1, \ldots, x_{r_k}) f(x_1) \ldots f(x_{r_k}) dx_1 \ldots dx_{r_k}$$

Here $u_n^{(0)}$ is a number and $u_n^{(r_k)}(x_1,\ldots,x_{r_k})$ is a continuous function
depending only on U and the indices (but not on f); it may be taken
to be a polynomial in $x_1,\ldots,x_{r_k}$. The convergence to U(f) of the
sum is uniform on sets in C[a,b] which are compact (in the sense of
the term "compact" in use at that time as introduced by Fréchet).
The functional $u_n^{(r_k)}$ is "entier, d'ordre r_k."

In this paper Fréchet also shows that if U is entire and of order
n, then $U(y_1 f_1 +\ldots+ y_p f_p)$ is a polynomial of degree at most n in
the real variables $y_1,\ldots,y_p$. Here $f_1,\ldots,f_p$ are members of C[a,b].
He also observes that U is representable (uniquely) in the form

$$U(f) = U_0 + U_1(f) +\ldots+ U_n(f)$$

where U_0 is constant and U_k is entire and of order k as well as
homogeneous of degree k.

In the last part of this 1910 paper Fréchet turns to a definition
of holomorphism. Suppose $g \in C[a,b]$. Then a functional U is called
holomorphic at f = g if there is a representation (necessarily
unique)

$$U(f) = U_0 + U_1(f - g) + U_2(f - g) +\ldots+ U_n(f - g) +\ldots$$

converging suitably under certain restrictions, where U_n is entire,
of order n, and homogeneous of degree n. The representation is to
be valid when $\max|f(x) - g(x)| < \varepsilon$, for a certain positive ε, and
the convergence is to be uniform with respect to f when f is confined
to a compact subset of the functions which satisfy the inequality.
Fréchet observes that Volterra had already considered particular
instances of series representations of this type.

Fréchet observes that if U is holomorphic at f = g, then U(g + tf)
is a holomorphic function of t at t = 0. Here he comes close to
an alternative approach to the establishment of a theory of analytic
functionals. He notes, however, that U can have the property that

U(tf) is holomorphic in t at t = 0 without U being holomorphic at
f = 0. The example he gives is this: U(f) = maxf(x) + minf(x).
This functinoal U is homogeneous of the first degree: U(tf) ≡ tU(f).
But U is not entire of order one, and hence is not holomorphic at
f = 0. Fréchet is here considering real scalars exclusively.

The next stage in the development is apparently due to R. Gateaux,
who was clearly strongly influenced by both Fréchet and Hadamard.
Gateaux was killed in September 1914, soon after the beginning of
World War I, and his manuscript work remained unpublished until 1919.
He made decisive contributions to the theory of analytic functions
in the framework of functional analysis. At this point I take note
only of his work in polynomial operations. This work is presented
in two different contexts (see [17] and [18]). In [17], the manuscript of which dates from March 1914, Gateaux considers the space
(which he calls E'_ω) of all complex sequences $(x_1, x_2, \ldots)$ with écart

$$E(x,x') = \sum_{n=1}^{\infty} \frac{1}{n!} \frac{\left| x_n - x'_n \right|}{1 + \left| x_n - x'_n \right|}$$

A "polynomial of degree n" is a functional P which is defined and
continuous on a certain set D in E'_ω and such that $P(\lambda z + \mu t)$ is an
ordinary polynomial of degree n in λ and μ for each z and t in D.
Here Gateaux is using *as a defining property* something which Fréchet
had observed as a property possessed by his entire functionals of
order n. In another posthumous paper [18] we find Gateaux considering functionals defined on a space of continuous functions. Citing
Fréchet, he speaks of a functional as entire of order n if its
differences of order n + 1 vanish identically while some difference
of order n does not so vanish. He then points out that this definition is not satisfactory in the complex case. As an example he
cites U(z) = x, where z = x + iy (x and y real) for which $\Delta_2 U \equiv 0$,
$\Delta_1 U \not\equiv 0$. Here U is continuous, but it is not suitable to regard it
as a "polynomial" in z. In view of this, Gateaux uses the following
definition: a continuous functional U is called entire, of order n,

if $U(\lambda z + \mu t)$ is a polynomial of degree n in λ and μ whenever z and t are in the given space of continuous functions.

At this point Gateaux introduces the concepts of variations for a functional:

$$\delta U(z,t) = \left[\frac{d}{d\lambda} U(z + \lambda t)\right]_{\lambda=0}$$

$$\delta^2 U(z,t) = \left[\frac{d}{d\lambda} \delta U(z + \lambda t, t)\right]_{\lambda=0}$$

These are what have subsequently become known as Gateaux differentials. We know that $\delta U(z,t)$ is not necessarily continuous or linear in t, but Gateaux does not discuss these points carefully. Instead, he says with reference to $\delta U(z,t_1)$, "C'est une fonctionnelle de z et de t_1 qu'on suppose habituellement linéaire, en chaque point z, par rapport à t_1." He is discussing the case of real scalars at this point. Again, citing Fréchet, Gateaux obtains the following formulas when U is an entire functional which is homogeneous and of order n:

$$U(z + \lambda t) = U(z) + \ldots + \frac{\lambda^k}{k!}\delta^k U(z,t) + \ldots + \frac{\lambda^n}{n!}\delta^n U(z,t)$$

and

$$U(t) = \frac{1}{n!}\delta^n U(z,t)$$

In a 1915 paper Fréchet [10] points out that the problem of determining the representation of a functional of the second order is reducible to the problem of representing a bilinear functional. This is an early allusion to the relation (which emerges in later studies by various authors) between entire functionals of order n and multilinear functionals.

The next major step, and a definitive one, was taken by Fréchet. In 1925 he published a note [13], later expanded into a paper [15] published in 1929, entitled "les polynomes abstraits." The paper is a natural and direct outgrowth of his papers of 1909 and 1910, but the setting is more general. Fréchet is now considering functions from one abstract space to another. His spaces are of a general type which he calls "espaces algèbrophiles." These form a class more extensive than the class of normed linear spaces, but less extensive than the class of topological linear spaces as currently defined. The scalars are real, and the characterization of a function from one such space to another as an abstract polynomial is by means of continuity and the identical vanishing of differences, just as in the 1910 paper. The principal result is expressed thus: "Tout polynome abstrait d'ordre entier n'est la somme de polynomes abstraits d'ordre h = 0,1,...,n, chaque polynome d'ordre h étant homogène et de degré h, et la décomposition étant unique." By 1929, of course, the theory of abstract linear spaces was much further advanced than in 1910. The difference between the 1910 and 1929 papers is not in the basic characterization of abstract polynomials, but in the development of concepts and a framework and technique for dealing suitably with linearity and continuity as they enter into this particular problem.

The next six years saw the completion of the theory of abstract polynomials. This completion consisted in tidying up the proofs, putting them into their ultimate general form, and completing the linkage between several different ways of founding the theory. In particular, the difference between the complex and the real case was clarified.

In his 1932 California Institute of Technology doctoral dissertation R. S. Martin [34] used the following defintion: A function f(x) from one normed vector space to another is a polynomial of degree n if it is continuous and if $p(x + \lambda y)$ is a polynomial of degree at most n in λ (with vector coefficients) and of degree exactly n for some x,y. This definition applies for the case of either real or complex scalars. Martin was interested in a general theory of analytic functions. He was the student of A. D. Michal. At this time Michal was giving lectures on abstract linear spaces, and he developed an approach to polynomials

by using multilinear functionals. These lectures were not published,
but references to the work of this period are found in Michal [37].

In 1935 the Polish mathematicians S. Mazur and W. Orlicz published
an important paper [36] systematizing the theory quite completely
for the case of linear spaces with real scalars. The first of their
results was announced to the Polish Mathematical Society in 1933.
Their work was independent of that of Martin, Michal, and his group.
They separated out the part of the development of the theory which
is purely algebraic. In the algebraic portion of the theory they
begin with multi-additive operations. If $U^*(x_1,\ldots,x_k)$ is defined
for $x_1,\ldots,x_k$ in X, with values in Y, and is additive in each x_i,
$U(x) = U^*(x_1,\ldots,x_k)$ is said to be rationally homogeneous of degree
k (provided it does not vanish identically). An operation U of mth
degree is one which has a representation

$$U(x) = U_0(x) + U_1(x) + \ldots + U_m(x)$$

where $U_m(x) \not\equiv 0$ and each U_k is either identically zero or is rationally homogeneous of degree k. It is then shown how an operation U
which is rationally homogeneous of degree k determines and is
uniquely determined by a symmetric k-additive operation U* such that
$U^*(x_1,\ldots,x_k)$ becomes U(x) if one puts $x_1 = x_2 = \ldots = x_k$. U* is
obtained from U by differencing, and there is a "multinomial theorem"
expressing $U(t_1 x_1 + \ldots + t_k x_k)$ as a sum of terms in which the coefficient of each term of the form $t_k^{v_1} \ldots t_k^{v_k}$ is a multiple of

$$U^*(\underbrace{x_1,\ldots,x_1}_{v_1},\ldots,\underbrace{x_k,\ldots,x_k}_{v_k})$$

Finally, operations of degree m are characterized by both the Fréchet
approach and the Gateaux approach. The case of complex scalars is
not considered.

Passing beyond the purely algebraic part of the theory, Mazur and Orlicz deal with operations mapping one space of type (F) into another. (Here the nomenclature is that of Banach (see [1], p. 35). It is important for some but not for all of the results that the spaces be complete as metric spaces.) They add the requirement of continuity: a continuous k-additive operation is called k-linear, a continuous operation which is rationally homogeneous of degree k is called a homogeneous polynomial of degree k, and a continuous operation of degree m is called a polynomial of degree m. It is then shown that this concept of a polynomial coincides, for mappings from one (F)-space to another, with the concepts as variously introduced by Fréchet, Gateaux, and Martin.

Finally, for spaces with complex scalars, I. E. Highberg [25], another student of A. D. Michal, showed that for a mapping f from one complex algèbrophile space to another the following two sets of conditions are equivalent:

(i) f is continuous; $f(x + \lambda y)$ is a polynomial of degree at most n in λ for each x,y and of degree exactly n for some x,y.

(ii) f is continuous; f has a Gateaux differential at each point; $\Delta_{n+1} f \equiv 0$ and $\Delta_n f \not\equiv 0$.

The requirement of Gateaux differentiability in the second condition is superfluous in the real case but not in the complex case.

3. Analytic functions of a complex variable

As I pointed out earlier, Fréchet's work of 1910 includes consideration of the notion of a functional U defined on a set in the space C[a,b] and holomorphic at a point g of that space. In particular he noted that, for such a U, U(g + tf) is a holomorphic function of t at t = 0. Fréchet was dealing with the case of real scalars, and thus in this situation t and U(g + tf) are real. The essence of the situation is that U(g + tf) is expressible as a power series in t. The concept of an analytic function of a complex variable, with values in a function space, does not appear to come in for consideration

at this period either in Fréchet's work or in that of Gateaux. In
this essay I cannot deal fully with the question of just how this
concept developed. Analytic dependence on a complex parameter appears
at many places in the study of differential and integral equations.
In Ivar Fredholm's famous 1903 paper [16], for instance, the solution
of the equation

$$f(s) - \lambda \int_a^b k(s,t) f(t) dt = g(s)$$

is presented in the form

$$f(s) = g(s) + \frac{\lambda}{d(\lambda)} \int_a^b D(s,t;\lambda) g(t) dt$$

provided $d(\lambda) \neq 0$. This is in a context where f and g are members
of C[a,b]. Here $d(\lambda)$ is an entire analytic function of λ and $D(s,t;\lambda)$
is a continuous function of s,t as well as being an entire analytic
function of λ. However, there is no suggestion at this stage of
regarding the analytic dependence on λ in terms of the conceptuali-
zation of an analytic mapping from the complex plane into a function
space.

If we turn to the work of F. Riesz [43] ten years later, we find
explicit recognition of what was latent in the work of Fredholm,
though not in precisely the same context. Riesz was studying
"linear substitutions" in the theory of system of linear equations
in an infinite number of unknowns. He considered, in particular,
substitutions of the type called *completely continuous*, acting in
the class of infinite sequences which later came to be denoted by
l^2 (the classical prototype of a Hilbert space). If A is such a
substitution and E is the identity substitution, Riesz's studies
led him to the assertion (see [43], p. 106) that the inverse substitution $(E - \lambda A)^{-1}$ is a meromorphic function of λ. In this conclusion we perceive Reisz's conception of $(E - \lambda A)^{-1}$ as a function
of the complex variable λ with values in the class of bounded linear

substitutions on l^2. However, there is no explicit discussion here
of exactly what it means in general for such a function to be analytic
at a particular point or to have a pole at a particular point. There
is only a discussion of the particular function $(E - \lambda A)^{-1}$. This is
couched in terms of what occurs when one looks at certain related
linear substitutions (and their inverses) involving only a finite
number of unknowns. These systems are dealt with by using determi-
nants of finite order. A bit further on (see [43], pp. 114-119)
Riesz examines $(E - \lambda A)^{-1}$ for the case of an arbitrary bounded linear
substitution A (i.e., not merely the completely continuous case).
Riesz shows that the class of regular points λ (those for which E -
λA is appropriately invertible) form an open set in the complex plane
and that $(E - \lambda A)^{-1}$ is analytic at each regular point μ in the
following sense: $(E - \lambda A)^{-1}$ can be represented as a power series
in $\lambda - \mu$ with certain coefficients which are bounded linear substitutions. The series converges in a well defined sense when λ is suffi-
ciently close to μ. The mode of convergence is that of what is
sometimes called the uniform topology of the bounded linear substitutions.

Further references to the work of Riesz will be made later when
we come to the consideration of the role of analyticity as a concept
in spectral theory. However, it should be mentioned at this point
that Riesz observed (see [43], pp. 117-119) that it is possible to
integrate $(E - \lambda A)^{-1}$ and other such functions along contours in the
complex plane and to make effective use of the calculus of residues
in the study of linear substitutions.

In a paper published in 1923 Norbert Wiener [60] pointed out that
Cauchy's integral theorem and much of the classical theory of analytic
functions of a complex variable remain valid for functions from the
complex plane to a complex Banach space (which did not then regularly
carry that name). Wiener made a few observations about applications.
In particular, he pointed out that some of the work of Maxime Bôcher
[2], on complex-valued functions $f(x,z)$ of a real variable x and a
complex variable z, can be regarded as a study of an analytic function

of z with values in the function space C[a,b]. (See Taylor [51],
p. 655, for a slight refinement of this.) In another paper Taylor
[50] extended Wiener's work to obtain some rather surprising things
in connection with an analytic function of z with values in $L^p(a,b)$.

In the 1930's A. D. Michal and his students began to use Wiener's
observations in the study of analytic mappings from one complex
Banach space to another. In 1935 A. E. Taylor discovered that the
following definition leads to a satisfactory general theory of such
mappings: a function f with values in a complex Banach space Y is
called analytic in a neighborhood N of a point x_o in the complex
Banach space X if f is continuous in N and if, for each x in N and
each u in X, $f(x + \lambda u)$ is analytic (i.e., differentiable) as a
function of the complex variable λ in some neighborhood of $\lambda = 0$.
This definition and the theory flowing from it are natural generalizations of the work of Gateaux [18]. The details of Taylor's work
are given in [45] and [47]. That Gateaux's work could be generalized
in this manner was also observed by L. M. Graves (see [20], pp. 651-
653). Taylor also studied some of the divergences from the classical
theory which appear in the case of vector-valued analytic functions
of a complex variable.

During the 1930's the research of American mathematicians was
increasingly directed to the study of Banach spaces. The theory of
analytic functions began to play a systematic role in spectral-
theoretic studies of linear operators. This will be dealt with
separately in a later part of this essay. Another interesting and
important development occurred as a result of studying the concept
of analyticity in the context of various alternative modes of convergence. In 1937 Nelson Dunford and Taylor independently discovered
very closely related results, both of which depend essentially on
what has come to be known as the principle of uniform boundedness.
Dunford's result, published in 1938 (see [3], p. 354) is that, if f
is a function from an open set D in the complex plane to a complex
Banach space X, then f is analytic on D, in the sense of being
differentiable at each point of D with respect to the strong topology
of X, provided that x*(f(z)) is analytic on D for each continuous

linear functional x* defined on X. This result, which has enormous usefulness, was initially very surprising because this is a case in which the weak convergence of difference quotients implies their strong convergence. The result of Taylor relates to the dependence of a bounded linear operator on a complex parameter. If A_λ is such an operator (from one complex Banach space X to another such space Y) for each complex λ in the open set D, Taylor showed (see [50], p. 576) that A_λ is differentiable on D as an operator-valued function provided that $A_\lambda x$ is differentiable on D as a vector-valued function for each vector x in X. That is, convergence of the difference quotients in the strong topology of Y implies convergence in the uniform topology of the space of operators. This result was also unexpected and surprising. The result of Dunford can be deduced from that of Taylor, and vice versa.

The use in functional analysis of analytic functions from the complex plane to a function space or an abstract space has developed and ramified enormously since the 1930s. Since this essay treats only the early stages of the subject I forego further details and examples. For a report on the subject as of 1943 see Taylor [51]. For a striking application to obtain a famous classical theorem by recognizing it as a functional analysis instance of another classical theorem see [52] (this is also dealt with in [54], pp. 211-212).

4. Analytic functions of a vector variable

In a 1909 paper David Hilbert sketched the start of a theory of numerical analytic functions of a countable infinity of complex variables (see [26], pp. 67-74). He used the Weierstrassian approach; that is, he dealt with functions represented by a "power series" the successive terms of which are: a constant, a linear form, a quadratic form, a ternary form, and so on. A definition of analytic functions based on this notion of a power series appears in the work of Helge von Koch [30] in 1899. This work deals with infinite systems of differential equations; von Koch imitates the classical pattern of argument used to obtain analytic solutions of differential equations.

Hilbert considers analytic continuation and the composition of
analytic functions. He gives an example to show that analytic
continuation can give rise to uncountably many branches of a function.

As was pointed out earlier in this essay, Fréchet in 1910 introduced a definition of what it means for a functional to be holomorphic at a point in the space C[a,b]. The basic idea is that of an
expansion in a series of homogeneous polynomials of ascending degree.
Fréchet's work is much clearer than that of Hilbert, perhaps mainly
because the questions surrounding proper definitions of convergence
and of domains of definition of forms and power series are obscure
when one deals with a countable infinity of complex variables without
an adequate consideration of the structure of the space composed of
the sequences $(x_1, x_2, \ldots)$. Fréchet's 1910 paper marked out quite
clearly the lines along which the theory was to develop, and the work
of Gateaux exploited Fréchet's beginning in brilliant fashion.
Doubtless there were other forerunners of the definitive work of
these pioneers. For example, Volterra had considered infinite
series in which integrals of the form

$$\int_a^b \cdots \int_a^b K(x_1, \ldots, x_n) f(x_1) \ldots f(x_n) dx_1 \ldots dx_n$$

play the role of the homogeneous polynomial of degree n on C[a,b].

At this stage there was lacking to Fréchet and Gateaux something
fully comparable to the Cauchy point of view, according to which a
function is characterized as analytic on a suitable set if it is
differentiable on each point of the set. The early workers in
functional analysis borrowed the concept of a "variation" from the
calculus of variations. Gateaux used variations in his theory of
analytic functionals (see [18]) to characterize a functional U as
being holomorphic on a certain domain of definition if it is continuous and admits a first variation $\delta U(z,t)$ at each point z of the domain.
(The definition of this variation was given earlier in this essay,
in the discussion of polynomials.)

The variation $\delta U(z,t)$ came to be called a Gateaux differential. However, it lacked certain properties which are desirable in a differential. Fréchet made a definitive contribution with his 1925 paper [12] in which he shows how to define the concept of a differential of a function which maps one abstract space (of suitable type) on another. Fréchet's differential has strong properties; it has become the standard instrument of differential calculus in modern Banach space theory. In his paper Fréchet ascribes to Hadamard a basic principle: the fundamental fact about the differential of a function in calculus is that it depends linearly on the differentials of the independent variables. He refers to Hadamard's book [22] on the calculus of variations, where Hadamard asserts that the methods of differential calculus can be extended to functionals $U(y)$ for which the first variation is a linear functional of the variation of y. Actually, Fréchet's concept of the differential is a direct extension to the abstract situation of the well known definition introduced into calculus by O. Stolz. The line of Fréchet's thinking as early as 1912 with respect to the Stolz definition is clearly shown in [24] and in other papers mentioned therein. However, the tools were not yet available to formulate the definition adequately in the abstract space context which Fréchet came to in 1925.

The Fréchet concept of the differential is stronger than the Gateaux concept. More precisely, if f is a function from a normed linear space X to a normed linear space Y, if f is defined in a neighborhood of the point x_0 and has a Fréchet differential at x_0, then f has a Gateaux differential at x_0 and coincides with the Fréchet differential. This proposition becomes false, however, if we interchange the names Fréchet and Gateaux in it. Some of the disparity between the two concepts is revealed by the observation that f may have a Gateaux differential and yet be discontinuous, while f must be continuous at a point if it has a Fréchet differential at that point. Because of this great difference between the two differentials it is remarkable that the following theorem is true. (It was discovered independently by L. M. Graves and Angus E. Taylor, but Graves has priority in the time of discovery. (See [20], as well as [45]

and [46]. Both Graves and Taylor knew the work of Gateaux.) Suppose
that X and Y are complex Banach spaces and that f is a function with
values in Y which is defined on an open set D in X. Then f has a
Fréchet differential at each point of D if and only if it is continuous and has a Gateaux differential at each point of D. Moreover,
under these conditions f has Fréchet differentials of all orders at
each point of D, f is holomorphic at each point of D, and the power
series expansion of f in the neighborhood of a point x_0 in D is
convergent within the largest spherical neighborhood centered at
x_0 which lies wholly in D. The homogeneous polynomials in $x - x_0$
which form the terms of the power series are expressible as multiples
of the Fréchet differentials of f at x_0 according to the appropriate
generalization of the Taylor series. The proof of this theorem
utilizes and exposes the core of the theory of analytic mappings
from one complex Banach space to another. Gateaux's work, building
on the original power series conception of Fréchet, set the whole
development in motion, but Gateaux's theory of analytic functionals
made no reference to or use of the Fréchet concept of a differential.

One consequence of all this is that there is a simple definition
of analyticity completely analogous to the classical definition from
the Cauchy standpoint: f is analytic on an open set if it has a
Fréchet differential at each point of the set (both domain and
range in complex Banach spaces).

A few years prior to the discoveries by Graves and Taylor the
power series approach to abstract analytic functions was under in-
tensive study by Michal and his students, especially Clifford and
Martin. They did not use the methods based on complex variables
and Cauchy's formulas, as Gateaux had done, but relied entirely on
the power series approach. (See [34], [38], and [39], especially
p. 71 in [39].)

In a posthumously published monography by Michael are given some
historical notes about his involvement with the theory of polynomials
and abstract analytic functions (see [37], pp. 35 - 39). Taylor was
Michal's student. His work, originally motivated by an interest in

a long paper of Luigi Fantappiè [7] on a rather different theory of analytic functionals, led into the consideration of $f(x + \lambda y)$ as a function of the complex variable λ, and from this, using Cauchy's integral formulas, Taylor discovered the linkage between the Gateaux and Fréchet differentials. It was apparent that Fantappiè's work did not lend itself to treatment in the framework of Banach spaces. It has subsequently become clear that a more general theory of topological linear spaces is necessary, and there have been extensive developments based on the foundations laid by Fantappiè. (See, for instance, [31], pp. 375-381, and references therein to Grothendieck, Sebastiao e Silva, and others.)

The state of subsequent development of the theory of analytic functions of a vector variable is partially indicated, with many references, in the massive book of Hille and Phillips ([28], Chapter 3, especially pp. 109 - 116). See also Dunford and Schwartz ([6], pp. 520 - 526) for an important application. Significant contributions were made in three papers by Max Zorn [61], [62], [63]. There are great untouched areas in the theory of analytic functions of a vector variable, especially in the study of such things as manifolds of singularity, convergence sets of power series, and envelopes of holomorphy. This is not surprising, since these subjects are so deep and complicated in the case of functions of a finite number of complex variables.

5. The role of analytic functions in spectral theory

I made reference earlier in this essay to the work of Riesz in showing that $(E - \lambda A)^{-1}$ depends analytically (and in some cases meromorphically) on λ. This work of Riesz foreshadows a wealth of important developments, but the full scope of what was latent in Riesz' book did not become evident for many years. Taylor, in a paper [49] published in 1938, showed that if T is a closed linear operator in a complex Banach space, the inverse $(\lambda I - T)^{-1}$, if it exists in a suitable sense for at least one λ, is an operator-valued analytic function on the (necessarily open) set of λs for which it is defined. He also showed that if T is bounded and everywhere defined there must be some value of λ for which $(\lambda I - T)^{-1}$ does not exist; that is, the spectrum of T is not empty. This result hinges on the use of Liouville's theorem for vector-valued analytic functions. The special case of this theorem on the spectrum, for operators in a Hilbert space, had been known earlier. Marshall Stone [44] obtained the result without using vector-valued analytic functions. He did this by applying linear functionals to $(\lambda I - T)^{-1}x$.

At about this same time and a few years later work was going on in the study of complete normed rings (later to be known as Banach algebras) with the use of analytic functions as a tool. Ring elements $\lambda e - a$ and their inverses were investigated by Gelfand [19], Lorch [33], Mazur [35], and Nagumo [40]. It is interesting to note that, although Gelfand defined the concept of an analytic function of the complex variable λ with values in a complete normed ring, he developed the theory of such functions by reducing the arguments to the numerical case through the use of linear functionals. All of this work in normed rings has many parallels with the more general spectral theory of linear operators.

Spectral theory of linear operators is the functional analysis counterpart of the theory of eigenvalues of linear transformations in spaces of finite dimension. The spectrum of a linear operator T acting in a Banach space corresponds to the set of eigenvalues of the linear transformation in the finite-dimensional case. The spectrum of T consists of all values of λ for which $\lambda I - T$ fails to have an inverse in a suitable sense. When the Banach space is finite-dimensional, the spectrum is a finite set of eigenvalues and $(\lambda I - T)^{-1}$ is a rational function of λ. Much of spectral theory, for a particular operator acting in a Banach space, may be viewed as the study of the operator-valued function $(\lambda I - T)^{-1}$, called the *resolvent* of T, and of the behavior of this function of λ both globally and locally. A great deal of information about T itself can be extracted from this study.

The calculus of residues and Cauchy's integral formula play important roles in spectral theory. Integrals of the form

$$\frac{1}{2\pi i} \int_C f(\lambda)(\lambda I - T)^{-1} d\lambda$$

over suitable closed contours C in the plane, where the complex-valued function f is analytic on a neighborhood of the spectrum of T, are used to define operators which can be used in significant ways. The heuristic basis for the definition of an operator denoted by f(T) is the symbolic "Cauchy's formula"

$$f(T) = \frac{1}{2\pi i} \int_C \frac{f(\lambda)}{\lambda - T} d\lambda$$

For the finite-dimensional case (i.e., for the study of finite square matrices) there are nineteenth-century applications of this sort of thing in the work of Frobenius and others, most explicitly in an 1899 paper by Poincaré. For references to this early work see Dunford and Schwartz ([6], pp. 606 - 607) and Taylor ([51], p. 662, [53], p. 190). Systematic exploitation of the calculus of residues and of the symbolic operational calculus based on the Cauchy integral formula, as applied to general spectral theory, starts with Riesz and picks up again in the years around 1940. The work of investigators of normed rings, referred to earlier, is part of this general development. Dunford [4], [5] and Taylor [51], [53], [54] dealt explicitly with operator theory. (See also Elinar Hille [27]. For further references see Hille and Phillips [28], pp. 164 - 183.)

The subsequent development and application in functional analysis of the ideas and methods of classical analytic function-theory have been varied and rich. I conclude this essay by mentioning just one example of such development. Analytic function-theory methods have been used to deal with perturbations of operators and their spectra. Perturbation theory goes back a long way in mathematics, of course.

The pioneering work in studying perturbations of operators in Hilbert space seems to be that of F. Rellich, dating from 1937. Numerous subsequent investigators have made use of the symbolic operational methods of Dunford and Taylor. For references to the work of Rellich, B. v. Sz. Nagy, F. Wolf, and others see Kato [29].

UNIVERSITY OF CALIFORNIA
BERKELEY

REFERENCES

1. Banach, Stefan, *Théorie des operations linéaires,* Warsaw, 1932.
2. Bôcher, Maxime, "On semianalytic functions of two variables," *Annals of Mathematics,* (2), *12* (1910), 18-26.
3. Dunford, Nelson, "Uniformity in linear spaces," *Transactions of the American Mathematical Society,* 44 (1938), 305-356.
4. ———, "Spectral theory," *Bulletin of the American Mathematical Society, 49* (1943), 637-651.
5. ———, "Spectral theory, I: Convergence to projections," *Transactions of the American Mathematical Society, 54* (1943), 185-217.
6. Dunford, N., and J. T. Schwartz, *Linear Operators, Part I: General Theory.* New York: Interscience Publishers, Inc., 1958.
7. Fantappiè, Luigi, "I funzionali analitici", *Memorie della R. Accademia Nazionale dei Lincei,* (6) 3 fasc., *11* (1930), 453-683.
8. Fréchet, Maurice, "Une définition fonctionnelle des polynomes," *Nouvelles Annales de Mathématiques,* (4) *9* (1909), 145-162.
9. ———, "Sur les fonctionnelles continues," *Annales Scientifiques de l'École Normale Superieure,* (3) 27 (1910), 193-216.
10. ———, "Sur les fonctionnelles bilinéaires," *Transactions of the American Mathematical Society, 16* (1915), 215-234.
11. ———, "Sur la notion de différentielle totale," *Comptes Rendus du Congrès des Sociétés Savantes en 1914,* Sciences, 5-8, Paris, 1915.
12. ———, "La notion de différentielle dans l'analyse générale," *Annals Scientifiques de l'Ecole Normale Supérieure,* (3) *42* (1925), 293-323.
13. ———, "Les transformations ponctuelles abstraites," *Comptes Rendus Acad. Sci. Paris, 180* (1925), 1816.
14. ———, *Les Espaces Abstraits et leur théorie considérée comme introduction à l'analyse générale.* Paris: Gauthier-Villars, 1928.
15. ———, "Les polynomes abstraits," *Journal de Mathématiques Pures et Appliquées,* (9) *8* (1929), 71-92.

16. Fredholm, Ivar, "Sur une classe d'équations fonctionnelles," *Acta Mathematica, 27* (1903), 365-390.

17. Gateaux, R., "Fonctions d'une infinité des variables indépéndantes," *Bullétin de la Societé Mathématique de France, 47* (1919), 70-96.

18. ——, "Sur diverses questions de calcul fonctionnel," *Bullétin de la Societé Mathématique de France, 50* (1922), 1-37.

19. Gelfand, I. M., "Normierte Ringe," *Matem. Sbornik, 9* (51) (1941), 3-24.

20. Graves, L. M., "Topics in the functional calculus," *Bulletin of the American Mathematical Society, 41* (1935), 641-662.

21. Hadamard, Jacques, "Sur les opérations fonctionnelles," *Comptes Rendus Acad. Sci. Paris, 136* (1903), 351.

22. ——, *Leçons sur le calcul des variations.* Paris: Hermann, 1910.

23. ——, "Le calcul fonctionnel," *l'Enseignement Mathématique, 14* (1912), 1-18.

24. ——, "Le développement et le rôle scientifique du calcul fonctionnel," in *Atti del Congresso Internazionale dei Matematici,* Bologna, 3-10 Settembre 1928 (VI) Tomo I, pp. 143-161.

25. Highberg, Ivar, *Polynomials in Abstract Spaces,* unpublished doctoral dissertation, California Institute of Technology, 1936.

26. Hilbert, David, "Wesen und Ziele einer Analysis der Unendlichvielen unabhängigen Variabeln," *Rendiconti del Circolo Matematico di Palermo, 27* (1909), 59-74.

27. Hille, Einar, "Notes on linear transformations, II: Analyticity of semigroups," *Annals of Mathematics,* (2) *40* (1939), 1-47.

28. Hille, Einar, and R. S. Phillips, *Functional Analysis and Semigroups,* rev. ed. Providence, R.I.: American Mathematical Society, 1957.

29. Kato, Tosio, *Perturbation Theory for Linear Operators.* Berlin: Springer-Verlag, 1966.

30. von Koch, Helge, "Sur les systèmes d'ordre infini d'équations differentielles," *Ófversigt af Kongl. Svenska Vetenskaps - Akademiens Forhandlingar, 61* (1899), 395-411.

31. Köthe, Gottfried, *Topologische Lineare Räume,* I. Berlin: Springer-Verlag, 1960.

32. Lévy, Paul, "Jacques Hadamard, sa vie et son oeuvre - Calcul fonctionnel et questions diverses," in Monographie №16 de *l'Enseignement Mathématique - La Vie et l'oeuvre de Jacques Hadamard.* Genève, 1967, pp. 1-24.

33. Lorch, E. R., "The theory of analytic functions in normed Abelian vector rings," *Transactions of the American Mathematical Society, 54* (1943), 414-425.

34. Martin, R. S., *Contributions to the Theory of Functionals,* unpublished doctoral dissertation, California Institute of Technology, 1932.

35. Mazur, S., "Sur les anneaux linéaires," *Comptes Rendus Acad. Sci. Paris, 207* (1938), 1025-1027.

36. Mazur, S. and W. Orlicz, "Grundlegende Eigenschaften der poly-
    nomischen Operationen" (erste Mitteilung), *Studia Mathe-
    matica, 5* (1935), 50-68.
37. Michal, Aristotle D., *Le Calcul Différential dans les espaces
    de Banach.* Paris: Gauthier-Villars, 1958.
38. Michal, A. D. and A. H. Clifford, "Fonctions analytiques
    implicites dans des espaces vectoriels abstraits," *Comptes
    Rendus Acad. Sci. Paris, 197* (1933), 735-737.
39. Michal, A. D. and R. S. Martin, "Some expansions in vector
    space," *Journal de Mathématiques Pures et Appliquées, 13*
    (1934), 69-91.
40. Nagumo, M., "Einige analytische Untersuchungen in linearen
    metrischen Ringen," *Japanese Journal of Mathematics, 13*
    (1936), 61-80.
41. Pincherle, Salvatore, "Mémoire sur le calcul fonctionnel
    distributif," *Mathematische Annalen, 49* (1897), 325-382.
42. ———, "Funktional-Gleichungen und Operationen," in *Encyklopädie
    de Mathematischen Wissenschaften mit Einschluss ihrer
    Anwendungen, II$_1$,* Heft 6, Leipzig, 1906, pp. 761-817.
43. Riesz, F., *Les systémes d' équations linéaires à une infinité
    d' inconnues.* Paris: Gauthier-Villars, 1913.
44. Stone, M. H., *Linear Transformations in Hilbert Space.* New York:
    American Mathematical Society, 1932.
45. Taylor, Angus E., *Analytic Functions in General Analysis,*
    unpublished doctoral dissertation, California Institute of
    Technology, 1936.
46. ———, "Sur la théorie des fonctions analytiques dans les
    espaces abstraits," *Comptes Rendus Acad. Sci. Paris, 203*
    (1936), 1228-1230.
47. ———, "Analytic functions in general analysis," *Annali della R.
    Scuola Normale Superiore di Pisa, 6* (1937), 277-292.
48. ———, "On the properties of analytic functions in abstract
    spaces," *Mathematische Annalen, 115* (1938), 466-484.
49. ———, "The resolvent of a closed transformation," *Bulletin of
    the American Mathematical Society, 44* (1938), 70-74.
50. ———, "Linear operations which depend analytically on a param-
    eter," *Annals of Mathematics, 39* (1938), 574-593.
51. ———, "Analysis in complex Banach spaces," *Bulletin of the
    American Mathematical Society, 49* (1943), 652-669.
52. ———, "New proofs of some theorems of Hardy by Banach space
    methods," *Mathematics Magazine, 23* (1950), 115-124.
53. ———, "Spectral theory of closed distributive operators,"
    Acta Mathematica, 84 (1951), 189-224.
54. ———, *Introduction to Functional Analysis.* New York: John
    Wiley & Sons, Inc., 1958.
55. Volterra, Vito, "Sopra le funzioni che dipendono da altre
    funzioni," Nota I. *Rendiconti della R. Accademia dei Lincei,*
    Series IV, vol. III (1887), pp. 97-105.
56. ———, Nota II, *ibid.,* pp. 141-146.
57. ———, Nota III, *ibid.,* pp. 153-158.
58. ———, "Sopra le funzioni dipendenit da linee," Nota I, *ibid.,*
    pp. 225-230.

59. Volterra, Vito, Nota II, *ibid.*, pp. 274-281.
60. Wiener, Norbert, "Note on a paper of M. Banach," *Fundamenta Mathematica, 4* (1923), 136-143.
61. Zorn, Max, "Characterization of analytic functions in Banach spaces," *Annals of Mathematics,* (2) *46* (1945), 585-593.
62. ———, "Gateaux differentiability and essential boundedness," *Duke Mathematical Journal, 12* (1945), 579-583.
63. ———, "Derivatives and Fréchet differentials," *Bulletin of the American Mathematical Society, 52* (1946), 133-137.

Bibliography

1. Abraham, R. and Marsden, J.E. (1978). *Foundations of Mechanics* (2nd ed.), Benjamin/Cumming Publishing, Reading, Massachusetts.

2. Alencar, R., Aron, R. and Dineen, S. (1984). A reflexive space of holomorphic functions in infinite many variables, *Proc. Amer. Math. Soc.* 90, 407-411.

3. Alexander, H. (1968). Analytic functions on Banach spaces, Thesis, University of California, Berkeley.

4. Alexiewicz, A. and Orlicz, W. (1953). Analytic operations in real Banach spaces, *Studia Math.* 14, 57-81.

5. Aron, R.M. (1972). Holomorphic functions on balanced subsets of a Banach space, *Bull. Amer. Math. Soc.* 78, 624-627.

6. ———(1973). Holomorphic types for open subsets of Banach spaces, *Studia Math.* 45, 273-289.

7. ———(1973a). The bornological topology on the space of holomorphic mappings of a Banach space, *Math. Annalen* 202, 265-272.

8. ———(1974). Entire functions of unbounded type on a Banach space, *Bull. U. M. It.* 4(9), 28-31.

9. Aron, R.M. and Schottenloher, M. (1974). Compact holomorphic mappings on Banach spaces and the approximation property, *J. Functional Analysis* 21(1), 7-30.

10. Averbukh, V.I. and Smolyanov, O.G. (1968). The various definitions of the derivative in linear topological spaces, *Russian Math. Surveys* 23(4), 67-113.

11. Banach, S. (1932). *Theorie des Operations Lineaires*, Monografje Matematyczne 1, Warszawa.

12. Banach, S. (1938). Uber homogene Polynome in (L^2), *Studia Math.* 7, 36-44.

13. Barroso, J.A. (1971). Topologias nos espacos de aplicacoes holomorfas entre espacos localmente convexos, *Anais. Acad. Bras. de Ciencas* 43, 527-546.

14. ———(1976). *Introduccion a la Holomorfia entre Espacios Normados,* Universidad de Santiago de Compostela, Spain.

15. ———(1979). *Introduccion a la Holomorfia entre Espacios Localmente Convexos,* Universidad de Valencia, Spain; in English, to appear in North-Holland Math. Studies.

16. ———Editor (1979). *Advances in Holomorphy,* North-Holland Math. Studies 34, Amsterdam.

17. ———Editor (1982). *Functional Analysis, Holomorphy and Approximation Theory,* North-Holland Math. Studies, 71, Amsterdam.

18. Bartle, R. (1974). *Elements of Real Analysis,* John Wiley, New York.

19. Bayoumi, A. (1980). Bounding subsets of some metric vector spaces, *Arkiv for Math.* 18(1), 13-17.

20. Bierstedt, K-D. and Meise, R. (1979). Aspects of inductive limits in spaces of germs of holomorphic functions on locally convex spaces and applications to a study of $(H(U),\tau_\omega)$. In *Advances in Holomorphy,* J.A. Barroso (ed.), North-Holland Math. Studies 34, 111-178.

21. Bochnak, J. (1970). Analytic functions in Banach spaces, *Studia Math.* 35, 273-292.

22. Bochnak, J. and Siciak, J. (1971). Polynomials and multi-linear mappings in topological vector spaces, *Studia Math.* 39, 59-76.

23. ———(1971a). Analytic functions in topological vector spaces, *Studia Math.* 39, 77-112.

24. Boland, P.J. (1974). Some spaces of entire and nuclear functions on a Banach space, Part I and Part II, *Journal fur die Reine und Angewandte Math.* 270, 38-60; 271, 8-27.

25   ———(1975). Holomorphic functions on nuclear spaces, *Trans. Amer. Math. Soc.* 209, 275-281.

26. ———(1976). *Holomorphic Functions on Nuclear Spaces,* Universidad de Santiago de Compostela, Spain.

27. Boland, P.J. and Dineen, S. (1974). Convolution operators on G-holomorphic functions in infinite dimensions, *Trans. Amer. Math. Soc.* 190, 313-323.

28. Bremermann, H.J. (1956). Complex convexity, *Trans. Amer. Math. Soc.* 82, 17-51.

29. ――――(1957). Holomorphic functionals and complex convexity in Banach spaces, *Pacific J. Math.* 7, 811-831.

30. ――――(1960). The envelopes of holomorphy of tube domains in Banach spaces, *Pacific J. Math.* 10, 1149-1153.

31. ――――(1965). Pseudo-convex domains in linear topological spaces. In *Proceedings of Conference on Complex Analysis,* Minneapolis, 1964, Springer-Verlag, 182-186.

32. ――――(1965a). Several complex variables. In *Studies in Real and Complex Analysis,* I.I. Hirschman, Jr. (ed.), Math. Assoc. Amer. Studies in Math. 3, 3-33.

33. Cartan, E. (1935). Sur les domaines bornes d'espaces n variables complexes, *Abhandlungen Math. Sem. Univer. Hamburg* 11, 116-162.

34. Cartan, H. (1971). *Differential Calculus,* Hermann/Houghton Miffin, New York.

35. ――――(1961). *Theorie Elementaire des Fonctions Anaytiques d'une ou plusieurs Variables Complexes,* Hermann, Paris; in English, *Elementary Theory of Analytic Functions of One or Several Variables,* Addison-Wesley, Reading, Massachusetts.

36. Cartan, H. and Thullen, P. (1932). Zur Theorie der Singularitaten der Funktionen mehrerer Veranderlichen: Regularitats und Konvergenzbereiche, *Math. Annalen* 106, 617-647.

37. Chae, S.B. (1970). Sur les espaces localement convexes de germes holomorphes, *C. R. Acad. Sci. (Paris)* 271, 990-991.

38. ――――(1971). Holomorphic germs on Banach spaces, *Ann. Institut Fourier* 21, 107-141.

39. ――――(1977). A holomorphic characterization of Banach spaces with basis. In *Infinite Dimensional Holomorphy and Applications,* M.C. Matos (ed.), North-Holland Math. Studies 12, 139-146.

40. ――――(1980). *Lebesgue Integration,* Monographs and Textbooks in Pure and Applied Math. 58, Marcel Dekker, New York and Basel.

41. Choquet-Bruhat, Y., Dewitt-Morette, C. and Dillard-Bleick, M. (1977). *Analysis, Manifolds, and Physics,* North-Holland, Amsterdam and New York.

42.   Coeuré, G. (1970). Fonctions plurisousharmoniques sur les
      espaces vectoriels topologiques et applications a l'etude des
      fonctions analytiques, *Ann. Institut Fourier* 20, 361-432.

43.   ———(1974). *Analytic Functions and Manifolds in Infinite
      Dimensional Spaces*, North-Holland Math. Studies 11, Amsterdam.

44.   ———(1977). Sur le rayon de bornologie des fonction holo-
      morphes. In *Journees de Fonctions Analytiques, Toulous*, Mai
      1976, P. Lelong (ed.), Springer-Verlag Lecture Notes in Math.
      578, 183-194.

45.   Colombeau, J.F. (1974). On some various notions of infinite
      dimensional holomorphy. In *Proceedings of Infinite Dimen-
      sional Holomorphy*, T.L. Hayden and T.J. Suffridge (eds.),
      Springer-Verlag Lecture Notes in Math. 364, 145-149.

46.   ———(1982). *Differential Calculus and Holomorphy*, North-
      Holland Math. Studies 64, Amsterdam.

47.   Davis, W.J., Garling, D.J.H. and Tomczak-Jaegermann, N. (1984).
      The complex convexity of quasi-normed linear spaces, *J.
      Functional Analysis* 55, 110-150.

48.   Dias, C.L.S. (1952). Espacos vectoriais topologicis e sua
      aplicacao nos espacos funcionais analiticos, *Bol. Soc. Mat.
      Sao Paulo* 5, 1-58.

49.   Dieudonne, J. (1953). Recent developments in theory of
      locally convex spaces, *Bull. Amer. Math. Soc.* 59, 495-512.

50.   ———(1960). *Foundations of Modern Analysis*, Academic Press,
      New York.

51.   Dineen, S. (1970). Holomorphic functions on a Banach
      space, *Bull. Amer. Math. Soc.* 76, 883-886.

52.   ———(1970a). Holomorphic types on a Banach space, *Studia
      Math.* 39, 241-288.

53.   ———(1970b). The Cartan-Thullen theorem for Banach spaces,
      Ann. Scuola Normale Superiore Pisa 24(4), 667-676.

54.   ———(1971). Bounding subsets of a Banach space, *Math.
      Annalen* 192, 61-70.

55.   ———(1971a). Runge's theorem for Banach spaces, *Proc.
      Royal Irish Acad.* 71(a), 85-89.

56.   ———(1972). Holomorphic functions on (c_o, X_b)-modules, *Math.
      Annalen* 196, 106-116.

57. Dineen, S. (1972a). Unbounded holomorphic functions on a Banach space, *Journal London Math. Soc.* 4(3), 461-465.

58. ———(1973). Holomorphic functions on locally convex topological vector spaces: I. Locally convex topologies on H(U); II. Pseudoconvex domains, *Ann. Institut Fourier* 23, 19-54; 155-185.

59. ———(1975). Surjective limits of locally convex spaces and their applications to infinite dimensional holomorphy, *Bull. Soc. Math. France* 103, 441-509.

60. ———(1981). *Complex Analysis in Locally Convex Spaces*, North-Holland Math. Studies 57, Amsterdam.

61. ———(1983). Entire functions on c_o, *J. Functional Analysis* 52, 205-218.

62. Dineen, S., Noverraz, P. and Schottenloher, M. (1976). Le probleme de Levi dans certains espaces vectoriels topologiques localement convexes, *Bull. Soc. Math. France* 104, 87-97.

63. Douady, A. (1966). Le probleme des modules pour les sous espaces analytiques compacts d'un espace analytique donne, *Ann. Institut Fourier* 16, 1-95.

64. Dunford, N. (1938). Uniformity in linear spaces, *Trans. Amer. Math. Soc.* 44, 305-356.

65. Dunford, N. and Schwartz, J.T. (1964). *Linear Operators, Part I: General Theory,* John Wiley and Sons, New York.

66. Enflo, P. (1973). A counter example to the approximation problem in Banach spaces, *Acta. Math.* 130, 309-317.

67. Earle, C.J. and Hamilton, R.S. (1970). A fixed point theorem for holomorphic functions, *Global Analysis, Proceedings of Symposia in Pure Mathematics,* Amer. Math. Soc. 16, 61-65.

68. Exbrayat, J.M. (1970). Fonctions analytiques dans un espaces de Banach (d'apres Alexander), *Seminaire Pierre Lelong 1969,* Springer-Verlag Lecture Notes in Math. 116, 30-38.

69. Fantappie, L. (1930). I funzionali analitici, *Memorie della Reale Accademia dei Lincei* 6,3,11, 453-683.

70. Franzoni, T. and Vesentini, E. (1980). *Holomorphic Mappings and Invariant Distances,* North-Holland Math. Studies 40, Amsterdam.

71. Frechet, M. (1925). La notion de differentielle dans l' analyse generale, *Ann. Sci. de l'Ecole Norm. Sup.* 3,42, 293-323.

72. Gâteaux, R. (1919). Fonctions d'une infinite des variables independants, *Bull. Soc. Math. France* 47, 70-96.

73. Glicksfield, B. (1970). Meromorphic functions of elements of a commutative Banach algebra, *Trans. Amer. Math. Soc.* 151, 293-307.

74. Globevnik, J. (1975). On complex strict and uniform convexity, *Proc. Amer. Math. Soc.* 47, 175-178.

75. Graves, L.M. (1935). Topics in the functional calculus, *Bull. Amer. Math. Soc.* 412, 641-662.

76. Greenfield, S.J. and Wallach, N. (1971). The Hilbert ball and biball are holomorphically inequivalent, *Bull. Amer. Math. Soc.* 77, 261-263.

77. Grothendieck, A. (1953). Sur les certains espaces de fonctions holomorphes, *Journal fur die Reine und Angewandte Math.* 192, 35-44; 192, 77-95.

78. ——(1954). Sur les espaces (F) et (DF), *Summa Brasil Math.* 3, 57-123.

79. ——(1958). *Espaces Vectoriels Topologiques* (2nd Ed.), Soc. Mat. Sao Paulo.

80. ——(1973). *Topological Vector Spaces,* Gordon & Breach, New York.

81. Gruman, L. (1974). The Levi problem in certain infinite dimensional vector spaces, *Illinois J. of Math.* 18, 20-26.

82. Gruman, L. and Kiselman, C.O. (1972). Le probleme de Levi dans les espaces de Banach a base, *C. R. Acad. Sci. (Paris)* 274, 821-824.

83. Gunning, R. and Rossi, H. (1965). *Analytic Functions of Several Variables,* Prentice-Hall, Englewood Cliffs, New Jersey.

84. Gupta, C.P. (1968). *Malgrange Theorem for Nuclearly Entire Functions of Bounded Type on a Banach Space,* Notas de Matematica 37, IMPA, Rio de Janeiro.

85. ——(1970). On Malgrange theorem for nuclearly entire functions of bounded type on a Banach space, *Indag. Math.* 32, 356-358.

86. Hagler, J. and Johnson, W.B. (1977). On Banach spaces whose dual balls are not weak* sequentially compact, *Israel J. Math.* 28, 325-330.

87. Harris, L. (1969). Schwarz's lemma and the maximum principle in infinite dimensional spaces, Thesis, Cornell University.

88. ———(1969a). Schwarz's lemma in normed linear spaces, *Proc. Nat. Acad. Sci. U.S.A.* 62, 1014-1017.

89. ———(1974). Bounded symmetric homogeneous domains in infinite dimensional spaces. In *Proceedings on Infinite Dimensional Holomorphy*, T.L. Hayden and T.J. Suffridge (eds.), Springer-Verlag Lecture Notes in Math. 364, 13-40.

90. ———(1975). Bounds on the derivatives of holomorphic functions of vectors. In *Colloque d'Analyse, Rio de Janeiro, 1972*, L. Nachbin (ed.), Hermann Paris Act. Sc. et Ind., 1367, 145-163.

91. ———(1979). Schwarz-Pick systems of pseudometrics for domains in normed linear spaces. In *Advances in Holomorphy*, J.A. Barroso (ed.), North-Holland Mathematics Studies 34, 345-406.

92. Hayden, T.L. and Suffridge, T.J. (1971). Biholomorphic maps in Hilbert space have a fixed point, *Pacific J. of Math.* 38, 419-422.

93. ———Editors (1974). *Proceedings on Infinite Dimensional Holomorphy*, Springer-Verlag Lecture Notes in Math. 364.

94. ———(1976). Fixed points of holomorphic maps in Banach spaces, *Proc. Amer. Math. Soc.* 60, 95-105.

95. Herve, M. (1963). *Several Complex Variables: Local Theory*, Tata Institute of Fundamental Research, Bombay and Oxford University Press, London.

96. ———(1968). *Analytic and Plurisubharmonic Functions*, Springer-Verlag Lecture Notes in Math. 198.

97. ———(1971). Analytic continuation on Banach spaces. In *Several Complex Variables II*, J. Horvath (ed.), Springer-Verlag Lecture Notes in Math. 185, 63-75.

98. ———(1977). Some properties of the images of analytic maps. In *Infinite Dimensional Holomorphy and Applications*, M.C. Matos (ed.), North-Holland Mathematics Studies 12, 217-229.

99. Hille, E. (1948). *Functional Analysis and Semigroups*, Amer. Math. Soc. Colloquium Publications 31.

100. ———(1972). *Methods in Classical and Functional Analysis*, Addison-Wesley, Reading, Massachusetts.

101. Hille, E. and Phillips, R.S. (1957). *Functional Analysis and Semigroups,* Amer. Math. Soc. Colloquium Publications $\underline{31}$.

102. Hirschowitz, A. (1969). Sur le non-plongement de varietes analytiques Banachiques reelles, *C. R. Acad. Sc. (Paris)* $\underline{269}$, 844-846.

103. ———(1970). Bornologie des espaces de fonctions analytiques en dimension infinite, *Seminaire Pierre Lelong 1969/70,* Springer-Verlag Lecture Notes in Math. $\underline{205}$, 21-33.

104. Hörmander, L. (1967). *An Introduction to Complex Analysis in Several Variables,* Van Nostrand, Princeton, New Jersey.

105. Horvath, J. (1966). *Topological Vector Sapces and Distributions I,* Addison-Wesley, Reading, Massachusetts.

106. Isidro, J.M. (1979). Topological duality on the functional space $(H_b(U;F),\tau_b)$, *Proceedings, Royal Irish Academy* $\underline{79A}$, 115-130.

107. Istratescu, V.I. (1981). *Introduction to Linear Operator Theory,* Monographs and Textbooks in Pure and Applied Math. $\underline{65}$, Marcel Dekker, New York and Basel.

108. Josefson, B. (1974). A counterexample in the Levi problem. In *Proceedings on Infinite Dimensional Holomorphy,* T.L. Hayden and T.J. Suffridge (eds.), Springer-Verlag Lecture Notes in Math. $\underline{364}$, 168-177.

109. ———(1975). Weak sequential convergence in the dual of a Banach space does not imply norm convergence, *Arkiv for Math.* $\underline{13}$, 79-89.

110. Kakutani, S. (1943). Topological properties of the unit sphere of a Hilbert space, *Proc. Imp. Acad. Tokyo* $\underline{19}$, 269-271.

111. Kakutani, S. and Klee, V. (1963). The finite topology of a linear space, *Arch. Math. Basel* $\underline{14}$, 55-58.

112. Kantorovich, L.V. and Akilov, G.P. (1964). *Functional Analysis in Normed Spaces,* A Pergamon Press Book, MacMillan Co., New York.

113. Katz, G. (1977). Domains of existence in infinite dimensions. In *Infinite Dimensional Holomorphy and Applications,* M.C. Matos (ed.), North-Holland Math. Studies $\underline{12}$, 239-247.

114. Kaup, W. (1983). A Riemann mapping theorem for bounded symmetric domains in complex Banach spaces, *Math. Zeit.* $\underline{183}$, 508-529.

115. Kaup, W. and Upmeier, H. (1976). Banach spaces with bi-holomorphically equivalent unit balls are isomorphic, *Proc. Amer. Math. Soc.* 58, 129-133.

116. Kelley, J.L. (1955). *General Topology*, D. van Nostrand, New York.

117. Kiselman, C.O. (1977). Geometric aspects of the theory of bounds for entire functions in normed spaces. In *Infinite Dimensional Holomorphy and Applications*, M.C. Matos (ed.), North-Holland Math. Studies, 12, 249-275.

118. Kobayashi, S. (1970). *Hyperbolic Manifolds and Holomorphic Mappings*, Monographs and Textbooks in Pure and Applied Math. 2, Marcel Dekker, New York and Basel.

119. Köthe, G. (1969). *Topological Vector Spaces I*, Springer-Verlag, Berlin and New York.

120. Krantz, S.G. (1982). *Function Theory of Several Complex Variables*, John Wiley and Sons, New York.

121. Lang, S. (1962). *Introduction to the Theory of Differential Manifolds*, John Wiley and Sons, New York.

122. ———(1969). *Analysis II*, Addison-Wesley, Reading, Massachusetts.

123. Lelong, P. (1971). Recent results on analytic mappings and plurisubharmonic functions in topological linear spaces. In *Several Complex Variables II*, J. Horvath (ed.), Springer-Verlag Lecture Notes in Math. 185, 97-124.

124. Lindenstrauss, J. and Tzafriri, L. (1977). *Classical Banach Spaces*, Springer-Verlag, Berlin.

125. Machado, S., Editor (1981). *Functional Analysis, Holomorphy and Approximation Theory*, Springer-Verlag Lecture Notes in Math. 843.

126. Martineau, A. (1963). Sur les fonctionnelles analytiques et la transformation de Fourier-Borel, *J. Anal. Math.* 11, 1-164.

127. ———(1966). Sur le topologie des espaces de fonctions holomorphes, *Math. Annalen* 163, 62-88.

128. ———(1967). Equations differentielles d'ordere infinie, *Bull. Soc. Math. France* 95, 109-154.

129.   Martin, R.S.   (1932).   Contributions to the theory of
       functionals.   Thesis, California Institute of Technology.

130.   ———(1974).   On the Cartan-Thullen theorem for some sub-
       algebras of holomorphic functions in a locally convex space,
       Journal fur die Reine und Angewandte Math. 270, 7-14.

131.   ———Editor.   (1977).   *Infinite Dimensional Holomorphy and
       Applications,* North-Holland Math. Studies 12, Amsterdam.

132.   Mazur, S. and Orlicz, W.   (1934).   Grundlegende Eigenschaften
       der polynomischen Operationen, I, II, *Studia Math.* 5, 50-68;
       179-189.

133.   McArthur, C.   (1972).   Development in Schauder basis theory,
       Bull. Amer. Math. Soc. 78, 877-908.

134.   de Moraes, L.A.   (1979).   Theorems of the Cartan-Thullen type
       and Runge domains.   In *Advances in Holomorphy,* J.A. Barroso
       (ed.), North-Holland Math. Studies 34, 521-561.

135.   Mujica, J.   (1975).   On the Nachbin topology in spaces of
       holomorphic functions, *Bulletin Amer. Math. Soc.* 81, 904-906.

136.   ———(1978).   *Germenes Holomorfos y Funciones Holomorfas en
       Espaces de Frechet,* Universidad de Santiago Compostela, Spain.

137.   ———(1979).   Spaces of germs of holomorphic functions.   In
       Studies in Analysis, Advances in Math. Suppl. Studies 4,
       G.C. Rota (ed.), Academic Press, 1-41.

138.   ———(1981).   Domains of holomorphy in DFC spaces.   In
       Functional Analysis, Holomorphy and Approximation Theory,
       S. Machado (ed.), Springer-Verlag Lecture Notes in Math.
       843, 500-533.

139.   Nachbin, L.   (1954).   Topological vector spaces of continuous
       functions, *Proc. Nat. Acad. Sci. U.S.A.* 40, 471-474.

140.   ———(1964).   *Lectures on the Theory of Distributions,*
       Lecture Note, University of Rochester.   Reprinted by the
       University Microfilms International (USA), 1980.

141.   ———(1967).   On the topology of the space of all holomorphic
       functions on a given open set, *Indag. Math.* 29, 366-368.

142.   ———(1967a).   On spaces of holomorphic functions of a given
       type.   In *Functional Analysis,* Irvine 1966, B.R. Gelbaum (ed.),
       Academic Press, 55-70.

143.   ———(1969).   *Topology on Spaces of Holomorphic Mappings,* Erg.
       der Math. Springer-Verlag 47, Berlin.

144. Nachbin, L. (1970). Sur les espaces topologiques d'application continues, *C. R. Acad. Sci. (Paris)* 271, 596-598.

145. ———(1970a). *Holomorphic Functions, Domains of Holomorphy and Local Properties*, North-Holland Math. Studies 1, Amsterdam.

146. ———(1970b). Concerning spaces of holomorphic mappings, Publications of Math. Dept., Rutgers University.

147. ———(1973). Recent developments in infinite dimensional holomorphy, *Bull. Amer. Math. Soc.* 79, 625-640.

148. ———(1974). A glimpse at infinite dimensional holomorphy. In *Proceedings on Infinite Dimensional Holomorphy,* T.L. Hayden and T.J. Suffridge (eds.), Springer-Verlag Lecture Notes in Math. 364, 69-79.

149. ———(1976). Some holomorphically significant properties of locally convex spaces. In *Functional Analysis,* J. Figuereido (ed.), Marcel Dekker, New York and Basel.

150. ———(1979). Some problems in the application of functional analysis to holomorphy. In *Advances in Holomorphy,* J.A. Barroso (ed.), North-Holland Math. Studies 34, 577-583.

151. ———(1980). Why holomorphy in infinite dimension? *Ens. Math.* 26, 257-269.

152. ———(1981). *Introduction to Functional Analysis: Banach Spaces and Differential Calculus,* Monographs and Textbooks in Pure and Applied Math. 60, Marcel Dekker, New York and Basel.

153. Nashed, M.Z. (1971). Differentiability and related properties of non-linear opeartors: some aspects of the role of differentials in non-linear functional analysis. In *Nonlinear Functional Analysis and Applications,* L.B. Rall (ed.), Academic Press, New York, 103-309.

154. Nissenzweig, A. (1975). W* sequential convergence, *Israel J. Math.* 22, 266-272.

155. Noverraz, Ph. (1969). Fonctions plurisubharmoniques et analytiques dans les espaces vectoriels topologiques, *Ann. Inst. Fourier* 19, 419-493.

156. ———(1973). *Pseudo-convexite, Convexite Polynomiale et Domaines d'Holomorphie en Dimension Infinie,* North-Holland Math. Studies 3, Amsterdam.

157. ———(1979). *El problema de Levi-Oka en dimension infinita,* Universidad de Santiago de Compostela, Spain.

158.   Noverraz, Ph.   (1979a).   Topologies associated with Nachbin
       topology.   In *Advances in Holomorphy,* J.A. Barroso (ed.),
       North-Holland Math. Studies, 34, 609-627.

159.   Pelczynski, A.   (1971).   Any separable Banach space with
       the bounded approximation property is a complemented sub-
       space of a Banach space with a basis,   *Studia Math.* 40,
       239-243.

160.   Pisanelli, D.   (1966).   Funzionali analitici dello spazio
       H(0), *Bull. U. M. It.* 3,21, 377-384.

161.   ———(1972).   Applications analytiques en dimension infinie,
       Bull. Soc. Math. France 96, 181-191.

162.   Pomes, R.   (1974).   Solution du probleme de Levi dans les
       espaces Silva a base, *C. R. Acad. Sc. Paris* 278, Ser. A,
       707-710.

163.   Rado, T.   (1949).   *Subharmonic Functions,* Chelsea, New York.

164.   Ramis, J.P.   (1970).   *Sous Ensembles Analytiques d'une
       Variete Banachique,* Erg. der Math Springer-Verlag, Berlin.

165.   Rickart, C.E.   (1969).   Analytic functions of an infinite
       number of complex variables, *Duke Math. J.* 36, 581-597.

166.   ———(1979).   *Natural Function Algebras,* Springer-Verlag,
       New York.

167.   Robertson, A. and Roberson, W.   (1964).   *Topological Vector
       Spaces,* Cambridge Tracts in Math. 53, Cambridge University
       Press.

168.   Rogers, J.T. and Zame, W.R.   (1982).   Extension of analytic
       functions and the topology in spaces of analytic functions.
       Preprint.

169.   Rudin, W.   (1977).   Holomorphic maps of discs into F-spaces.
       In *Complex Analysis, Kentucky, 1976,* Springer-Verlag Lecture
       Notes in Math. 599, 104-108.

170.   Rusek, K.   (1974).   Remarks on H-bounded subsets in Banach
       spaces, *Zeszyty Nauk. Uniw. Jagiellonski. Proce Mat.* 356(16),
       55-59.

171.   Schaefer, H.H.   (1971).   *Topological Vector Spaces,* Graduate
       Texts in Math. 3, Springer-Verlag.

172.   Schottenloher, M.   (1974).   Analytic continuation and regular
       classes in locally convex Hausdorff spaces, *Port. Math.* 33,
       219-250.

173. Schottenloher, M. (1974a). Bounding sets in Banach spaces and regular classés of analytic functions. In *Functional Analysis and Applications*, Springer-Verlag Lecture Notes in Math. 384, 109-122.

174. ———(1977). Holomorphe Funktionen auf Gebieten uber Banachraumen zu vorgegebenen Konvergenzradien, *Manuscripta Math.* 21, 315-327.

175. ———(1977a). Richness of the class of holomorphic functions on an infinite dimensional space. In *Functional Analysis: New Results and Surveys*, K.D. Bierstedt and B. Fuchssteiner (eds.), North-Holland Math. Studies 27, 209-226.

176. Singer, I. (1970). *Bases in Banach Spaces I*, Grundlehren der Math. Wissenschaften 154, Springer-Verlag, Berlin and New York.

177. Taylor, A.E. (1937). Analytic functions in general analysis, *Ann. Sc. Normale Sup. Pisa (2)*,6, 277-292.

178. ———(1943). Analysis in complex Banach spaces, *Bull. Amer. Math. Soc.* 49, 652-669.

179. ———(1970). Historical notes on analyticity as a concept in functional analysis. In *Problems in Analysis*, R.C. Gunning (ed.), Princeton University Math. Series 31, 325-343, and in this book.

180. ———(1974). The differential: nineteen and twentieth century developments, *Arch. Hist. of Exact Sciences* 12(4), 355-383.

181. ———(1980). *Introduction to Functional Analysis* (2nd ed.), Wiley, New York.

182. Thorp, E. and Whitley, R. (1967). The strong maximum modulus theorem for analytic functions into a Banach space, *Proc. Amer. Math. Soc.* 18, 640-646.

183. Titchmarsh, E.C. (1939). *The Theory of Functions* (2nd ed.), Oxford University Press, London.

184. Willard, S. (1968). *General Topology*, Addison-Wesley, Reading, Massachusetts.

185. Zame, W.R. (1975). Extendability, boundedness and sequential convergence in spaces of holomorphic functions, *Pacific J. of Math.* 57, 619-628.

186. Zapata, G.I. Editor (1983). *Functional Analysis, Holomorphy and Approximation Theory,* Lecture Notes in Pure and Applied Math. 83, Marcel Dekker, New York and Basel.

187. Zorn, M.A. (1945). Gateaux differentiability and essential boundedness, *Duke Math. J.* 12, 579-593.

188. ———(1945a). Characterization of analytic functions in Banach spaces, *Ann. of Math.* 46, 585-593.

189. ———(1946). Derivatives and Frechet differentials, *Bull. Amer. Math. Soc.* 52, 133-137.

Notation

Index